PLANETARY NEBULAE

PLANETARY NEBULAE

G. A. Gurzadyan

Burakan Astrophysical Observatory
Armenian Socialist Soviet Republic, USSR

Revised by the author,
Translated and edited by
D. G. HUMMER
with the assistance of
C. M. VARSAVSKY
and
Z. LERMAN

D. REIDEL PUBLISHING COMPANY

DORDRECHT-HOLLAND

Copyright © 1970 by GORDON AND BREACH, Science Publishers, Inc.
Softcover reprint of the hardcover 1st edition 1970

150 Fifth Avenue, New York, N.Y. 10011
Library of Congress Catalog Card Number: 69-11664

ISBN 978-94-010-3062-5   ISBN 978-94-010-3060-1 (eBook)
DOI: 10.1007/978-94-010-3060-1

Preface to the English Edition

The publication in English of this monograph seems to me to indicate the ever-increasing interest of astrophysicists in the physical and dynamical problems of planetary nebulae—one of the most interesting and fruitful branches of theoretical astrophysics. Their interest in part arises from the fact that the methods of identifying the physical processes occurring in planetary nebulae, as well as the many theoretical results, are now acquiring a degree of universality as their sphere of application increases. Finally, the special cosmic significance of planetary nebulae is becoming apparent.

The English edition of *Planetary Nebulae* differs considerably from the Russian version published in 1962, primarily because of the new results included in it, but also because of numerous editorial revisions.

The problems of magnetic fields and hydrodynamics in planetary nebulae are beginning to occupy an important place in the study of the dynamics of these objects. Recent studies by D. H. Menzel confirm the idea advanced in the present monograph as to the existence of magnetic fields in planetary nebulae. New light is being cast on the dynamics of planetary nebulae by the hydrodynamic investigations of F. D. Kahn, W. G. Mathews and others. Unfortunately I was not able to include these and other interesting results in the present edition.

I would like to acknowledge my appreciation to the translators of this book for their arduous work and especially thank Dr. Hummer for assuming the role of editor as well. It seems to me that the translators have succeeded in preserving in the English version the style of the original, and that skillful editing has in many respects improved the monograph.

G. A. Gurzadyan
29 January 1969

Translation Editor's Preface

Gurzadyan's *Planetary Nebulae*, published in 1962, gives a comprehensive survey of the observational material available at that time and of the earlier theoretical work on planetary nebulae. A substantial portion of the book is devoted to Dr. Gurzadyan's own research, especially on the origin, evolution and morphology of these objects. While some of the results of this work run counter to currently accepted ideas, the enormous number of new suggestion he makes must, in the long run, be beneficial to our understanding of planetary nebulae. That he has raised more questions than he has answered cannot be held a fault of the book.

The present translation incorporates nearly eighty pages of new material prepared by Dr. Gurzadyan. The most important of the work published subsequently is referred to in editorial footnotes. The most recent account of research on planetary nebulae is given in the Proceedings of the IAU Symposium No. 37, held in Tatranska Lomnica, Czechoslovakia in September 1967.

Draft translations of four chapters have been prepared by Dr. Carlos Varsavsky and of two chapters by Mr. Z. Lerman, of the Israel Program for Scientific Translations. I have translated the remaining material and edited the whole work, so that responsibility for any errors rests with me. To increase the utility of the list of references, many of which are to work published in USSR, I have added the titles, when available, to the citations. References to English translations have also been included. The transliteration system recommended by the Royal Society has been used throughout.

D.G. Hummer

Foreword

The theoretical study of those extremely interesting and mysterious celestial objects — the planetary nebulae — over a period of more than thirty years of persistent search and laborious investigations, has resulted in important discoveries.

In 1927 Bowen identified the nebular lines with the forbidden lines of different atoms and ions and developed a theory for the emission of light from the nebulae arising from electron collisions. In the same year, Zanstra established the fluorescent origin of the nebular light emission and shortly afterwards, in 1930, proposed a method for determining the temperature of the nuclei of planetary nebulae. In 1933 V. A. Ambartsumyan developed the mathematical theory of radiative equilibrium in planetary nebulae and pointed out the important role played by radiation pressure in their dynamics.

The preceding works were concerned with the basic *physics* of planetary nebulae. Interest in these studies spread rapidly, leading to a new series of investigations. Within a short time the physics of planetary nebulae became one of the most active branches of theoretical astrophysics. This situation was favoured, in particular, by the extremely simple physical conditions in the nebulae: both the density of matter and the radiation density are small. In this respect the development of atomic physics also played a significant role. The most important results obtained concerning the physics of planetary nebulae were the following: in visual light the planetary nebulae are ten to a hundred times brighter than their exciting stars; the nuclei of planetary nebulae belong to the group of hottest stars in the Galaxy; planetary nebulae represent gigantic, strongly ionized and continuously expanding gas formations of small mass; the dynamics of planetary nebulae are tied to the action of radiation pressure and electromagnetic forces, Newtonian forces playing almost no role; the chemical composition of planetary nebulae is similar to that of the ordinary stars.

The first serious observations of planetary nebulae are associated with the names of Curtis, who obtained photographs of a large number of planetary nebulae, and of Wright, who made the first observations of their spectra during the years 1916–1917. Later on the most interesting photometric and spectrophotometric observations of planetary nebulae were made by Aller, Minkowski and Page. Wilson carried out important work on the subject of the internal motions of planetary nebulae.

On the theoretical side very important results were obtained: the theory of diffusion of radiation in nebulae by V. V. Sobolev; the theory of the luminosity of nebulae in the hydrogen lines by Cillié, Menzel and their associates; the calculation of atomic parameters of astrophysical interest by M. J. Seaton; the theory of the continuous spectrum of nebulae by A. Ya. Kipper, J. Greenstein and L. Spitzer. The first catalogue of planetary nebulae was compiled by B. A. Vorontsov-Vel'yaminov.

The present monograph is devoted to an account of the basic results obtained in the fields of the physics and the dynamics of planetary nebulae. The initial development of the *dynamics* of planetary nebulae dates from a later epoch. This is explained, in particular, by the fact that the solution of the dynamical problem depends to a large extent upon the physical conditions in planetary nebulae. Also of great significance was the accumulation of observational material of the highest quality which, in turn, is tied in with the introduction of new and more powerful telescopes. The most important conclusions that can be drawn at the present state in the investigation of the dynamics of planetary nebulae are based mainly upon the results of the work of the author, namely: each nebulae is born with one envelope and ends its life with two envelopes; in the planetary nebulae there exist regular, relatively strong, magnetic fields that control their internal structure; there is no direct genetic tie between planetary nebulae and novae or supernovae; the life times of planetary nebulae are a hundred thousand times shorter than those of the common stars; planetary nebulae appear to be the remains of processes of stellar formation; each fiftieth star in our Galaxy was at sometime the nucleus of a planetary nebula. These conclusions, however, do not appear to be definitive and need to be verified further.

The author takes advantage of this opportunity to express his deep gratitude to V. A. Ambartsumyan, V. V. Sobolev and A. B. Severny for their observations, useful remarks and advice that were made upon the examination of the manuscript.

The Author

Contents

Chapter I

Basic Observational Data

1. Forms of Planetary Nebulae

The forms of planetary nebulae are characterized by their great variety. The brightest nebulae have disk-like or oval forms, and at the telescope they look like planets, with a rather greenish appearance (which explains their name). The majority of the planetary nebulae appear in the photographs as bright rings, in the center of which are located the stars that excite the light-emission of the nebulae – the so-called nucleus of the nebula. With longer exposures, however, the rings disappear and turn into continuous disks. This seems to indicate that the real forms of the nebulae have nothing in common with a ring but rather appear in space as shells of finite thickness. The observed ring-like forms are due to the projection effect of the nebulae on the sky.

The ideal round nebula is seldom observed. Usually they are somewhat flattened or stretched out. At one time it was assumed, mainly on the basis of the results of the spectrographic investigations of Campbell and Moore [5], that the flattening was due to the rotation of the planetary nebulae around their minor axis; measurements were even made of the magnitude of the rotational velocity. Later these results were not confirmed; more exact and reliable observations by Wilson, made with more precise equipment, did not detect any trace of nebular rotation to within the limit of observational error. On the other hand, to explain the observed flattening, the rotational velocity should be of the same order as the velocity of expansion of the nebula, but the latter is easily detected and its magnitude considerably exceeds the error in the measurements.

The flattening or stretching out of the planetary nebulae, as we shall show later, can be apparently explained as a consequence of the presence of magnetic fields. We should also point out that the shape or, rather, the external outline of the nebulae, differs strongly from an ellipsoid. On the other hand, if the flattening of the nebulae were actually caused by rotation, their shapes should be ellipsoidal with different ratios of semi-major to semi-major axes according to the magnitude of the angular velocity.

As examples of ring-like planetary nebulae we have NGC 6720, in the constellation of Lyra, one of the brightest of all planetary nebulae (the integrated stellar magnitude is $9^m.3$), and the gigantic—at least in apparent size—nebula NGC 7293 in the constellation of Aquarius ($12' \times 15'$).

Elongated nebulae appear as bodies rotated around their major axes. Therefore, generally speaking, it is impossible to draw conclusions about the real shapes of nebulae from only their apparent form, that is, without taking into account the orientation of the nebulae relative to the line of sight. For example, nebulae whose photographic images appear as disks or rings may actually be elongated along the line of sight.

2. The Structure of Nebulae

It has been established that the brightness at the end of the short axis of the nebulae is greater than at the end of the major axis. In some cases this difference becomes so pronounced that it is apparent in the photographs, without need of measurement. In these cases one talks of *bipolar* planetary nebulae.

There is a certain variation of bipolar nebulae. As a basic criterion one can take the brightness difference at the ends of the minor and major axes of the nebulae. In some objects this difference is relatively small, that is, the bipolar structure is shown only by an increase in brightness at the ends of the short axis. This is observed, for example, in the ring-like nebulae NGC 6720 and NGC 7293, where the brightness at the ends of the minor axis is one and a half to two times greater than at the end of the major axis. In this respect these nebulae seem also to be bipolar.

In the second group the difference in brightness is so great that it gives the impression that the nebulae are divided into two bright parts, so called "beads", whose shapes are half-moons placed at the ends of the short axes, symmetrically relative to the center of the nebulae. Sometimes these "beads" have a sharp boundary on the inside. As examples of such objects we can mention one of the most interesting nebulae, NGC 3587 (the "Owl"), nebula A 70 (that is, nebula No. 70 in Abell's [6] list), NGC 7354, 7662, 3195 (in the Southern sky) and others.

Finally, in the third group of bipolar nebulae the "beads" are strongly developed and one can almost completely isolate one from the other. Only with long exposures does a faint background appear between them. Such shape is shown, for example, by the nebulae A 66, A 17, A 19, the anonymous nebula (from now on we shall designate such nebulae briefly as "anon."—for more details see Section 11 of this chapter) at $16^h 10^m.5$, $-54°50'$, and others.

In a small number of cases the bipolar nebulae seem to have a so-called "rectangular" shape, which is observed in a few planetary nebulae. These nebulae differ from the common bipolar nebulae in the sense that they are strongly compressed along the minor axis and elongated along the major axis. A beautiful example of this type of

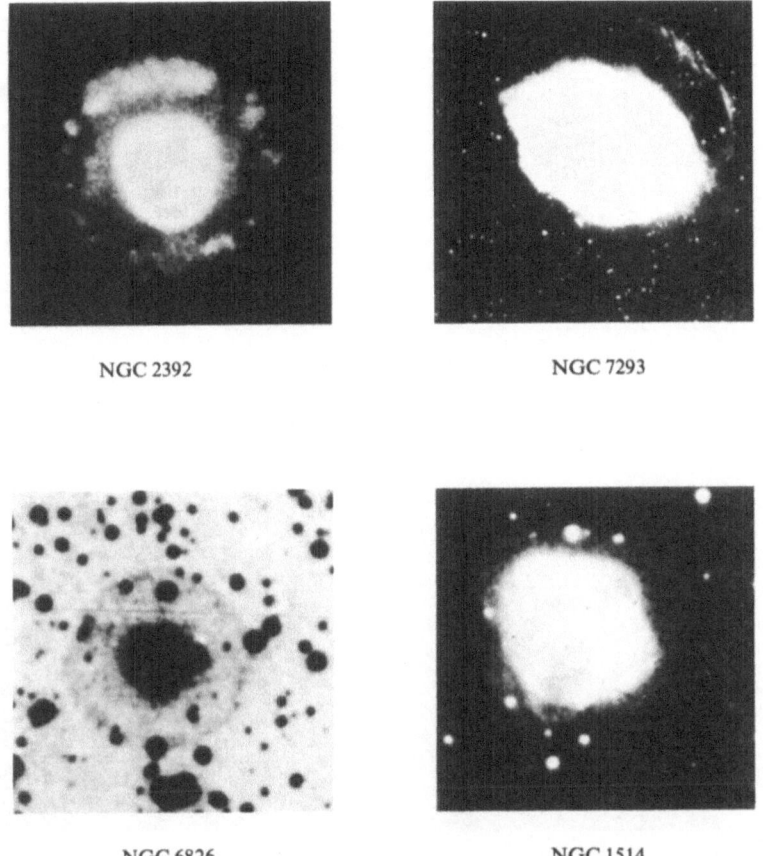

NGC 2392   NGC 7293

NGC 6826   NGC 1514

PLATE I.   Double-envelope nebulae.

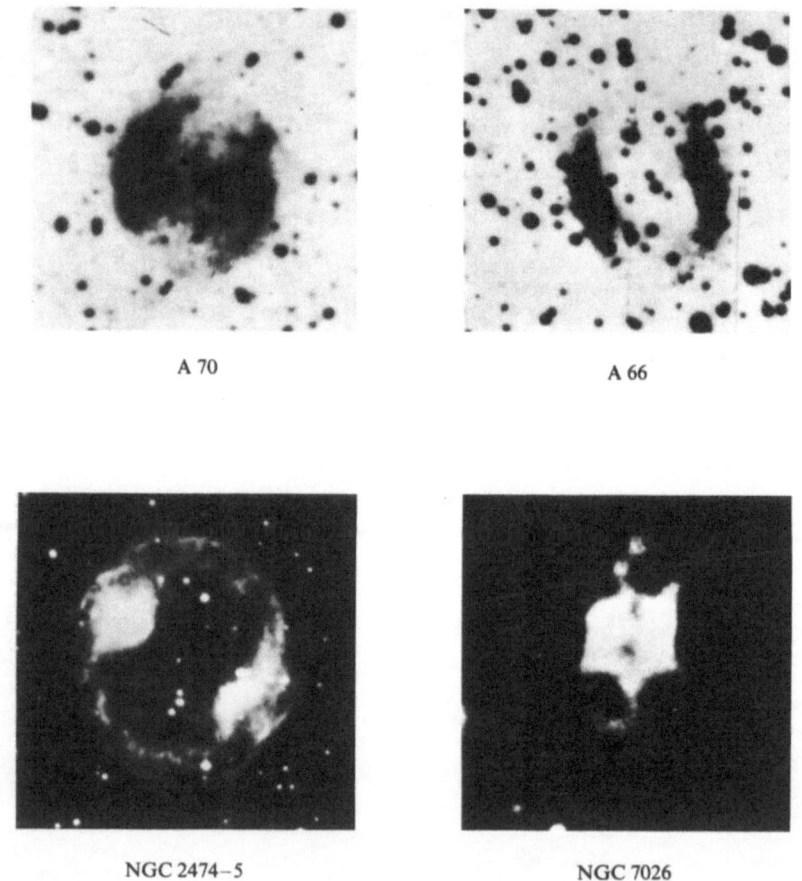

A 70

A 66

NGC 2474–5

NGC 7026

PLATE II. Bipolar nebulae.

Anon. 16^{h}10^m.5

NGC 3587

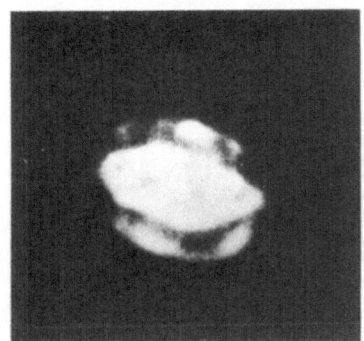

NGC 7009

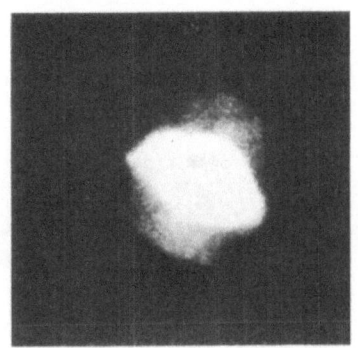

NGC 7662

PLATE III.   Bipolar double-envelope nebulae.

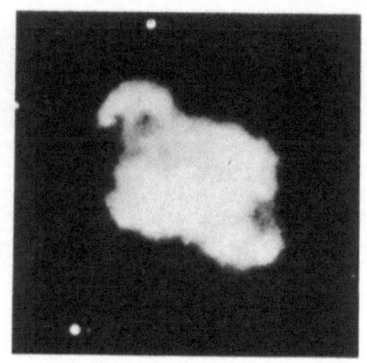

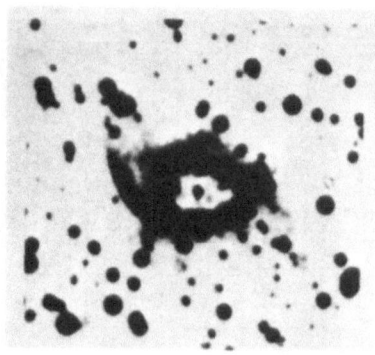

NGC 4361   A 65

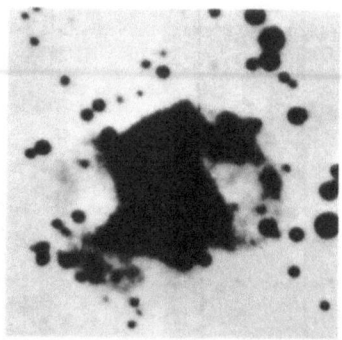

NGC 650–1

PLATE IV.   Spiral nebulae.

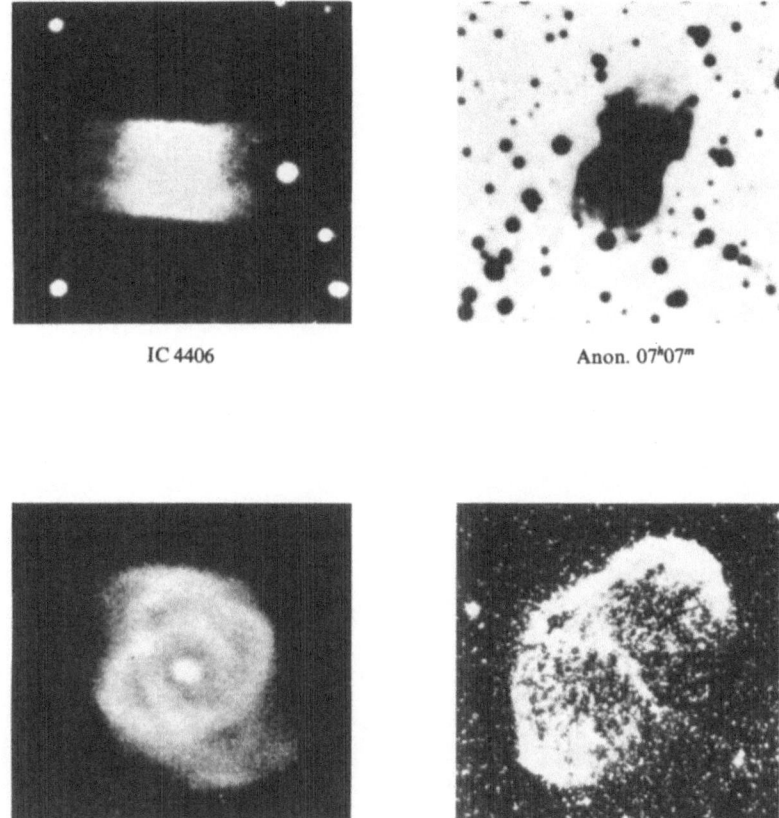

IC 4406

Anon. 07^{h}07^m

NGC 6543

NGC 6888

PLATE V. Anomalous nebulae.

nebula is the well known planetary nebula IC 4406 in the Southern sky. A variety of this type are the nebulae of the "hour-glass" type (anon. $07^h07^m-00°43'.3$).

Almost half of the planetary nebulae whose photographs permit the study of their structure are bipolar. On the other hand, as will be shown in the chapter on bipolar nebulae (Chapter IX), the "beads" can be observed only when the major axis of the nebula is almost perpendicular to the line of sight. It follows, then, that among the remaining planetary nebulae there are also bipolars, whose major axes are oriented almost parallel to the line of sight. This circumstance once again shows the decisive role played by the relative orientation of the observations on the apparent form of the nebulae.

Thus, the bipolarity of planetary nebulae is one of the important characteristics of their structure and is present, to a greater or lesser degree, in almost all planetary nebulae.

Of special interest are the planetary nebulae that consist of two concentric shells around the nucleus. In this case we shall speak of *double-envelope* planetary nebulae. The surface brightness of the second (outer) envelope is, as a rule, less than the surface brightness of the primary (inner) envelope. The ratio between the radii of the outer and inner envelopes covers a wide range; at minimum it differs slightly from unity, and it can reach several units; in one case it is even greater than ten.

Examples of double-envelope nebulae are NGC 2392, 1535, 6826, 7354, 7662 and others. In some cases the surface brightness of the second envelope is so small that it is revealed with difficulty, and only with very long exposures and very powerful telescopes. In this way the presence of a second envelope was established in the case of the nebulae NGC 6720 and NGC 7293. A very weak second envelope was discovered also in the nebulae NGC 1514 and 6853. The presence of a very weak and very extended second envelope was established in one of the most interesting planetary nebulae, NGC 6543.

The possibility of verifying the presence or absence of a second envelope in a given planetary nebula depends on the observing conditions, and, in particular, on the light-gathering power of the telescope. In recent years a second envelope was discovered in many planetary nebulae that earlier had been assumed to have only one envelope. One gets the impression that all those of the largest dimensions, that is, ring-like nebulae, should have two envelopes. About one out of every three planetary nebulae for which we have reasonably good photographs, seems to have two envelopes.

In Chapter VII, which is devoted to the theory of the formation of double-envelope nebulae, it will be shown that all planetary nebulae are born as single-envelope nebulae and end their life as double-envelope nebulae. In other words, the appearance of a second envelope around the first at a definite stage in the life of the nebula is a phase through which all planetary nebulae must pass.

A sparse group is made up by the so-called *spiral* planetary nebulae. In the majority of cases they are common ring-like or two-envelope nebulae, in which one observes two appendages that look like spiral arms, placed symmetrically at the two sides of the nebula. Typical representatives of such nebulae are NGC 4361, 6210, A 65. A variety of the spiral nebulae are the Z-shaped nebulae, of which there are very few (NGC 2452, 6778 and others). They appear as bipolar nebulae, whose "beads" are joined by a more or less bright, oblique arch. However, unlike the galactic spiral structure, in planetary nebulae spiral structure does not have a dynamical character. In particular, it does not show rotation of the nebula or of the spiral arms. The origin of the spiral arms and, like them, of the appendages can be simply explained, and it is related to the magnetic fields in the nebulae themselves.

The fundamental special shapes mentioned above sometimes can be present in a single nebula simultaneously in different combinations. For example, the nebula NGC 4361 is spiral and, at the same time, has two envelopes; the anonymous nebula $16^h10^m.5 - 54°50'$ is bipolar and also has two envelopes, and furthermore in the second envelope there are traces of spiral structure; in the nebula NGC 650-1 appear simultaneously bipolarity, spiral structure, and two envelopes. One can find many such examples. Most often double-envelope nebulae are also bipolar.

We should emphasize that, independently of the peculiarities in the shape and structure of the nebulae, all of them have three-dimensional structure. In other words, plane (two-dimensional) forms are not characteristic of planetary nebulae.

To obtain the true picture of the spatial structure of planetary nebulae, one must solve the problem of the distribution of matter inside the nebula. In turn, to answer this question one needs, first of all a map of the distribution of the surface brightness (isophotes) in the emission lines of different atoms and ions. This is a comparatively simple problem; good isophotes of several planetary nebulae have been constructed, for example, by means of the balancing microphotometer of the University of Michigan [7], as shown in Aller's book [8]. Secondly, one needs to find a way to go from the apparent distribution of surface brightness over the nebula to the spatial distribution of energy emitted per unit volume in the different spectral lines. In one special case, namely for a spherical nebula with a concentric distribution of material density, this question was tackled with a method analogous to that used for determining the distribution of spatial density in the chromosphere; in this case the problem was solved by Abell's equation.

This method was applied by a number of authors [9–11, 208] to the planetary nebulae NGC 6572, 6720, 7293 and IC 418 and on the basis of the results curves were constructed showing the spatial distribution of the concentration of different atoms and ions along different axes of the nebulae.

However, as has been shown above, a true spherical shape and, in particular, a concentric distribution of material density do not obtain in planetary nebulae. Therefore, the results obtained using a spherical model of the nebula cannot, in general, produce a true picture of the distribution of the concentration of atoms and ions in a real, nonspherical nebula. However, in certain cases (for example, when the axis of the nebula is almost perpendicular to the line of sight) this method can be used to find the distribution of the electron or ion concentration in the equatorial plane of the nebula.

A characteristic feature of *small* planetary nebulae is their straight-forward, clear-cut shape, with sharp boundaries. This indicates the dominant role of a force that controls the dynamics of the nebulae at a given period in their lives and prevents their disruption or destruction. As the size of the nebula increases the role played by this force decreases due to the expansion, and the outline of the nebula loses its sharpness. Thus we reach a transition type of planetary nebulae – the *diffuse* nebulae. The number of the latter (A 16, 56) is not great, because the decrease in surface brightness caused by the expansion of the nebula makes its discovery difficult.

The most numerous group is formed by the so-called *stellar* planetary nebulae. In this group one finds nebulae having very small apparent sizes. A fraction of them have, undoubtedly, small linear dimensions and, possibly, represent an early phase in the development of planetary nebulae. However, others can very well be nebulae of larger dimensions, of well-known types, whose star-like appearance is due just to the fact that they are very far away from us.

More than half of all known planetary nebulae are stellar. As a rule they are discovered either with objective prisms, through the characteristic nebular spectrum, or by direct photography with combinations of different filters. Many such nebulae were discovered by Minkowski [12], mainly in the direction of the center of the Galaxy. The apparent dimensions of these nebulae cannot be determined. Their study is, as a rule, restricted to the measurement of relative intensities in their spectra, and also to the determination of their integrated apparent brightness. In spite of this limitation, the observation and study of star-like nebulae can have a special significance since only among them can one discover nebulae in their earlier stages of formation and evolution.

The angular dimensions subtended by the majority of the planetary nebulae are smaller than one minute of arc. The dimensions of the most interesting objects fall in the range 20″–40″. It is practically impossible to determine the dimensions of nebulae smaller than 2″. A very small fraction of the nebulae have dimensions larger than 1′. A few have a gigantic extent – greater than 10′, and in one case (A 72) the diameter of the nebula is 40′! Table 1–1 presents data on the apparent dimensions of several interesting planetary nebulae.

As far as the linear extent of planetary nebulae is concerned, the problem is intimately connected with that of the determination of their distances from us, and we shall treat it in Chapter III.

TABLE 1-1

Apparent Dimensions of Several Planetary Nebulae

Nebula	Dimensions	Type
IC 4997	$2''$ (?)	stellar
NGC 7027	$11'' \times 18''$	?
NGC 2392	$\begin{cases} 15'' \times 19'' \\ 43'' \times 47'' \end{cases}$	double-envelope
NGC 7009	$\begin{cases} 26'' \times 30'' \\ 26'' \times 44'' \end{cases}$	double-envelope spiral
NGC 4361	$\begin{cases} 39'' \times 44'' \\ 81'' \end{cases}$	double-envelope spiral
NGC 1514	$1'.5 \times 2'$	double-envelope
NGC 3587	$3'.3$	bipolar
NGC 2447−5	$6'.7 \times 7'.5$	bipolar
NGC 7293	$\begin{cases} 12' \times 15' \\ 21' \end{cases}$	double-envelope
A 72		ring

3. Luminosity of Nebulae and their Nuclei

The luminosity of a nebula is caused by the radiation of its central star — the *nucleus* of the nebula. In spite of this, the total brightness of the nebula is many times that of the star. This is due to the fact that the nebula absorbs the invisible ultra-violet energy radiated by the nucleus and re-emits it in frequencies that correspond to visible light. Since the nuclei of planetary nebulae are very hot stars, the largest fraction of their energy is emitted in the ultra-violet region of the spectrum. As a result, the nebula appears brighter than the illuminating central star.

The observations often give the possibility of determining the integrated photographic stellar magnitude, m_n of the nebula, that is, the sum of the brightness of the nebula and the nucleus. Besides, one can also determine the apparent photographic stellar magnitude, m_*, of the nucleus. Since the nebulae are completely transparent in the frequencies corresponding to visual light, the magnitude m_n represents the total energy emitted by the nebulae in all directions and per unit time in the frequencies of the lines in the photographic region of the spectrum. The difference $\delta = m_* - m_n$ is always positive and it is an important parameter that does not depend on the distance to the nebula. This quantity determines the temperature and size of the nucleus, as well as the mass of the nebula.

The magnitudes m_n and m_* have been determined photographically for a large number (around 150) of planetary nebulae by V. A. Vorontsov-Vel'yaminov and P. P.

Parenago [13]. According to these measurements the integrated apparent photographic magnitudes of planetary nebulae vary between very wide limits – from $6^m.5$ (NGC 7293)* to 15^m and lower (NGC 6620, 6630). The majority of planetary nebulae have integrated photographic magnitudes in the range $11–14^m$, and the maximum number of nebulae corresponds to $12^m.5$. The quantity δ varies from $0^m.1$ (NGC 7635, anon. $18^h13^m.0$) to 7^m (NGC 7048). On the average, for the majority of planetary nebulae $\delta \sim 3–4^m$.

It should be pointed out that the magnitude determinations in [13] are not of very high accuracy. This is explained mainly by the fact that the spectral sensitivity curve of the photographic plates was not well known, so that the calibration of their stellar magnitude system to the international system could not be performed with sufficient accuracy. The question then arises of a new determination of m_n for all nebulae. Preference should be given to photoelectric photometry, since the finite dimensions of the planetary nebulae make the measurement of the integrated brightness by photographic means a very difficult task, and these difficulties lead to inevitable errors.

Such measurements were carried out by Liller [14] for relatively bright planetary nebulae (brighter than 12^m). Table 1–2 gives his results for 17 planetary nebulae. It also gives the photovisual stellar magnitude m_{pv} of the nuclei of a few of them. The discrepancies between the data given in [13] and [14] for m_n are in several cases quite considerable, and reach $2^m.5$.

If one knows m_n and the apparent size, or mean angular diameter D'', of the nebula (in seconds of arc), it is possible to determine H, the mean surface brightness of

TABLE 1-2

Integrated Photographic Magnitudes (m_n) of Nebulae and Photovisual Magnitudes (m_*) of the Nuclei According to the Photoelectric Measurements of Liller [14].

Nebula	m_n	$m_*(pv)$	Nebula	m_n	$m_*(pv)$
NGC 40	$10^m.7$	$11^m.0$	NGC 6572	9.0	—
NGC 1535	10.4	—	NGC 6720	9.7	—
IC 418	9.4	—	NGC 6803	12.4	—
IC 2149	11.2	—	NGC 6818	10.0	$12^m.0$
NGC 2392	10.4	10.4	NGC 6826	9.8	9.9
NGC 3242	8.8	11.3	NGC 7009	8.9	10.9
IC 4593	11.4	10.9	NGC 7027	9.6	11.3
NGC 6210	9.7	11.3	NGC 7662	9.2	11.1
NGC 6543	8.8	10.3	CD + 30° 3639	—	10.1

* According to new measurements, the integrated brightness of the nebula NGC 7293 in photovisual light is $m_n = 9.4$ [208].

the nebula. This quantity, of course, does not say anything about the brightness distribution over the visible disk of the nebula, which, in the majority of cases, has a specific character and depends on the type and structure of the nebula, but it describes the emitting power per unit surface (1 cm^2) of nebula. In particular, the use of the parameter H forms the basis of one of the methods of determining the electron concentration, and hence the total mass, of the nebula.

The mean value of H decreases from 4^m (NGC 6153) to $14^m.5$ (NGC 6772, 3587) in a circle of 1' diameter [13]. In one case $H = 19^m.4$ (NGC 2474–5). The largest surface brightness occurs with the envelopes of nebulae with small apparent dimensions, although sometimes one finds planetary nebulae of considerable size having high surface brightness. The planetary nebulae of largest apparent size have in general small surface brightness. Many of the planetary nebulae discovered by Abell belong to this type [6].

The concept of *relative exposure* introduced by Curtis [15] can serve as a criterion of the relative brightness of a nebula. He defined relative exposure (RE) as the time necessary to obtain a photograph of the brightest region of the nebula, taking as unit of time the length of exposure for several standard regions in the Orion nebula (under identical photographic conditions). The relative exposures of planetary nebulae vary between 0.1–0.2 (NGC 7027, 6543) to 250–300 (NGC 6778, 3587).

As far as the color index is concerned, in the case of planetary nebulae, it does not have the same physical significance as for the common stars. In the stars it depends basically on one parameter—the color temperature. In the nebula it represents the difference between the stellar magnitude summed over the intensity of all the emission lines (and part of the continuous spectrum) that lie in the photographic region of the spectrum, and that summed over the intensity of the emission lines that lie in the photovisual region. But this difference depends on many factors and can vary from one nebula to the next. The brightest lines—N_1, N_2 and H_β—occur where the photographic and photovisual regions meet. The spectral sensitivity curves of these systems intersect in the neighborhood of $\lambda = 5000$ Å. Another bright line, H_∞, falls practically beyond the limit of the photographic region. It follows that the ratio $\int P_\lambda E_\lambda d\lambda / \int V_\lambda E_\lambda d\lambda$, where P_λ and E_λ are the relative sensitivities of the photographic and photovisual systems, respectively, will be somewhat greater than unity (because of the Balmer lines beginning with H_y).

Therefore, the color index for planetary nebulae will be always positive (in the International System). Its numerical value can vary somewhat from one nebula to another, but on the average it should be of the order of $+ 0^m.7$.

The nuclei of planetary nebulae are very faint stars. Only two nebulae have nuclei brighter than tenth magnitude: NGC 7635 ($m_* = 8^m.5$) and NGC 1514 ($m_* = 9^m.7$). The photographic stellar magnitude of the nuclei of several of them reach $18^m.5$ (NGC 6439). On the average, the majority of the planetary nebulae have m_* (pg) = $= 14–15^m$. The central star has not been discovered in about 15 planetary nebulae.

Undoubtedly, the nuclei of these nebulae, which have small apparent size, are even weaker and their discovery against the bright background of the nebula becomes very difficult. One could try to discover these weak nuclei by means of photographs taken with specially chosen filters, so as to exclude the strong emission lines of the nebula, thus decreasing the influence of the background.

We know very little about the absolute brightness of planetary nebulae (M_n) and their nuclei (M_*) since we know little about their distances from us. Thus, for example, according to B.A. Vorontsov-Vel'yaminov's catalogue, the absolute integrated magnitude M_n varies within a relatively small interval—from $-1^m.5$ to $0^m.0$, with a mean value of $M_n = -0.64$ [16]. On the other hand, according to the data in Shklovskii's catalogue, the dispersion in the quantity M_n is very large, from -4 to $+5^m$. Even greater is the dispersion in the absolute magnitudes of the nuclei: The extreme values of M_* may differ by 10^m. The mean value of M_* is positive and approximately 4–6^m. According to Berman $\overline{M}_n \sim -2$ and $\overline{M}_* \sim +2$ [18]. No matter how inexact and uncertain the data may be, we can draw one conclusion: the nuclei of planetary nebulae are not giants and, most probably, are dwarfs of the usual brightness. This conclusion was reached long ago, in the early stages of the investigation of planetary nebulae.

4. The Spectra of the Nebulae

The most characteristic feature of the planetary nebulae is their spectra. The general structure of the spectrum is identical for all planetary nebulae. On the other hand, they differ strongly from the spectra of other celestial bodies. Because of this, it is always possible to distinguish unmistakably the spectrum of a planetary nebula from, say, that of a peculiar star, when the apparent size of the nebula is too small to allow its identification by direct photography.

The spectrum of planetary nebulae consists of a series of intense and weak emission lines, superimposed on a weak continuous background. The general structure of the spectrum of planetary nebulae is shown schematically in Fig. 1-1: In almost

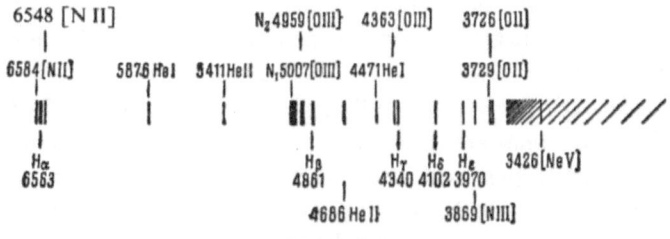

Figure 1-1

General structure of the spectra of planetary nebulae in the photographic and visual regions.

all nebulae the brightest lines are the so-called "nebulium" lines: $\lambda = 5007$ Å (N_1) and $\lambda = 4959$ Å (N_2). Bowen was the first to show [2] that these, and several other formerly mysterious lines, originate from forbidden transitions downwards from metastable states in several atoms and ions. As a rule, these transitions are not observable in the laboratory, since it is almost impossible to produce the conditions necessary for their study.

The lines N_1 and N_2 belong to doubly ionized oxygen. Sometimes they are re-referred to as the O III doublet and written: $\lambda5007$ Å [O III] and $\lambda4959$ Å [O III]; the square brackets indicate as a rule that the line in question corresponds to a forbidden transition. These lines also appear in the spectra of diffuse nebulae, in the spectra of envelopes ejected by novae, and also in the spectra of nonstable stars surrounded by gaseous envelopes. However, in planetary nebulae the lines N_1 and N_2 are considerably brighter. The intensity ratio of the lines N_1, N_2 and H_β is $10 : 3 : 1$, while in diffuse nebulae it varies from $2 : 0.6 : 1$ to $0.3 : 0.1 : 1$.

The ratio I_{N_1}/I_{N_2} is roughly equal to three. It is constant for all nebulae and it does not depend on the physical conditions in them. The ratio $I_{N_1+N_2}/I_H$ is different for different nebulae. In Table 1-3 the value of this ratio is given for a number of planetary nebulae, according to the photoelectric measurements of Liller and Aller [19]; the mean value is of the order of 13.

TABLE 1-3

The Ratio $I_{N_1+N_2}/I_{H_\beta}$ for Several Planetary Nebulae [19]

Nebula	$I_{N_1+N_2}/I_{H_\beta}$	Nebula	$I_{N_1+N_2}/I_{H_\beta}$
IC 418	1.78	NGC 1535	15.4
NGC 2149	5.4	NGC 6572	15.6
NGC 6543	8.86	NGC 7009	16.1
NGC 6826	9.32	NGC 7662	18.2
NGC 6818	11.6	NGC 7027	20.2
NGC 6210	15.1		

Doubly ionized oxygen also produces the considerably weaker forbidden line $\lambda4363$ Å [O III], found just at the side of the line H_γ—the third line in the Balmer series of hydrogen ($\lambda = 4340$ Å).

Singly ionized oxygen gives the well-known violet doublet $\lambda3726$ Å [O II] and $\lambda3729$ Å (O II). These lines are so close to each other that they are not resolved in the common spectrographs, and blend into one line with equivalent wavelength $\lambda = 3727$ Å. Unlike the ratio I_{N_1}/I_{N_2}, the ratio I_{3729}/I_{3726} depends on the physical conditions in the nebula and, in particular, on its electron concentration.

The intensity of the $\lambda3727$ Å [O II] line, as compared with the sum of the intensities of the N_1 and N_2 lines, is considerably smaller in planetary than in diffuse nebulae. This is explained by the high temperature of the nuclei of planetary nebulae in comparison with the temperature of the stars that excite the light emission of diffuse nebulae. Due to the high temperature a larger fraction of oxygen atoms is found in the doubly ionized state, and hence the lines N_1 and N_2 are strengthened.

A feature always present in the spectra of planetary nebulae is the red doublet of singly ionized nitrogen—$\lambda 6548\,\text{Å}$ [N II] and $\lambda 6584\,\text{Å}$ [N II]. These lines fall to the left and to the right of H_α ($\lambda = 6563\,\text{Å}$). In spectrograms obtained with common spectrographs all three lines blend into one. They can be resolved only by high dispersion spectrographs (as a rule with diffraction gratings). The ratio I_{6584}/I_{6548} is also equal to three and does not depend on the physical conditions in the nebulae. The magnitude of the ratio $I_{6584+6548}/I_{H_\beta}$ is of the order of unity, but is different for different nebulae.

In the nebula NGC 7027, for example, it is 3.2. The influence of the blend on the lines $\lambda 6584\,\text{Å}$ [N II] and $\lambda 6548\,\text{Å}$ [N II] can be deduced from the observed ratio I_{H_α}/I_{H_β}, whose magnitude normally lies in the region between 3 and 8.

The lines of neutral helium $\lambda 5876\,\text{Å}$ (He I) and $\lambda 4471\,\text{Å}$ (He I) are observed in the so-called low-excitation nebulae (and also in diffuse nebulae). The line of ionized helium $\lambda 4686\,\text{Å}$ (He II) appears in the spectra of several planetary nebulae, which indicates a high degree of excitation in the nebula. Other signs of a high degree of excitation which appear in the spectra of several nebulae are the forbidden lines of doubly and, in particular, four times ionized neon—$\lambda 3869\,\text{Å}$ [Ne III] and $\lambda 3426\,\text{Å}$ [Ne V]. This latter line falls in the region of the continuum beyond the Balmer limit ($\lambda < 3646\,\text{Å}$).

As far as hydrogen is concerned, the first four lines of the Balmer series (H_α, H_β, H_γ, H_δ) have relatively high intensities and can be easily measured, but the rest of the lines, beginning with H_ε, blend with lines of other elements and their measurement with the usual observational techniques is practically impossible. Hydrogen also produces a continuum: weak in the visual region of the spectrum, considerably stronger beyond the Balmer limit.

The infra-red region of the spectrum of planetary nebulae is poorly known. There are no particularly bright lines in this region of the spectrum. The most intense ones are the forbidden doublet of singly ionized oxygen—$\lambda 7320\,\text{Å}$ [O II] and $\lambda 7330\,\text{Å}$ [O II], whose total intensity is comparable with the intensity of H_β, and the forbidden line of doubly ionized sulphur $\lambda 9069\,\text{Å}$ [S III].

The emission lines enumerated above are characteristic of all planetary nebulae. In addition to these, the spectra of a number of nebulae show many weak lines, the majority due to forbidden transitions, which belong to various ions of helium, chlorine, sulphur, silicon, magnesium, aluminum, potassium, calcium and manganese. The identification of these lines in the spectrograms and, moreover, their measurement, present great difficulties, and it is not possible for all nebulae. In this respect we should mention the classic paper of Aller, Bowen and Minkowski [20], dedicated to a detailed study of the spectrum of the nebula NGC 7027. They obtained 9 spectrograms of this nebula using diffraction spectrographs of different dispersions, from 6.7 to 56 Å/mm; the observations were made with the 100″ and 200″ telescopes. The exposure times varied from 100 minutes to 21 hours. The results of this work include the wavelengths and relative intensities of 263 spectral lines. Almost all lines were identified, with only a few exceptions.

5. The Spectra of the Nuclei

In contrast to the spectra of planetary nebulae, the spectra of their nuclei show great differences. To a certain extent this is understood: it could be partially explained by a large dispersion in absolute brightness of the nuclei. On the other hand, the nuclei of the majority of planetary nebulae are still unknown—the faintness of the nuclei strongly hinders their spectrographic study. One can assume, however, that the real dispersion in spectral types of the nuclei must be quite large.

At the present time the spectra of forty nebulae are more or less well known. Most of the credit for this knowledge belongs to Aller, who obtained the largest number of spectrograms of nuclei of planetary nebulae and studied them thoroughly. Among the earlier investigations one should mention the work of Wright at Lick Observatory. References [21–32] contain data on the spectral types, characteristics and behaviour of the spectral lines—such as their intensity, width, and so forth—for the nuclei of a series of planetary nebulae.

According to their spectra, the nuclei of planetary nebulae can be classified in the following types:

1. Wolf-Rayet type (NGC 7026, 6803).

2. Of type; emission and absorption lines appear simultaneously in the spectra (NGC 2392, IC 4593).

3. O type; absorption lines appear in the spectra, but emission lines are absent (IC 2149, NGC 246).

4. Continuum type (C); the spectrum consists entirely of continuous radiation, without any traces of absorption or emission lines (NGC 1535, 6807, 7009).

It is well known that the classical Wolf-Rayet stars are divided into two subclasses. In one of them the lines (bands) of carbon predominate, while the nitrogen lines are almost absent. This subclass is represented with the symbol WC. In the stars of the second subclass, on the contrary, the nitrogen lines predominate and the carbon lines are almost absent. They are designated by the symbol WN.

It seems that there exists a certain similarity between the spectra of planetary nebulae and Wolf-Rayet (WR) stars, thus, for example, the nuclei of the nebulae NGC 40, IC 1747, NGC 7026 and others have spectra of type WC, while the nuclei of the nebulae J 320, anon. $22^h 29^m$ and others belong to the type WN. We should add that the majority of the nuclei are of type WC.

In several cases nitrogen and carbon occur simultaneously: in this case they have a certain combination-type nucleus—WC + WN. At the present time only four planetary nebulae are known whose nuclei belong to this combination-type—NGC 6543, 6572, 6826 and IC 4997.

The nuclei that belong to the stellar type Of offer special interest. These stars have been well studied by Oke [24], who determined the variation in intensity and shape of the emission lines in their spectra with time. In some cases it can amount to the almost complete disappearance of several emission lines and their subsequent reappearance. If it is true that the nuclei of planetary nebulae can belong to the type Of, then we can conclude that they are nonstationary objects and, in particular, that they belong to the category of young, recently formed stars.

The nuclei of type O (absorption) is encountered frequently in very extended planetary nebulae, *i.e.*, in those which are in the later phases of their life.

Of special interest are the nuclei with spectra of type C, *i.e.* consisting of just a continuous spectrum, without any lines. This type almost never appears among the common stars.* Obviously, the disappearance of the lines (absorption or emission) from the spectra of several nuclei is due to a very high radiation temperature, and also to the disappearance of the gaseous envelope around them. The distribution, in per cent, of the nuclei of planetary nebulae among the different types is as follows. The largest fraction, about 50%, corresponds to nuclei with spectra of the WR type. About 25% of the nuclei have spectra of type C. Nuclei of types Of and O are found approximately in the same numbers ($\sim 10\%$).

In isolated cases nuclei have been found that belong, although not with complete certainty, to later type spectra. Thus, for example, the nucleus of the nebula NGC 1514 corresponds, according to McLaughlin [33], to the stellar class B9; the nucleus of the nebula NGC 246 is a double system composed of a hot star 07 and a late type dwarf, class F or even G ([8], page 205).

The classical WR stars are, as a rule spectroscopic binaries. Up to the present time double systems have not been discovered among nuclei with WR spectra (because of their faintness). General brightness variability of the nuclei has not been discovered either, although it is suspected to occur in several nuclei.

When talking about nuclei that belong to types WR, Of or O one must keep in mind that all that is meant is that there exists a similarity between only the *spectral* characteristics of these stars and the nuclei of planetary nebulae. Actually, there exists a huge difference in absolute magnitude between these two types of objects. We have already referred to the absolute magnitudes of the nuclei of planetary nebulae. The mean absolute magnitude of classical WR stars is of the order of -5^m and in some cases it reaches $-9^m(!)$

---

* Continuous spectra without lines are observed in several white dwarfs.

However it may be, it is apparent that the difference in stellar magnitude between the nuclei of planetary nebulae and the common WR, Of and O stars of 3 to 4^m is real. It is important to point out that the nuclei of planetary nebulae, although they may resemble dwarfs, are, according to the physical processes that occur in their atmospheres and possibly in their interiors, in general identical with common hot giants.

We shall go deeper into the question of the nature of the nuclei in the last chapter, when discussing the formation of planetary nebulae.

6. Stratification of Radiation

Interesting results can be derived from the study of spectrograms of planetary nebulae obtained with objective prisms or with slitless spectrographs (spectrographs in which the slit jaws are very far apart). Such spectrograms consist of a series of *monochromatic* images of the nebula in the frequencies of the different lines. The photometric study of these images allows one to reach certain conclusions about the distribution of radiation of the nebula in the emission lines of different atoms and ions.

By means of these measurements it has been established that the sizes of the monochromatic images of the nebula are different for different lines and, on the average, that they decrease with increasing ionization potential of the atom that produces a given line. This is the phenonemon of stratification of radiation, first discovered by Wright [23]. In Table 1-4 are shown some interesting results for several nebulae, according to the observations of Wright and Wilson [34].

TABLE 1-4

Dimensions of Monochromatic Images of Several Planetary Nebulae

Line	H_β	$N_1 + N_2$	$\lambda3868$ Å [Ne III]	$\lambda4686$ Å He II	$\lambda3426$ Å [Ne V]
Ionization potential of ion in e.v.	13.5	35.0	40.9	54.2	96
NGC 2392	13"	14".2	13".6	12".8	11".8
NGC 6816	17.6	19.7	19.2	14.4	10.8
NGC 6886	3.9	5	—	3.4	3.1
NGC 7662	13.9	14.1	14.7	12.1	9.1
NGC 2440	5.7	6.4	6.5	5.3	3.1
NGC 7026	6.0	6.3	6.1	3.7	—
NGC 2165	5.3	5.7	5.9	4.4	3.4

As can be seen from the data in this Table, the above-mentioned phenomenon is particularly strong in the lines $\lambda4686$Å He II and $\lambda4326$Å [Ne V]. In several cases the monochromatic images in the line $\lambda3426$Å [Ne V] are one and a half to two times smaller than in the lines of [O III]. A different situation occurs in the case of the lines $N_1 + N_2$; the size of the monochromatic image of the nebulae in the hydrogen lines, although their ionization potentials differ by a factor of almost three.

Together with these facts it has also been established that the shape of the nebula differs from one wavelength to another. Thus, for example, the nebula IC 2165 has a well defined bipolar structure in the hydrogen lines and also in the lines $\lambda6548$Å [N II] and $\lambda6584$Å [N II]. This structure is very weak in the line $\lambda3869$Å [Ne IV] and it entirely disappears in $\lambda3426$Å [Ne V]. Summarizing: in the hydrogen lines the nebula is bipolar, in the $\lambda3869$Å [Ne III] line it is ring-like, and in the $\lambda3426$Å [Ne V] line its shape is that of a true disk whose diameter is approximately equal to that of the inner, less dense, region of the nebula.

A similar situation occurs with the nebula NGC 7662: it is both a bipolar and a double-envelope nebula in the lines of hydrogen and doubly ionized oxygen, ring-like in the lines 4686 He II and 3896 [Ne III], and quite disk-like in the 3426 [Ne V] line. In the case of the ring-like nebula NGC 6720 (Lyra) the images in the lines N_1, N_2 are ring-like, but the image in the 4686 He II line is considerably smaller (c.f. Table 3-2) and its shape is that of a uniformly illuminated disk, filling the inner region of the ring. Beautiful monochromatic images of these and other planetary nebulae have been obtained by Wilson and can be seen in the above mentioned work [34].

The stratification of radiation can be explained in two ways: either the different chemical elements have different distributions inside the nebula, or the conditions controlling their emission depend on their distances from the exciting star. If the first explanation were true, the sizes of all the images that belong to different ions of the same element should be identical. This, as follows from Table 1-4, does not occur. Therefore, one must accept the second possibility. As will be shown in the next chapter, the stratification of radiation in the nebulae is entirely explained by the different conditions under which the atoms and ions radiate.

7. Expansion of Planetary Nebulae

The internal motions in the nebulae can be studied only by measuring the radial velocities and analyzing the shapes of different spectral lines with slit spectrographs. The angular displacement of the edges of the nebulae, or of different details in the direction perpendicular to the line of sight, is too small to be detected. Since the majority of the bright nebulae have very small angular sizes ($10''-20''$) and, furthermore, the magnitude of the expansion velocity is relatively small, in order to obtain spectrograms suitable to determine the expansion of the nebula, it is necessary to work with telescopes of very long focal length (at the Coudé focus) and with spectrographs of very high dispersion. This explains the relative scarcity of data concerning the internal motions and the expansion of planetary nebulae.

The first investigations of the internal motions and expansion of nebulae were carried out by Wright [23] and Zanstra [35]. But the basic results on this subject were obtained by Wilson [34], working with the 100″ and 200″ telescopes. He successfully applied a method designed to obtain spectrograms of different parts of the disk of the nebula simultaneously by means of "many slits" ("multislit"), which allow him to gather complete information about the kinematics of all parts of the nebula.

The basic results on the internal kinematics of planetary nebulae are the following. The planetary nebulae are expanding gaseous envelopes whose expansion velocities are relatively small, of the order of 20 km/sec. This is shown by the shape of the spectral lines; they appear split into two components of approximately the same intensity, placed symmetrically with respect to the central frequency (Fig. 1-2).

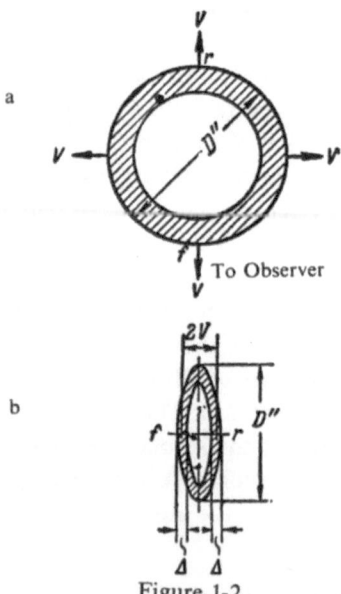

Figure 1-2

Expanding planetary nebula (*a*) and the geometrical form of its spectral lines (*b*).

The distance between the components is largest in the center of the image of the nebula and decreases with increasing distance from the center. At the edges of the nebula the two components merge into one. The general outward shape of the spectral line resembles an ellipse whose major axis corresponds to the radius of the nebula ($D''/2$) and the minor axis to the expansion velocity [$v = (\Delta\lambda/\lambda)c$]. The violet component of the spectral line, marked f in Fig. 1-2(b), corresponds, obviously, to radiation coming from the region of the nebula closer to us (front), and the red component—r—to that from the back part of the nebula. It must be noticed that this method for the determination of the expansion velocity of the nebula does not require the knowledge of its radial velocity of motion with respect to the observer.

The fact that both components of the spectral line have almost the same brightness shows that the radiation from the back part of the nebula can pass unimpeded through the whole nebula without suffering absorption. From this fact we can conclude that the nebulae are completely transparent at the frequencies of visual light. The several exceptions to this rule can be explained as asymmetry in the nebulae with respect to the meridian plane perpendicular to the line of sight.

One should add that the spectral lines would have exactly the same structure if the nebulae were contracting. However, this possibility must be rejected on dynamical and cosmogonical grounds.

In those cases when different layers of hydrogen nebulosity expand with different velocities (*i.e.* the expansion occurs with a velocity gradient) the components of the (split) spectral lines should be broadened accordingly. But in general these components will be broadened by the thermal motions of the atoms, corresponding to a nebular temperature of the order of $10,000°K$. The observations confirm that the components of the spectral lines are indeed broadened. However, at the present time it is difficult to establish how much of the broadening is due to thermal motions of the atoms and how much to a gradient in the expansion velocity. This question, of course, can be settled from an analysis of the structure of the lines near the center and at the edges of the nebular image. The difference in the amount of broadening of the components depends, obviously, on the effect of the gradient in the expansion velocity. But in order to obtain trustworthy results the observations should be made with very high dispersion spectrographs.

With the exception of the hydrogen and, to some extent, the helium lines, the widths of the components of the spectral lines [Fig. 1-2 (b)] which belong to other elements are almost identical and correspond to a velocity of 10 km/sec. Since the expansion velocity is two to three times greater and almost identical for all nebulae, it can be assumed that the *turbulent* velocities in planetary nebulae are small compared to the expansion velocities.

This conclusion is of particular interest. It happens that in several nebulae the brightness distribution over the disk is far from uniform. In different cases the fluctuations in brightness distribution reach very large values. To the intensity fluctuations must correspond, although to a lesser degree, fluctuations in electron (ion) concentration in the nebula. If we assume that the fluctuations in the distribution of electron density are tied to turbulent motions (an assumption that is, generally speaking, quite attractive), then we would expect the presence of fluctuations in the velocities of the internal motions. However, as was pointed out before, this is not observed. Therefore we must look for some other reason to explain the origin of the fluctuations in electron concentration (see Section 10, Chapter IX). One should only add that it is difficult to imagine from a hydrodynamical point of view, that a gas sphere of such dimensions as those of a planetary nebula can expand without showing turbulent motions. Obviously, there must exist forces that maintain a "nonturbulent" condition during the expansion of the nebula.

The following fact, established by Wilson, is of special interest. When he tried to determine the magnitude of the expansion of the nebulae, not only by means of the hydrogen lines, but using the lines of other atoms as well, he obtained different results. In this way a certain regularity was discovered, namely, that the magnitude of the splitting, which is equal to twice the expansion velocity, on the average decreases with increasing ionization potential of the ion or atom which produces a given line. This is observed in both high-excitation and low-excitation nebulae. Table 1-5 gives data on the value of twice the expansion velocity, determined from lines of different atoms, for several planetary nebulae.

TABLE 1-5

Twice the Expansion Velocity (in km/sec) of Several Planetary Nebulae,
Determined from Lines of Different Elements

Ion and ionization potentials, e.v.	H 13.5	[O II] 13.6	[O III] 35.0	[Ne III] 40.9	He II 54.2	[Ne V] 96
NGC 3242	40.8	—	39.6	39.0	33.6	—
NGC 6818	55.5	60.2	56.2	58.0	42.4	32.6
NGC 7027	42.4	47.2	40.9	44.7	39.6	38.2
NGC 7662	52	58	53	52	46	39

The explanation of this phenomenon can be found if we convert the relationship between the magnitude of the splitting and the ionization potential into a relationship between the magnitude of the splitting and the sizes of the monochromatic image of the nebula in a given line. Indeed, as was pointed out in the previous section, the size of the monochromatic image of the nebula decreases with increasing ionization potential. Comparing these two correlations, one reaches the conclusion that there exists a gradient in the expansion velocity inside the nebula, i.e. that the outer layers expand appreciably faster than the inner ones. Obviously, there has to be some force responsible for the origin of the difference in expansion velocities.

The slower expansion of the inner layers of the nebula relative to the expansion velocity of its outer layers means that with increasing size of the nebula its thickness

TABLE 1-6

Twice the Expansion Velocity $2v_0$ of Planetary Nebulae in km/sec (Wilson)

Nebula	$2v_0$, km/sec	Nebula	$2v_0$, km/sec	Nebula	$2v_0$, km/sec
IC 315	29.0	NGC 3242	39.8	CD + 30° 3639	52.8
NGC 1535	39.5	NGC 6210	42.2	NGC 6818	56.5
J 320	34.6	IC 4634	29.6	NGC 6884	44.9
IC 418	42.3	IC 6543	24*)	NGC 6886	37.1
IC 2149	40.3	NGC 6567	35.8	NGC 7009	41.2
IC 2165	40.0	NGC 6572	31.5	NGC 7026	81.7
J 900	35.4	NGC 6720	60	NGC 7027	42.7
NGC 2392	107.3	NGC 6741	41.4	NGC 7662	52.2
NGC 2440	44.6	NGC 6803	28.0		

* From the measurements of Zanstra [35].

also increases. As will be shown in Section 8, Chapter VI, the increase in thickness of the nebula with the increase of its size always proceeds more slowly than would be expected in the case of free expansion.

Table 1-6 gives the magnitude of twice the expansion velocity for about thirty planetary nebulae according to Wilson's [34] measurements. The minimum expansion velocity occurs for the nebula NGC 6803 (14 km/sec), the maximum for the nebula NGC 2392 (53.6 km/sec). No correlation has been found between the magnitude of the expansion velocity and the nebular type or the degree of excitation.

The fact that the dispersion in the expansion velocity of planetary nebulae is small cannot be assumed to be accidental. Undoubtedly it is directly related to the origin and formation of the nebulae.

We shall return in Chapter VI to the question of the expansion and, in particular, the deceleration of planetary nebulae.

8. Apparent Distribution of Planetary Nebulae

A total of 593 planetary nebulae are known up to the present time [209]. Almost two-thirds of them are stellar.* Figure 1-3 shows the apparent distribution of all known planetary nebulae in galactic coordinates. Their concentration towards the galactic plane is obvious, although not as marked as in the case of the hot giants

* In addition, Henize discovered 137 emission objects in the Southern part of the sky, the majority of which are identified with planetary nebulae [8].

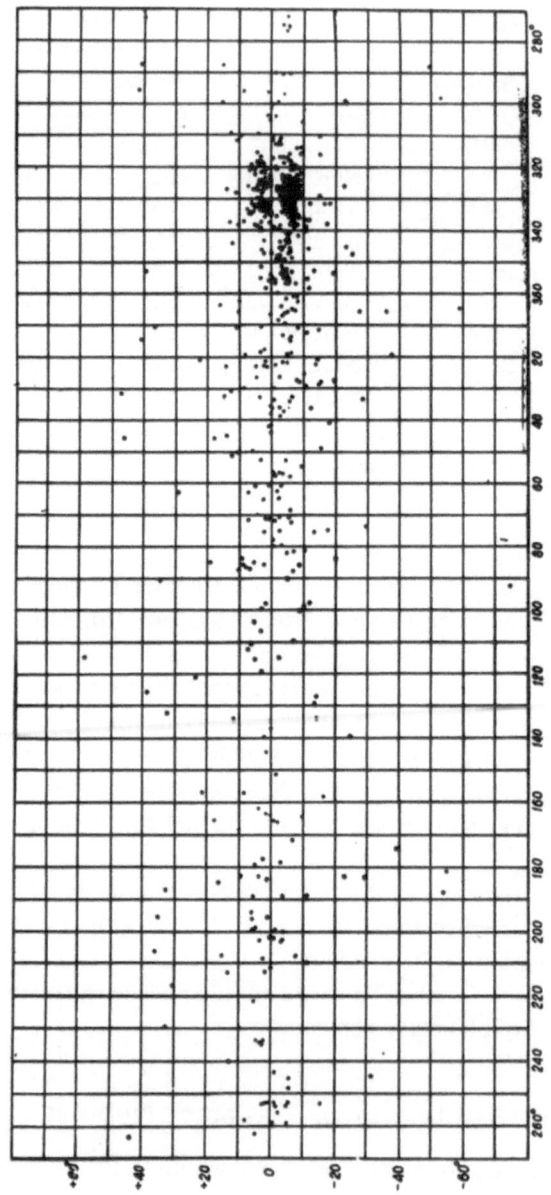

Figure 1-3

Apparent distribution of planetary nebulae.

of classes O–B. The galactic latitude of isolated planetary nebulae reaches as high as 60° and even higher. The strong concentration of planetary nebulae in the direction of the center of the galaxy ($l \sim 325°$) is quite apparent. More than 150 nebulae, *i.e.* about one seventh of all nebulae, are concentrated in the region of the sky between $l = 320°$ and $l = 340°$ and between $b = 0°$ and $b = -10°$.

Planetary nebulae with large apparent dimensions are distributed more or less uniformly over the sky, while small and stellar nebulae are concentrated near the galactic plane. The true galactic concentration of planetary nebulae must be, obviously, greater than the observed since, because of the interstellar absorption of light, weak nebulae near the plane of the galaxy cannot be discovered, while at high galactic latitudes they can be discovered easily.

The apparent distribution and the distances of planetary nebulae show that they form a system in the galaxy which is intermediate between the flat and the spherical subsystems, and more similar to the system of RR Lyrae stars.

The sharpness and narrowness of the emission lines in the spectra of planetary nebulae permit the measurement of their radial velocities with high accuracy. Campbell and Moore [5] made such measurement for 120 planetary nebulae. The radial velocities were quite high. Among these planetary nebulae eight had a velocity greater than 100 km/sec.

The maximum positive velocity observed is for the nebula NGC 6644 (+ 193 km/sec) and the maximum negative velocity is that of the nebula IC 4732 (− 145 km/sec).

Several investigators have tried to use the radial velocities of planetary nebulae for the study of galactic rotation [18], [37], [38], and the solar motion [39]. The velocity of the solar motion relative to the system of planetary nebulae was found to be 31 km/sec, which agrees with the velocity of the solar motion relative to class O stars (for which it is 30 km/sec), but it is considerably greater than the velocity of the Sun relative to stars of classes B and A.

9. Planetary Nebulae in Other Galaxies

The observation of planetary nebulae in other galaxies can provide very valuable information. Apart from purely statistical data, in this case we can also obtain some idea about the limiting values of the absolute brightness of planetary nebulae from their apparent brightness and the distance to the galaxy in question.

The first observation of this kind occurred in the nearest galaxy—the Small Magellanic Cloud. In 1933 Cannon drew up a list of emission objects discovered in the Small Magellanic Cloud [40]. This list contained thirty emission objects, two of which were O class stars. Henize and Miller working with a 15° objective prism and a 10.5″ telescope, discovered 120 new emission objects, of which 32 were identified with nebulae [41]. However, their observational method did not enable them to single out the planetary nebulae among them. A catalogue of emission objects in the Small Magellanic Cloud was compiled later [42]. Of the 152 objects in this catalog, about 40% have been identified with nebulae.

More conclusive results were obtained by Lindsey [43]. The presence of the N_1 and N_2 lines in the spectrum of nebular objects was used as the observational criterion. Of the 100 stellar-like H_α emission objects which he discovered, 33 also showed the N_1 and N_2 lines. The latter were then distributed in two groups. The members of the first group (13 objects) were hot stars, associated with gaseous nebulae and causing their luminosity; the mean absolute magnitude of this group was $- 4^m.9 \pm 0^m.6$, and the mean color index (photographic magnitude minus photovisual magnitude) was $- 0^m.47 \pm 0^m.19$. The members of the second group (17 objects) were planetary nebulae; the mean color index of the objects in this group was $+ 0^m.77 \pm 0^m.29$. (The data appear in Section 3 of this chapter). The absolute photographic magnitude of these objects extends over the interval from $- 5^m.2$ to $- 2^m.5$; the mean magnitude is $- 3^m.7 \pm 0^m.6$. In addition, Lindsey believes that eleven more emission objects may be planetary nebulae.

The limiting apparent photographic magnitude reached by Lindsey was $16^m.6$. Koelbloed, working with a more powerful telescope, pushed this limit to $18^m.1$ [44]. Working in seven areas of the Small Magellanic Cloud he discovered sixteen objects with the N_1 and N_2 lines. The absolute photographic magnitude of these ob-

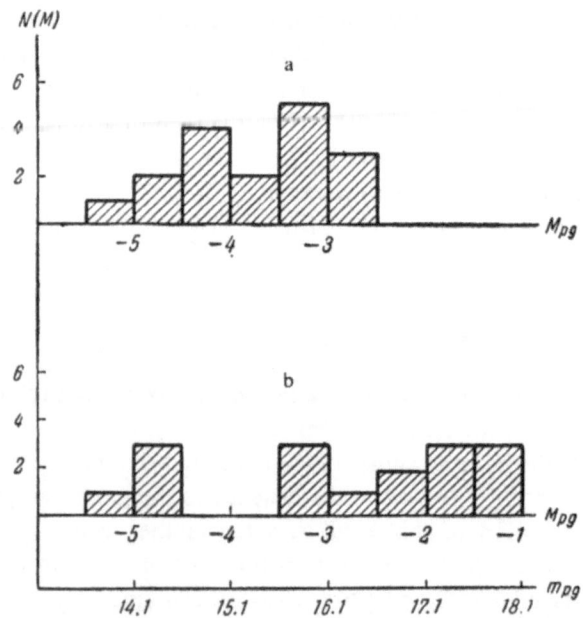

Figure 1-4

Distribution of absolute stellar magnitudes of the nucleus of planetary nebulae in the Small Magellanic Cloud (a—according to Lindsey, b—according to Koelbloed.).

jects, which can be assumed to be planetary nebulae (they are all stellar, except one whose diameter is 7″) lies in the interval from $-5^m.5$ to $-1^m.0$. Figure 1-4 presents a histogram of the distribution in absolute magnitude of the planetary nebulae discovered by Lindsey and Koelbloed.

For a long time, the efforts made to discover planetary nebulae in the Large Magellanic Cloud did not give positive results. According to Westerland's data [210] the relative number of planetary nebulae (per unit mass) in the Large Magellanic Cloud is considerably smaller than in the Small Magellanic Cloud or in our Galaxy.

Baade [45], with the 200″ telescope, discovered five planetary nebulae in a region of the Andromeda nebula (M 31) located at a distance of 96′ south of the center. The photographic magnitude of these nebulae lies in the range $21^m.7$ to $22^m.2$, which corresponds to a mean absolute photographic magnitude $\overline{M}_{pg} = -2^m.2$. But from these data it is not yet possible to draw any conclusions about the absolute brightness and distribution of the giant nebulae in this galaxy.

Planetary nebulae are apparently present in other galaxies too. A report concerning the discovery of a planetary nebula at the edge of the globular cluster in M 15, and which could be a member of it [46], has been confirmed.

10. Spectral and Morphological Classification of Nebulae

Classification of the spectra: Several classification systems have been proposed for the spectra of planetary nebulae [2, 47]. The most successful is the system of Page [48] and Aller ([8], page 65). We shall describe the basis of their classification and some modifications to it.

To choose the necessary criteria for classification one should point out that planetary nebulae fall into two basic groups: nebulae with a *low degree of excitation* and nebulae with a *high degree of excitation*. Obviously it would be difficult to find one general criterion for both groups. For the low-excitation nebulae a satisfactory criterion can be the intensity ratio $I_{N_1+N_2}/I_{H_\beta}$, and for the high-excitation nebulae ratio I_{4686}/I_{H_β} or the ratio $I_{3426[Ne\ V]}/I_{H_\beta}$.

The spectra are classified on a scale from 1 to 10. In class 1 are the nebulae in whose spectra the lines of [O III] are absent. In classes 2–5 are included the nebulae that do not show the He II line at $\lambda\,4686$ Å, and the ratio $I_{N_1+N_2}/I_{H_i}$ increases as we go from class 2 to class 5. Further classification is based mainly on the magnitude of the ratio I_{4686}/I_{H_β}. For the low-excitation nebulae a reliable criterion can be the ratio $I_{3727[O\ II]}/I_{H_\beta}$. In Table 1-7 we give examples of nebulae of classes 2–10.

Classification of shapes: The classification of the shapes of planetary nebulae is a considerably more difficult matter. In order to work out a classification system of nebular shapes, and to distribute the individual nebulae among the different classes, we depend upon the observational material (photographs) obtained with

TABLE 1-7

Classification of the Spectra of Planetary Nebulae

Nebula	Anon. 18^h15^m	IC 418	IC 2149	IC 4634	NGC 7026	J 900	NGC 6309	IC 2165	Anon. 21^h31^m
$I_{N_1+N_2}/I_{H_\beta}$	0.21	1.9	5.5	10.6	12.4	16.6	14.1	18.1	10.4
I_{3727}/I_{H_β}	32	3.1	0.20	0.03	0.04	0.03	0.03	0.02	0.04
I_{4686}/I_{H_β}	—	—	—	—	0.13	0.47	0.77	0.6	0.9
I_{3426}/I_{H_β}	—	—	—	—	—	0.28	0.5	0.8	2.4
Class	2	3	4	5	6	7	8	9	10

the existing observational means. But, as experience shows, each time that a larger telescope comes into use, it uncovers new, and frequently more important, traits in the structure or shape of a given nebula, and, as a consequence of this fact, our idea about it changes.

The success of any system of classification of nebular forms depends to a large degree on the state of the theory. Theory shows us which are the suitable criteria necessary for a correct classification, gives the proper weight to this or that particular nebular shape or structure and their different role in the dynamics and development of the nebulae.

Observational facilities are being constantly improved, theory changes and therefore any system of classification of the shapes of planetary nebulae (and we already have several [49–51]) cannot be "fixed". They must perfect and complete themselves, retaining from the existing systems the more valuable aspects and discarding the superfluous.

The classification of the shapes of planetary nebulae should rest mainly upon the most characteristic traits of their shape and structure. Since planetary nebulae, from a dynamical point of view, are nonstationary systems, it is desirable that this classification have at the same time an *evolutionary* meaning, *i.e.* that it reflect the stages of development.

As we shall see later, differences in the nature of the effective forces produce completely different nebular shapes and structures. By considering the influence of these forces as criteria for a morphological classification, we give to it a *physical* meaning, showing the dominating role of this or that force at a given stage of development of the nebulae.

Starting from these conditions we can complement and modify the classification previously given by the author [51] in the following way. First of all we should retain the class of *planet-like* nebulae, *i.e.* those which consist of only *one envelope*, with a brightness distribution more or less uniform over the disk. Such nebulae often have small dimensions (both apparent and absolute). We shall designate this class by the symbol I.

By means of the symbol II we shall designate the *double-envelope* nebulae (regardless of their relative size or other pecularities). These nebulae seem to be the evolutionary sequel of the planetoid nebulae.

The symbol III designates the *ring* nebulae, in which no second envelope is seen, but have an evident ring shape. Thus, the symbols I, II and III fully characterize certain nebular shapes and structures, and together they acquire an evolutionary meaning: the development of the nebulae goes in the direction I → II → III (planet-like → double-envelope → ring).

Another element that should be included in the classification of forms of planetary nebulae is the *bipolar* character of their structure. Bipolarity can be present in the three types of nebulae mentioned above. On the other hand, there exist different degrees of development of bipolarity, due to differences in the influence of the magnetic and the inertial forces. Therefore, the class of bipolar nebulae are designated by the letter B, to which we add the index a, b, c, d, which indicates the degree of development of bipolarity. Thus, in the nebulae of type Ba bipolarity shows only as an increase in brightness at the ends of its minor axis (NGC 6720), in type Bb two "caps" are well developed (NGC 3587), in type Bc the "caps" are quite clearly separated (A 66). The extreme case of bipolarity (and also of flatness) are the "rectangular" nebulae, and also the "hour-glass" type nebulae, which are called Bd (IC 4406).

By the symbol Sp we shall designate, as before, the class of *spiral* nebulae, and by the symbol Sz the class of Z-like nebulae, which are a variety of the spiral nebulae.

As far as the effect of flatness (stretching out) is concerned, *i.e.* a deviation from sphericity in the shape of the nebulae, no special symbol has been introduced to designate it, since *if a nebula is bipolar, then it is also flattened*.

To conclude, the symbol D stands for a transition type nebula (from planetary to diffuse) which corresponds to the later stages in the development of a planetary nebula. There are few of them, and this is accounted for by the observational conditions. In fact, the real number of nebulae of this type should be large, since the expansion velocity of the nebulae decrease as their dimensions increase [51], and hence the number of large nebulae should be high.

We shall retain the class of stellar planetary nebulae. Here we lump together all nebulae whose *apparent* dimensions are so small that they cannot be distinguished from stars from their outward appearance only.

In Fig. 1-5 the different types of planetary nebulae are depicted schematically, with their designations.

Thus, according to the proposed classification all planetary nebulae can be distributed among the following basic types (a typical example of each type is given

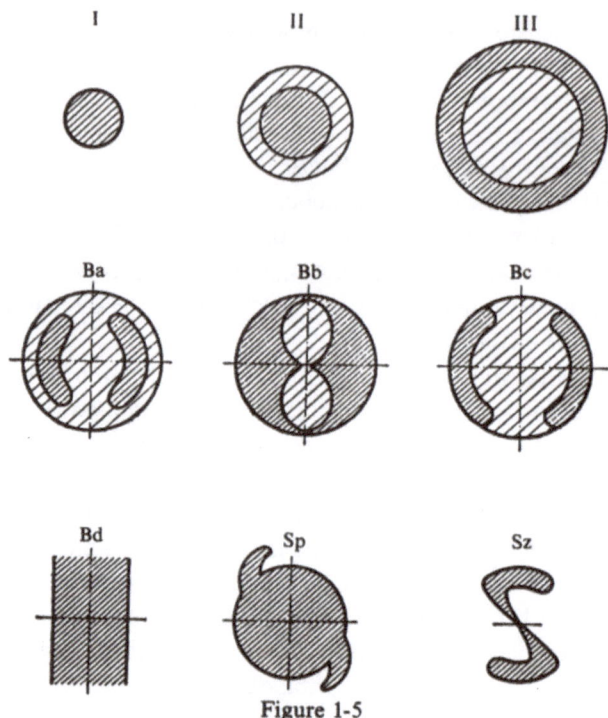

Figure 1-5

Schematic image forms of planetary nebulae of various classes.

in parentheses):

I   —planet-like (NGC 6803)
II   —double-envelope (NGC 1534, 7662)
III   —ring (NGC 2418, A 72)
Ba   —bipolar of first type (NGC 6720, 7293)
Bb   —bipolar of second type (NGC 3587, A 70)
Bc   —bipolar of third type (A 19, A 66)
Bd   —"rectangular" (IC 4406, anon. 07^h 07^m)
Sp   —spiral (NGC 4361, 6210)
Sz   —Z-like (NGC 6778)
S   —stellar (IC 4997)
D   —diffuse (A 16, A 50)

It is important to remember that in the majority of cases nebulae can belong to several types at the same time. In such cases one must resort to a combination system of classification. Thus, for example, the nebulae NGC 4361 is spiral (Sp), has a second envelope (II), and also shows an increase in brightness at the ends of the minor axis (Ba). Therefore, this nebula belongs to the class SpIIBa. The symbols are ordered according to the degree in which each aspect shows itself—the more important and stronger trait of the nebula is in the first place, the less important ones later. Below we give some examples of combination systems of classification:

Nebula	Class
NGC 7026	BcSp
NGC 7293	BaII
Anon. $16^h 10^m.5$	BeIISp
A 65	SpIII
NGC 6543	SpII
NGC 7662	IIBc

11. Designations and Catalogues

Planetary nebulae are designated by their number in Dreyer's New General Catalogue of Clusters and Nebulae (abbreviated NGC). This catalogue was published in 1887. Later on, it was followed by two supplements (in 1894 and 1908): First Index Catalogue and Second Index Catalogue. They are designated IC I and IC II. If, for example, the nebula appears in the second supplement it is called thus: IC II 2022, or NGC II 2022; if in the first, IC I 351 or NGC I 351.

From an examination of the Palomar Sky Atlas, Abell discovered on the order of sixty new planetary nebulae, mostly with small surface brightness and large apparent dimensions. His list appears in PASP 67, 258, 1955. Later on, these nebulae were designated by the letter "A"; thus, for example, A 17 means the planetary nebula No. 17 in Abell's list.

In recent years many planetary nebula with very small apparent dimensions (stellar) were discovered by means of objective prisms. Minkowski, for example, discovered in this way about 80 planetary nebulae. These and other nebulae still have no designation, *i.e.* remain anonymous. Up to now they are referred to by their coordinates (often only the right ascension α). For example, anon. $17^h 45^m.8$ means the planetary nebulae with coordinates $\alpha = 17^h 45^m.8$ and $\delta = -34°38'$ (for the epoch 1900).

A list of about 600 planetary nebulae (NGC and anonymous) with their equatorial and galactic coordinates is given by B.A. Vorontsov-Vel'yaminov in *Reports of the Shternberg Astronomical Institute*, No. 118, 1962. Here he also gathers probably the most complete data on the apparent magnitudes of the nebulae and their nuclei, surface brightness, radial velocity, and so forth. Catalogues with distances and dimensions of planetary nebulae in different systems are given in [16, 17, 18, 36, 82].

[Ed. note] The most recent and complete catalogue is that of L. Perek and L. Kohoutek, *Catalogue of Galactic Planetary Nebulae*, (Prague, 1967).

Chapter II

The Origin of the Emission Lines

1. The Nature of Nebular Emission

The spectra of the nebulae are completely different from the spectra of their central stars. In particular, the energy emitted by a nebula in the different lines is considerably greater than the energy of the corresponding regions in the spectrum of the star. Nevertheless, the luminosity of a nebula is entirely due to the radiation of its central star, or nucleus. It follows, then, that the light-emission mechanism of the nebulae is not at all a simple, or even resonant, reflection of the light of the nuclei.

Furthermore, the integrated brightness of a nebula (without its nucleus) is ten to many hundred times greater than the brightness of the central star. Therefore we have to accept that the nebulae are made luminous by energy radiated by the nuclei in invisible regions of the spectrum. Since the nuclei of planetary nebulae are high temperature stars, and since most of their energy is emitted in the far ultraviolet region of the spectrum, one can assume that the nebulae derive their energy from this ultraviolet radiation of the stars. The problem is to elucidate by what means the nebula "transforms" this energy into visible light.

Before answering this question let us see what properties the light of the nuclei will have after travelling a distance many times greater than the radius of the star.

Let us assume that the nucleus emits as a blackbody at the temperature T_*. The radiation density ρ_v^* per unit frequency interval in thermodynamic equilibrium at temperature T_* is given by Planck's law

$$\rho_v^* = \frac{8\pi h v^3}{c^3} \frac{1}{e^{(hv/kT_*)} - 1}. \tag{1}$$

At the surface of the star the radiation density should be half the above value, since emission occurs from only one half of the sphere. As the distance to the star increases this quantity decreases, and at a distance r the radiation density ρ_v will be given by the equation

$$\rho_v = W_v \rho_v^*, \tag{2}$$

where W_v is the so-called dilution coefficient. The physical meaning of W_v is obvious from (2): it represents the ratio of the radiation density at the point of observation to the radiation density at the surface layers of the star. In general, W_v depends also on the opacity of the medium and, hence, on the frequency v, and has the following form:

$$W_v = \frac{1}{2}\left[1 - \sqrt{1 - \left(\frac{r_*}{r}\right)^2}\right]e^{-\tau_v},\tag{3}$$

where r_* is the radius of the star, and τ_v the optical depth at the frequency v. In general, at the inner edge of the nebula one takes $\tau_v \ll 1$ and then

$$W = \frac{1}{2}\left[1 - \sqrt{1 - \left(\frac{r_*}{r}\right)^2}\right]\tag{4}$$

In this case the dilution coefficient takes on a geometrical meaning: it represents the ratio of the solid angle subtended by the central star at the point of observation to 4π. Then, when $r_*/r \ll 1$, we have, instead of (4):

$$W = \frac{1}{4}\left(\frac{r_*}{r}\right)^2\tag{5}$$

In planetary nebulae the ratio r_*/r is of the order of 10^{-6} to 10^{-7}. Therefore, the radiation density given by (2) shows, in the case of planetary nebulae, a weakening by a factor of 10^{13} in comparison with the radiation density at the surface of the star. In spite of this the composition of the spectrum does not change, *i.e.* as it was pointed out before, the emission maximum falls in the far ultraviolet region of the spectrum. This would not occur if the radiation in the nebula were in equilibrium. Actually, if we call T_1 the temperature that would obtain in the nebula in the presence of thermodynamic equilibrium, the integrated radiation density can be written: $\rho = aT_1^4$, where a is Stefan's constant. On the other hand, for the integrated radiation density, we have from equation (2): $\rho = WaT_*^4$. Equating these two values we find: $T_1 = T_*W^{1/4}$ which leads to a value of T_1 of the order of a few tens of degrees (by taking T_* of the order of tens of thousands of degrees). This means that the maximum of the energy distribution over the spectrum would be displaced to the very far infrared which, as we have seen, does not occur.

Thus, thermodynamic equilibrium does not obtain in the nebulae. From the incompatibility shown above between the spectral distribution and the density of the radiation coming from the nucleus of the nebula, it follows that interaction between radiation and matter must lead to a redistribution in frequency of the radiation in the direction of a more probable distribution, *i.e.* to a redistribution of energy over the spectrum that strengthens the long-wavelength region and weakens the short-wavelength region. Quantitatively this process can be expressed by means of Rosseland's theorem.

2. Rosseland's Theorem

The essence of Rosseland's theorem is the following: under nebular conditions, *i.e.* in the presence of dilute radiation, short-wavelength radiation energy emitted by the star is transformed into long-wavelength radiation energy.

Let us consider an atom having three energy levels—1, 2, and 3—with energies $\varepsilon_1 < \varepsilon_2 < \varepsilon_3$. As the result of the absorption of a light quantum, we can have a simple transition, related to resonant scattering, of the type $1 \to 2 \to 1$, or more complex cyclic processes. The most important among these will be the mutually opposite processes of type

$$1 \to 2 \to 3 \to 1 \quad \text{and} \quad 1 \to 3 \to 2 \to 1.$$

In the first of these processes a sort of synthesis occurs when, as a result of the absorption of two quanta of lower frequency ν_{12} and ν_{23} a quantum of higher frequency ν_{13} is emitted. In the second case, on the contrary, there occurs a process of subdivision of a quantum when, as a result of the absorption of one high-frequency quantum ν_{13}, two quanta of lower frequency, ν_{12} and ν_{23} are emitted. The problem consists in establishing which of these two cyclic transitions predominate under nebular conditions, *i.e.* under conditions of highly diluted radiation. To solve this problem let us compute the ratio of the number of transitions $1 \to 2 \to 3 \to 1$ to the number of transition $1 \to 3 \to 2 \to 1$ per unit time.

The number of transitions from the first state to the second per unit time and per unit volume is $n_1 B_{12} \rho_{12}$, where n_1 is the number of atoms per unit volume in the first state, B_{12} is the Einstein absorption coefficient and ρ_{12} is the radiation density at the frequency corresponding to the transition between the first and the second states. Of this number of atoms a fraction will undergo spontaneous transitions downwards to the first state, and a fraction will go to the third state by absorption of a quantum $h\nu_{23}$. The fraction of atoms that make a transition from the second state to the third is, obviously:

$$\frac{B_{23}\rho_{23}}{A_{21} + B_{12}\rho_{12} + B_{23}\rho_{23}}.$$

From the third state spontaneous transitions are possible to both the first and the second states. We are only interested in transitions to the first state. The relative number of these transitions is:

$$\frac{A_{31} + B_{31}\rho_{13}}{A_{31} + B_{31}\rho_{13} + A_{32} + B_{32}\rho_{23}}. \tag{6}$$

Therefore, for the number of atoms that follow the path $1 \to 2 \to 3 \to 1$ per unit time we have

$$N_{(1 \to 2 \to 3 \to 1)} =$$

$$= n_1 B_{12} \rho_{12} \frac{B_{23} \rho_{23}}{A_{21} + B_{21}\rho_{12} + B_{23}\rho_{23}} \cdot \frac{A_{31} + B_{31}\rho_{13}}{A_{31} + B_{31}\rho_{13} + A_{32} + B_{32}\rho_{23}} . \quad (7)$$

In the same way we can find that the number of atoms, following the path $1 \to 3 \to 2 \to 1$ per unit time is

$$N_{(1 \to 3 \to 2 \to 1)} = n_1 B_{13}\rho_{13} \frac{A_{32} + B_{32}\rho_{23}}{A_{32} + B_{32}\rho_{23} + B_{31}\rho_{13}} \cdot \frac{A_{21} + B_{21}\rho_{12}}{A_{21} + B_{21}\rho_{12} + B_{23}\rho_{23}} . \quad (8)$$

From (7) and (8) we can write down the ratio of the number of processes of type $1 \to 2 \to 3 \to 1$ to the number of processes of type $1 \to 3 \to 2 \to 1$:

$$\frac{N_{(1 \to 2 \to 3 \to 1)}}{N_{(1 \to 3 \to 2 \to 1)}} = \frac{B_{12}\rho_{12}B_{23}\rho_{23}(A_{31} + B_{31}\rho_{13})}{B_{13}\rho_{13}(A_{32} + B_{32}\rho_{23})(A_{21} + B_{21}\rho_{12})} \quad (9)$$

To simplify this expression we can introduce the Einstein relations:

$$A_{ki} = B_{ik}\frac{g_i}{g_k}\sigma_{ik} \; ; \; B_{ki} = \frac{g_i}{g_k}B_{ik}, \quad (10)$$

where

$$\sigma_{ik} = \frac{8\pi h v_{ik}^3}{c^3} , \quad (11)$$

and g_i and g_k are the statistical weights of the levels. In addition we shall call

$$\bar{\rho}_{ik} = (e^{h v_{ik}/kT*} - 1)^{-1}, \quad (12)$$

and using (2) we can write

$$\rho_{ik} = W\sigma_{ik}\bar{\rho}_{ik} . \quad (13)$$

By means of the above definitions we can write instead of (9)

$$\frac{N_{(1 \to 2 \to 3 \to 1)}}{N_{(1 \to 3 \to 2 \to 1)}} = W \frac{\bar{\rho}_{12}\bar{\rho}_{23}(1 + W\bar{\rho}_{13})}{\bar{\rho}_{13}(1 + W\bar{\rho}_{12})(1 + W\bar{\rho}_{23})} . \quad (14)$$

In the photospheres of the stars we find $W = 1$ and the ratio (14) is almost exactly equal to 1. In the gaseous envelopes of different stars $W < 1$, and therefore in those cases (14) will be smaller than 1. Finally, in nebulae where $W \sim 10^{-13}$, and the factor $\bar{\rho}_{12}\bar{\rho}_{23}/\bar{\rho}_{13}$ is of the order of unity, we have

$$\frac{N_{(1 \to 2 \to 3 \to 1)}}{N_{(1 \to 3 \to 2 \to 1)}} \approx W. \quad (15)$$

i.e. the number of transitions of type $1 \rightarrow 2 \rightarrow 3 \rightarrow 1$ is extremely small in comparison with the number of transitions of type $1 \rightarrow 3 \rightarrow 2 \rightarrow 1$. This is Rosseland's theorem.

Thus, according to Rosseland's theorem, the nebulae, which absorb high frequency (ultra-violet) quanta emitted by the central star, transform them into quanta of lower frequencies, including quanta of visible light. Therefore, since the number of ultraviolet quanta emitted by the star is very high in comparison with the number of visual quanta, the total number of visual quanta emitted by the nebulae will also be very high. Therefore, the visual brightness of the nebulae will considerably exceed the visual brightness of the stars. Thus is explained one of the mysteries concerning the nature of the nebulae. At the same time, Rosseland's theorem explains the properties of the emission spectra of the nebulae.

3. Fluorescence—Zanstra's Theory

The process of transformation of short-wave into long-wave radiation, *i.e.* the process by means of which high frequency quanta are transformed into low frequency quanta, is none other than that known as *fluorescence*. When applied to real nebulae with definite optical properties and real atoms with many, not just three, energy states, it leads to interesting results.

Let us consider a hydrogen nebula that surrounds the nucleus on all sides. In spite of a small density of ionizing radiation (it is $1/W$ times smaller than at the surface of the star), the degree of ionization in the nebula, as we shall see later, will be very high—of the order of one thousand (*i.e.* $n^+/n_1 \approx 10^3$, where n^+ and n_1 are, respectively, the ion and neutral atom concentrations). This is explained by the small material density in the nebula, *i.e.* the small electron concentration (the degree of ionization is inversely proportional to the electron concentration). From this it follows that almost all hydrogen atoms will be ionized. As far as the small number of neutral atoms is concerned, practically all of them will be found in the ground state; excited atoms will be almost nonexistent since the density of the radiation necessary for excitation is much too small.

Thus, in the nebulae the atoms will be found practically in only two states: ground and ionized. If the mass of the nebula is sufficiently large it will be opaque at the frequencies absorbed by the ground state of hydrogen (beyond the Lyman limit). It can be shown that this condition will obtain with a density as low as 1 atom cm^{-3} if the radius of the nebula is of the order of 10,000 a.u. In this case the optical depth of the nebula τ_c at the frequencies beyond the Lyman limit will be of the order of one. At the same time, the nebula will be completely transparent in the subordinate lines of hydrogen (Balmer, Paschen, *etc.*).

On the other hand, the ratio of the absorption coefficient of one neutral hydrogen atom in the first lines of the Lyman series (L_α, L_β and so forth) to the absorption coefficient of the same atom beyond the series limit is of the order of 10^4 to 10^5. Therefore, the optical depth in these lines, under the assumed conditions ($\tau_c \approx 1$), is of the order of 10^4 to 10^5, *i.e.* very high. Let us see what will happen to the ultraviolet quanta absorbed by the nebula under these conditions. In what follows, by ultraviolet or L_c-quanta we shall mean quanta whose frequencies are higher than that of the Lyman series limit.

An L_c-quantum that reaches the nebula will sooner or later be absorbed by a hydrogen atom, as a result of which the atom is ionized. This process is called photoionization. After a certain time the electron will be captured by a proton. There now exist two possibilities: 1) the electron falls directly to the first (ground) level, and 2) the electron falls into one of the higher levels. In the first case an L_c-quantum is emitted and everything goes back to the beginning. In the second case the electron, according to Rosseland's theorem, has to make a series of cascade transitions, thus emitting quanta in the subordinate series which escape from the nebula since the nebula is completely transparent at the frequencies of these series. In the nebulae the radiation is so diluted, and the material density is so low, that this chain of cascade transitions goes uninterrupted in the vast majority of cases. The last link in the cascade transitions is a transition to the first level, accompanied by the emission of some quantum in the Lyman series. Now two things can happen.

Let us assume that the electron finds itself in the second level. This obviously must be preceded by a transition of the atom from some higher state (discrete or continuous) to the second and, hence, by the emission of one, and only one, quantum in the Balmer series or Balmer continuum, which escapes from the nebula. Afterwards a $2 \rightarrow 1$ transition takes place with emission of one L_α-quantum. However, according to our assumptions, the optical depth of the nebula in the frequencies of the Lyman lines and, among them, of L_α is very high. Therefore, this quantum, after travelling a short distance inside the nebula, will be absorbed by some neutral hydrogen atom in the ground state. The atom is excited to the second state and again, due to the lack of collisions, it will make a spontaneous transition to the ground state after a very short time (of the order of 10^{-8} sec), emitting a L_α-quantum. Thus, under nebular conditions the L_α-quanta cannot be modified further: they will experience many scatterings until they finally reach the outer limit of the nebula and escape.

Let us assume now that the electron has been captured to the third state. This capture is accompanied by the emission of a quantum beyond the Paschen limit, which escapes from the nebula. Now the electron has two possibilities: it can go either directly to the first level with emission of an L_β-quantum, or first to the second level and then to the first with emission of two quanta, H_α and L_α. In the first case, the emitted L_β-quantum will be absorbed again due to the large optical depth of the nebula in the frequencies of the Lyman lines, and later on, it will again excite an atom to the third state. This process will continue until the second possibility is realized. Then the H_α-quantum leaves the nebula and the L_α-quantum remains and is scattered many times. The same reasoning can be applied to the cases where, instead of an L_β-quantum, one has L_γ-quanta, L_δ-quanta, and so forth.

Thus, one L_α-quantum and one Balmer quantum result from each L_c-quantum absorbed in the nebula. If the optical depth τ_c of the nebula is of the order of unity or higher, it will absorb all the L_c-quanta emitted by the star per unit time. In that case the total number of Balmer quanta emitted by the nebula should equal the number of L_α-quanta.

This is how the Balmer quanta emitted by the nebulae originate. The fluorescent mechanism just described was first proposed by Zanstra [4]. This mechanism can be extended to explain the luminosity of the nebulae, not only in the hydrogen lines but in the lines of other elements, helium, oxygen, and so forth, as well.

4. Excitation of the Forbidden Lines

The excitation of many emission lines, including the most intense, of the nebulae cannot be attributed to fluorescent mechanisms. Bowen [2] was the first one to show that in the nebulae there exists still another mechanism, in principle quite different from that of fluorescence, for the excitation of emission lines.

Bowen began by identifying the important nebular lines N_1 and N_2 ($\lambda 5007$ and $\lambda 4959$ Å respectively). It turned out that they are due to forbidden transitions in doubly ionized oxygen. Another pair of nebular lines—$\lambda 3726$ Å and $\lambda 3729$ Å—are also due to forbidden transitions, but this time of singly ionized oxygen; the lines $\lambda 6548$ Å and $\lambda 6584$ Å correspond to forbidden transitions in singly ionized nitrogen, and so forth. A schematic representation of these and other forbidden transitions is given in Figure 2-1. The correctness of these identifications is confirmed by many different facts. For example, the theoretical intensity ratio of the components of the green oxygen doublet (N_1 and N_2), as well as that of the components of the red nitrogen doublet ($\lambda 6548$ Å [N II] and $\lambda 6584$ Å [N II]), should be constant for all nebulae and equal to three; the observations agree with this prediction. This is explained by the fact that the upper level of the N_1 and N_2 lines is the same ($^1D_2$). Therefore, the intensity ratio depends simply on the ratio of the spontaneous emission coefficients $A(^1D_2 \rightarrow {}^3P_1)/A(^1D_2 \rightarrow {}^3P_2)$ which, in this case, is equal to three. Exactly the same argument applies to the red doublet of ionized nitrogen.

Forbidden lines are not observed under usual laboratory conditions. Therefore we must accept that in the nebulae there exist special conditions that favor the formation of such lines.

First of all let us remember that the usual "selection rules" for atomic transitions are connected to some approximation or other made in the calculation of transition probabilities. In the first approximation, when only dipole radiation is taken into account, one finds that the probabilities of certain transitions should be equal

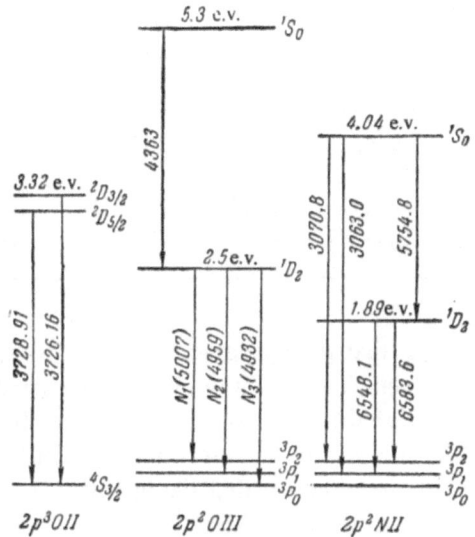

Figure 2-1

Representation of the transitions for several forbidden lines of the ions O II, O III and O IV.

to zero. Actually, when the problem is treated more exactly, *i.e.* when quadrupole and other radiations are included, as well as other effects connected with the magnetic moment of the electron, the magnetic moment of the nucleus, and so forth, one finds that the probabilities of those transitions are different from zero, although their values are a million or a billion times smaller than the probabilities of allowed transitions. Tables 2-1 and 2-2 show the spontaneous emission coefficient A for several transitions according to Garstang [53]. Let us remember that for the usual transitions these coefficients are of the order of $10^8 - 10^7$ sec^{-1}.

TABLE 2-1

Transition Probabilities A for Forbidden Transitions in $2p^2$ Configurations

Transition	[O III]		[N II]	
	λ (Å)	A (sec^{-1})	λ (Å)	A (sec^{-1})
$^1D_2 - {}^3P_2$	5007 (N$_1$)	2.10×10^{-2}	6584	0.30×10^{-2}
$^1D_2 - {}^3P_1$	4959 (N$_2$)	0.71×10^{-2}	6548	0.103×10^{-2}
$^1D_2 - {}^3P_0$	4932 (N$_3$)	1.90×10^{-6}	6528	4.2×10^{-7}
$^1S_0 - {}^1D_2$	4363	1.6	5755	1.08

TABLE 2-2

Transition Probabilities A for Forbidden Transitions in $3p^2$ Configurations

Transition	[O II]		[N I]	
	λ (Å)	A (sec^{-1})	λ (Å)	A (sec^{-1})
$^4S_{3/2}$—$^2D_{3/2}$	3726	43.15×10^{-5}	5198	16.25×10^{-6}
$^4S_{3/2}$—$^2D_{5/2}$	3729	4.08×10^{-5}	5201	6.95×10^{-6}

However, the simple fact that the magnitude of the transition probability is different from zero is not a sufficient condition for the production of an observable line. The *intensity* of the forbidden line is very significant. If it is not sufficiently great, the forbidden line still will be unobservable. Evidently, a necessary condition for observation is the requirement that the intensity of the forbidden line be comparable to the intensity of the usual lines.

In connection with this we can add that if the forbidden lines are emitted by a transition whose upper level can also give rise to a permitted line, then the intensity of the forbidden will never, under any conditions, be comparable with the intensity of the permitted line, since for the given upper level permitted transitions, which give rise to the permitted line, will occur a million times faster than the forbidden transitions. From this argument follows the first condition: *for the emission of a forbidden line of observable intensity it is necessary that the upper level, from which the forbidden transition originates, be metastable* (i.e. one that is connected to the lower levels only by forbidden transitions).

However, even if they satisfy this condition, not all forbidden lines can be observed. If several forbidden transitions are possible from a given metastable state then, obviously, those which are more strongly forbidden cannot give a sufficiently strong line. We can illustrate this point with the example of the O III ion. In this ion the level $^1D_2$ is metastable. Below this level we have three which are very close to each other: $^3P_2$, $^3P_1$ and $^3P_0$. Therefore, in addition to the N_1 and N_2 lines which arise from transitions from the level $^1D_2$ to the levels $^3P_2$ and $^3P_1$, one could expect the emission of still another line, N_3, with wavelength 4932 Å, originating from the transition $^1D_2 \rightarrow {}^3P_0$ (see Figure 2-1). However, as follows from the date in Table 2-1, the probability of this transition is ten thousand times smaller than the probability of the $^1D_2 \rightarrow {}^3P_2$ jump which gives the N_1 line. It follows, then, that the N_3 line will be ten thousand times less intense than N_1, i.e. it will be observed only with great difficulty and in special cases.* The same can be said of the λ 6528 Å [N II] line.

---

* Aller, Bowen, and Minkowski were able to observe the N_3 [O III] line in the spectrum of NGC 7027. According to their measurements the ratio E_{N_1}/E_{N_3} was greater than 2000.

From the transition probabilities of forbidden lines shown in Table 2-1 it follows that the mean life of the atom or ion in a metastable state is very large, of the order of a fraction of a second or greater. If during this interval of time the excited atom undergoes a collision with some other particle, then the excitation energy of the metastable atom will be transformed into kinetic energy of this particle instead of radiation energy in the forbidden line. The second condition follows: *for a forbidden line to appear it is necessary to have a relatively low material density.* This condition can be formulated quantitatively in the following way. We shall designate by n_2 the number of atoms per unit volume in the metastable state. The number of spontaneous transitions $2 \rightarrow 1$, per unit time which give origin to the forbidden line will be obviously, $n_2 A_{21}$. The number of induced transitions $2 \rightarrow 1$ caused by collisions of the second kind is $n_2 a_{21}$, where a_{21} is the transition probability $2 \rightarrow 1$ by means of collisions.

Therefore, the condition for the appearance of a forbidden line can be written $n_2 A_{21} \gg n_2 a_{21}$ or

$$a_{21} \ll A_{21}. \tag{16}$$

When the collision occurs with free electrons we have: $a_{21} = n_e \psi_{12}(T_e)$ where n_e is the electron concentration in the nebula and $\psi_{12}(T_e)$ is some function which depends on the electron temperature in the nebula. Then, from (16) we have:

$$n_e \ll \frac{A_{21}}{\psi_{12}(T_e)} \tag{17}$$

This condition must be satisfied by the electron concentration and, consequently, by the material density in the nebula, to avoid the "extinction" of the forbidden line. Let us denote the right hand side of (17) by n_e^0, the "limiting concentration". When $n_e < n_e^0$ the forbidden line will be observed with full intensity. But this still does not mean that the line cannot appear when the electron concentration is of the order of the "limiting concentration", *i.e.* when $n_e \approx n_e^0$. Forbidden lines can be observed even when $n_e > n_e^0$ but they will be very weak. The quantitative calculation of the intensity of a forbidden line in the latter case ($n_e \gtrsim n_e^0$) presents special problems which shall be considered in Chapter IV.

Table 2-3 gives the limiting values of the electron concentration n_e^0 for several forbidden lines. The function $\psi_{12}(T_e)$ is the coefficient of n_e in Eq. (27) of Chapter IV; the data on the magnitude of A_{21} are given in Tables 2-1 and 2-2. For the calculation it was assumed that $T_e = 10^4$.

For the majority of planetary nebulae $n_e \lesssim 10^4 \text{ cm}^{-3}$. If we compare this value with those which appear in Table 2-3, we see that the requirement of small density (more exactly, of small electron concentration) in planetary nebulae is fully satisfied for the N_1 and N_2 lines, and partially for $\lambda 3727 \text{ Å}$ [O II]. As far as the $\lambda 4363 \text{ Å}$ [O III] line is concerned, it can be observed even in the very dense planetary nebula

TABLE 2-3

Maximum Possible Electron Concentrations
for the Excitation of Forbidden Lines

Line	$n_e^0\,(\mathrm{cm}^{-3})$
N_1, N_2 [O III]	10^6
$\lambda\,4363\,\text{Å}$ [O III]	10^8
$\lambda\,6548\,\text{Å}$ [N II], 6584 [N II]	0.7×10^6
$\lambda\,3727\,\text{Å}$ [O II]	10^4

($n_e \approx 10^8\,\mathrm{cm}^{-3}$), and also during early phases of the development of nova envelopes, long before the appearance of the nebular lines N_1 and N_2.

However, an atom in a metastable state can, under the influence of the external
radiation, make a transition to a higher state. In this case the forbidden transition
will not be emitted. Therefore we require that no light quantum be absorbed during
the lifetime of the metastable state. From here follows the third condition, formulated by Eddington [52], necessary for the appearance of a forbidden line: *the radiation density in the nebula must be small.* As we shall see in the following section,
this condition is also satisfied in the nebulae.

5. Accumulation of Atoms in Metastable Levels

The intensity E_{ik} of a forbidden transition is proportional to the spontaneous
transition probability A_{ik} and to the number of atoms in the level n_i

$$E_{ik} = n_i A_{ik} h\nu_{ik}. \qquad (18)$$

The spontaneous transition probability from a metastable level is, as we saw before,
extremely small. Nevertheless, the intensity of the forbidden lines observed in the
nebulae is very high. Obviously, this is possible only if there is a large number of
atoms in the metastable state. As we have seen, under nebular conditions the atom
can remain in the metastable state for a long time (with respect to the possibility
of a spontaneous transition downwards). Therefore, a huge number of atoms can
accumulate in such a state. The problem of the accumulation of atoms in a metastable state was first discussed theoretically by V. A. Ambartsumyan [54]. Below
we give the basic results of his work, retaining to a large extent the original treatment.

Let us consider a simple atom consisting of three energy levels, of which the middle
one is metastable. Assuming, to start with, that the excited atom can only make
radiative transitions, we can write the equilibrium conditions for the number of

atoms per unit volume found in the first and third states:

$$n_1 B_{12}\rho_{12} + n_1 B_{13}\rho_{13} = n_2 A_{21} + n_3 A_{31},$$

$$n_1 B_{13}\rho_{13} + n_2 B_{23}\rho_{23} = n_3 A_{31} + n_3 A_{32}. \tag{19}$$

Here we neglect induced transitions since the radiation field is strongly diluted $(W \ll 1)$.

From Eq. (19) we get, eliminating n_3:

$$\frac{n_2}{n_1} = \frac{B_{12}\rho_{12} + (1 - p)B_{13}\rho_{13}}{A_{21} + pB_{23}\rho_{23}}, \tag{20}$$

where $p = A_{31}/(A_{31} + A_{32})$. Using the relation between the Einstein coefficients, and also expression (13) for the quantity ρ_{ik} from (20) we find:

$$\frac{n_2}{n_1} = W \frac{A_{21}(g_2/g_1)\bar{\rho}_{12} + (1 - p)A_{31}(g_3/g_1)\bar{\rho}_{13}}{A_{21} + pA_{32}(g_3/g_2)W\bar{\rho}_{23}}. \tag{21}$$

We also have the condition that the second state is metastable, *i.e.* $A_{21} \ll A_{31}, A_{32}$. Therefore, the first term in expression (21) can be neglected in comparison with the second. However, in the denominator each one of the two terms can dominate depending on the circumstances. We have to consider, the, two possibilities:

I. $W \ll A_{21}/A_{31}$. In this case the second term in the denominator can be neglected, and we get:

$$\frac{n_2}{n_1} = W(1 - p) \frac{A_{31}}{A_{32}} \frac{g_3}{g_1} \bar{\rho}_{13} \approx W \frac{A_{31}}{A_{21}} \bar{\rho}_{13}, \tag{22}$$

where we have taken $(1 - p)g_3/g_1 \approx 1$.

II. $W \gg A_{21}/A_{32}$. In this case we can neglect the first term in the denominator, and we get

$$\frac{n_2}{n_1} = \frac{g_2}{g_1} \frac{\bar{\rho}_{13}}{\bar{\rho}_{23}} \approx \bar{\rho}_{12}. \tag{23}$$

Let us examine, for comparison, the situation that would arise if the second level were not metastable, *i.e.* when the transition $2 \to 1$ is permitted. If we designate the Einstein coefficient for the permitted transition by A_{21}^0, we can write: $A_{21} \approx A_{31}, A_{32}$ and from formula (21) we find:

$$\frac{n_2^0}{n_1} = W \frac{g_2}{g_1} \bar{\rho}_{12} + W(1 - p) \frac{A_{31}}{A_{21}^0} \frac{g_3}{g_1} \bar{\rho}_{13} \approx W \frac{A_{31}}{A_{21}^0} \bar{\rho}_{13}, \tag{24}$$

since both terms on the right hand side of (21) are of the same order.

Comparing Eqs. (22) and (24) we see that when we have a metastable state in the first of the two cases considered, the ratio n_2/n_1 exceeds the value of this ratio for a normal excited state by a factor of the order of the ratio of the probabilities for the allowed and the forbidden transitions $(n_2/n_2^0 \approx A_{21}^0/A_{21})$. This means that the product of the number of atoms times the probability of spontaneous emission, *i.e.* the intensity of the forbidden line, is in this case of the same order as the intensity of the permitted line (since $n_2 A_{21} \approx n_2^0 A_{21}^0$).

In the second of the two cases the ratio n_2/n_1 is determined approximately by Boltzmann's formula and, as shown by comparing Eqs. (23) and (24), it is of the order of W^{-1} times greater than in the case of a normal excited state. It follows that, again in this case, there is a strong accumulation of atoms in the metastable state. However, the number of transitions from the metastable state is not as large as the number of transitions from the normal state. Indeed, we have:

$$n_2 A_{21} \approx n_1 \bar{\rho}_{12} A_{21} \approx n_2^0 \frac{A_{21}}{W} \ll n_2^0 A_{21}^0.$$

This means that in case II the intensity of the forbidden line will be small in comparison with the intensity of the permitted line.

The physical meaning of cases I and II is as follows. The first term in the denominator of Eq. (21) corresponds to the forbidden transition $2 \rightarrow 1$, and the second to upwards transitions from the metastable state $2 \rightarrow 3$. In the first case the transitions $2 \rightarrow 1$ predominate and the forbidden line is strong. In the second case there is a relatively strong radiation field which induces the transition $2 \rightarrow 3$ and the forbidden transition is weak.

Since in planetary nebulae W is of the order of 10^{-13}, almost all metastable states correspond to case I. A possible exception is the $2^3 S$ state of helium, where the electron can remain a rather long time (of the order of days and weeks). On the contrary, in envelopes of small radius (*i.e.* in the envelopes of Wolf-Rayet, B_e and other similar stars) almost all metastable states correspond to case II.

Thus, the third condition necessary for the appearance of a forbidden line—discussed at the end of the previous section—can be expressed by the relation:

$$W \ll A_{21}/A_{21}^0. \tag{25}$$

The ratio A_{21}/A_{21}^0 is of the order of 10^{-8} for the $\lambda 4363$ Å [O III] line; 10^{-10} for N_1 and N_2; 10^{-12} for $\lambda 3727$ Å [O II].

Equations (16) and (25) determine the upper limit of the material and the radiation densities. These conditions must be fulfilled by the mechanism of excitation of the atoms.

6. The Role of Electron Collisions

In Sec. 4 it was pointed out that various lines, among them the important nebular lines N_1 and N_2, cannot arise as the result of the photoionization of the ions and their subsequent recombination. This conclusion is confirmed by the following facts.

Let us assume that the luminosity of the nebula in the N_1 and N_2 lines is due to stellar energy beyond the limit of the principal series of doubly ionized oxygen. Then the number of quanta emitted by the star beyond the limit of this series ($\lambda < 226$ Å) cannot be smaller than the number of quanta emitted by the nebula in the N_1 and N_2 lines. In order to satisfy this condition we require a very high stellar temperature—of the order of a million degrees.

The following argument is more convincing; since the ionization potentials of O III and He II are almost identical (54.5 and 54.2 e.v. respectively) these two ions absorb stellar energy in the same region of the spectrum. Therefore, if the luminosity of the nebula in the N_1 and N_2 lines were due only to this energy (fluorescence), then the intensity of the N_1 and N_2 lines could not exceed that of the He II lines, since the number of helium atoms is not smaller than the number of oxygen atoms. However, in a number of planetary nebulae the helium lines can hardly be seen (and in diffuse nebulae they disappear altogether), while the N_1 and N_2 lines appear with great intensity.

All this shows that in the nebulae there exists an excitation mechanism other than recombination. Such a mechanism was suggested by Bowen [2] and, according to it, the luminosity of the nebulae in the "nebulium" lines comes from the *kinetic energy of the free electrons*.

The excitation potential of the metastable levels that give rise to forbidden lines, among them the nebular lines N_1 and N_2, is not large. The excitation potentials of several metastable levels are given in the last column of Table 4-5. For example, for the $^1D_2$ level of O III, from which the N_1, N_2 doublet originates, this potential is 2.5 e.v. For N II it is even smaller—1.89 e.v, and so forth. On the other hand, the residual energy of the ultraviolet quanta after the photoionization of hydrogen is transformed into kinetic energy of the free electrons. This energy is sufficiently high to excite the O III ions to the metastable state through inelastic collisions with the electrons. These collisions of course occur only rarely since the density of matter in the nebulae is very low—low enough so that the ion once in a metastable state will not undergo another collision with a free electron before making a spontaneous transition downwards. Nevertheless, the number of collisions is high enough to excite the ground state ions to the metastable levels since the time that the ion (for example, O III) spends in the ground state is very large (of the order of a week).

Thus, collisions with free electrons raise the ions from the ground to the metastable state at the expense of the kinetic energy of the electrons, and then spontaneous transitions downwards produce the forbidden lines. In other words, the nebulae are almost entirely "deprived" of the kinetic energy of their electrons in order to emit the forbidden lines.

Let us consider a two-level atom, 1 and 2, of which the higher (2) is metastable.

Let us call $n_1 b_{12}$ the number of transitions from the normal to the metastable level by electron collisions of the first kind; $n_2 a_{21}$, the number of transitions from the metastable to the ground state by means of collisions of the second kind, and $n_2 A_{21}$ the number of spontaneous transitions from the metastable to the ground state. The equilibrium condition gives:

$$n_1 b_{12} = n_2(a_{21} + A_{21}). \tag{26}$$

The coefficients b_{12} and a_{12} depend on the concentration and density distribution of the free electrons. One makes the basic assumption that the free electron velocities are, under nebular conditions, given by a Maxwellian distribution [55]. In that case, the relation between b_{12} and a_{21} is given by Eq. (26) of Chapter IV:

$$b_{12} = a_{21} \frac{\omega_2}{\omega_1} e^{-\varepsilon_{12}/kT_e}, \tag{27}$$

where T_e is the temperature of the electron gas and ε_{12} the excitation energy of the metastable level. Then, from (26) and (27) we get:

$$\frac{n_2}{n_1} = \frac{a_{21}}{A_{21} + a_{21}} \frac{\omega_2}{\omega_1} e^{-\varepsilon_{12}/kT_e}. \tag{28}$$

If $a_{21} \gg A_{21}$ Eq. (28) goes into Boltzman's law for n_2/n_1; however, this inequality indicates a preponderance of induced over spontaneous transitions $2 \to 1$. Under such conditions the nebular lines would be weakened. Actually, as we have seen before, in the nebulae condition (16) obtains and therefore

$$n_2 = n_1 \frac{a_{21}}{A_{21}} \frac{\omega_2}{\omega_1} e^{-\varepsilon_{12}/kT_e}. \tag{29}$$

For the intensity of the forbidden line we get:

$$E_{21} = n_2 A_{21} h\nu_{12} = n_1 a_{21}(\omega_2/\omega_1)h\nu_{12}e^{-\varepsilon_{12}/kT_e}, \tag{30}$$

or, using (27)

$$E_{21} = n_1 b_{12} h\nu_{12}, \tag{31}$$

i.e. the number of forbidden transitions is equal to the number of excitations by electron collisions.

7. The Stratification of Radiation

The fact that in the nebulae there exist at least two different excitation mechanisms for the emission should lead to monochromatic images of the nebulae in the various lines of different sizes, i.e. to the presence of stratification of radiation, to which reference was made in the first chapter. Let us analyze this question somewhat further following Bowen [56], assuming that the basic premises of the theory of ionization of the nebulae (Sec. 10 of this chapter) are known to the reader.

Suppose we have a flattened spherical nebula with a very hot star in its center. Let us now investigate which lines will be emitted by, for example, oxygen atoms at different distances from the center of the nebula. Obviously, near the nucleus the oxygen will be very highly ionized, say, in the fourth stage of ionization. After ionization these same atoms will undergo recombination with electrons, together with cascade transitions downwards from the upper states (according to Rosseland's theorem) and as a result different lines of the O IV spectrum will be emitted. Thus, in the central regions of the nebula we shall observe such lines as $\lambda 3412 \text{ Å}$ O IV, $\lambda 3737 \text{ Å}$ O IV, and so forth. However, these lines will be unobservable beyond this (first) zone, where quanta with wavelength less than 160 Å (necessary for the fourth ionization of oxygen) cannot reach. In the second zone the oxygen will be found in the third ionization stage, by absorbing quanta with wavelength shorter than 225 Å. The capture of an electron by an O^{+++} ion gives rise to the permitted lines in the spectrum of O III. Therefore, the dimensions of the region of the nebula where the O III lines are emitted (for example, the lines at $\lambda 3340 \text{ Å}$, $\lambda 3429 \text{ Å}$, and so forth) will be greater than the dimensions of the region that corresponds to O IV. This argument can be applied to the succeeding zones: with each lower ionization stage the dimensions of the zone where the permitted lines of the spectrum of a given ion are emitted, increase.

As regards the excitation of forbidden lines, in addition to the ions one must have a sufficient number of free electrons.

The main source of free electrons in the nebulae is hydrogen, which is fully ionized in the central regions. Therefore, in these regions we should also observe the forbidden lines of the doubly ionized atoms.

However, it is easy to see that the size of the image in the forbidden lines of O III should be greater than in the permitted lines of O III. The permitted lines of O III are emitted when an electron is captured by an O^{+++} ions. On the other hand, the forbidden lines of O III (N_1, N_2, $\lambda 4363 \text{ Å}$) are emitted when an O^{++} ion is excited by a free electron.

From the preceding arguments we can understand why, for example, the size of the monochromatic image of the nebula in the $\lambda 3426 \text{ Å}$ [Ne V] line, the strongest and most easily accessible in this region of the spectrum, is always smaller than the size of the image in the $\lambda 3869 \text{ Å}$ [Ne III] line.

The smallest ionization potential is that of the hydrogen atom. Therefore, the largest image size should correspond to the hydrogen lines. Neutral oxygen has an ionization potential almost identical to that of hydrogen. However, because of the abundance of hydrogen atoms in the nebula the oxygen atoms are usually left with an insufficient fraction of the ionization energy. It follows, then, that the size of the region containing O I lines should be smaller than the size of the region containing hydrogen lines. It is also clear that none of the forbidden lines can occupy a zone greater than that of the hydrogen lines.

As far as helium is concerned, the $\lambda4686$ He II lines are emitted when an electron is captured by a doubly ionized helium atom which has an ionization potential four times greater than that of hydrogen. Doubly ionized helium can be found only in the inner regions of the nebula, and therefore the size of the image in the $\lambda4686$ Å He II line should be considerably smaller than the H II region.

The problem of stratification of radiation has been treated here in a very schematic way: the picture given by actual photographs differs to a certain degree from the description given here. The complexity of the problem increases because in reality we have large number of different atoms with different abundances. Nevertheless, the basic ideas in this interpretation of the stratification of radiation in the nebulae are correct and, on the whole, they agree with the observed facts.

8. The Intensity of the Balmer Lines

The fact that the nebula is transparent to the radiation in the subordinate lines of hydrogen and, in particular, the Balmer series, opens the possibility of solving the important question of the distribution of the atoms over the different excited levels. The solution of this problem gives, in particular, the possibility of calculating the relative intensity of the Balmer lines, or, as it is generally called, the *Balmer decrement*. The sole process involved is that of ionization from the ground state followed by recombination and cascade transitions downwards. Ionizations from excited states can be ignored because of the negligible number of excited atoms.

The starting point in calculating the population of the excited states is the equilibrium condition, according to which the number of atoms that arrive in a given state must be exactly equal to the number of atoms leaving it.

Let us write the equilibrium conditions for several levels in hydrogen, characterized by the principal quantum number n and the azimuthal quantum number l. To do so we have to write down the total number of atoms entering the level nl, then the total number of atoms leaving the same level, and equate the two expressions.

The total number of atoms arriving in the level nl is given by the sum of the following three quantities:

1. The number of free electron captures directly to the level nl, which is equal to

$$N_e N^+ C_{nl}(T_e),$$

where $C_{nl}(T_e)$ is the recombination coefficient of the electron to the level nl; N_e and N^+ are, respectively, the free electron and proton concentrations.

2. The number of spontaneous transitions from higher discrete states $n'l'$ to

the states nl, which is equal to

$$\sum_{n'=n+1}^{\infty} \sum_{l'=0}^{n'-1} N_{n'l'} A_{n'l',nl}$$

where N_n is the total number of atoms per unit volume in the level n, i.e.
ficient for the spontaneous transition $n'l' \to nl$.

3. The number of transitions from the ground state to the level nl under the influence of radiation, which is equal to

$$N_1 B_{1,nl} \rho_{1,nl} ,$$

where $B_{1,nl}$ is the Einstein coefficient for the induced transition $1 \to nl$; $\rho_{1,nl}$ is the
radiation density corresponding to this transition.

Thus, the total number of atoms arriving at level nl per unit time and per unit
volume will be

$$N_e N^+ C_{nl}(T_e) + \sum_{n'=n+1}^{\infty} \sum_{l'=0}^{n'-1} N_{n'l'} A_{n'l',nl} + N_1 B_{1,nl} \rho_{1,nl}. \tag{32}$$

On the other hand, the total number of atoms leaving the level nl will be

$$N_{nl} \sum_{n''=1}^{n-1} \sum_{l''=0}^{n''-1} A_{nl,n''l''}, \tag{33}$$

since from the state nl only the spontaneous transitions downwards are possible
(to the levels $n''l''$); N_{nl} is the number of atoms per unit volume in the state nl.

If we equate (32) and (33) we get:

$$N_e N^+ C_{nl}(T_e) + \sum_{n'=n+1}^{\infty} \sum_{l'=0}^{n'-1} N_{n'l'} A_{n'l',nl} + N_1 B_{1,nl} \rho_{1,nl} = N_{nl} \sum_{n''=1}^{n-1} \sum_{l''=0}^{n''-1} A_{nl,n''l''}. \tag{34}$$

In the case of thermodynamic equilibrium the distribution of atoms over the
levels nl (or $n'l'$) is given by the relation

$$N_{nl} = \frac{2l+1}{n^2} N_n, \tag{35}$$

where N_n is the total number of atoms per unit volume in the level n, i.e.

$$N_n = \sum_{l=0}^{n-1} N_{nl}, \tag{36}$$

and $2(2l+1)$ and $2n^2$ are, respectively, the statistical weights of the levels nl and n.

The solution of the system (34) cannot satisfy the relation (35). However, taking
(35) into account we can simplify the system (34) quite considerably and in this way
obtain an approximate solution of the problem. Substituting (35) into (34) and summing over all the terms with a given value of n and all possible values of l (from $l = 0$

to $l = n + 1$), we get the following system of algebraic equations:

$$N_e N^+ C_n(T_e) + \sum_{n'=n+1}^{\infty} N_{n'} A_{n',n} + N_1 B_{1n} \rho_{1n} = N_n \sum_{n'=1}^{n-1} A_{n,n'}, \qquad (37)$$

where

$$C_n(T_e) = \sum_{l=0}^{n-1} C_{nl}(T_e),$$

$$A_{n',n} = \frac{1}{n^2} \sum_{l=0}^{n-1} \sum_{l'=0}^{n'-1} (2l' + 1) A_{n'l',nl}, \qquad (37a)$$

$$B_{1n} = \sum_{l=0}^{n-1} B_{1,nl} \ , \ \rho_{1n} = \sum_{l=0}^{n-1} \rho_{1,nl},$$

and N_n is given by (36).

Written in the most general way the system (37) is independent of the optical properties of the nebula. One often considers two nebular models. In the first one (case A), we assume that the nebula is thin in the frequencies of the Lyman series. In this case the last member of the left hand side of (37) will be equal to zero ($\rho_{1n} = 0$) and the equations take the form:

$$N_e N^+ C_n(T_e) + \sum_{n'=n+1}^{\infty} N_{n'} A_{n',n} = N_n \sum_{n'=1}^{n-1} A_{n,n'}. \qquad (38)$$

In the second case (case B) we assume that the nebula is thick in the Lyman lines; this corresponds to Zanstra's model. In this case all quanta in the lines of the Lyman series emitted in the nebula will be also absorbed there. In other words, the number of $n \to 1$ transitions must be exactly balanced by the number of $1 \to n$ transitions, i.e.

$$N_n A_{n1} = N_1 B_{1n} \rho_{1n}.$$

Therefore, in case B, equation (37) becomes

$$N_e N^+ C_n(T_e) + \sum_{n'=n+1}^{\infty} N_{n'} A_{n',n} = N_n \sum_{n'=2}^{n-1} A_{n,n'}. \qquad (39)$$

Thus, in both cases we are led to a system of linear algebraic equations which depend on the parameter $N_n/N_e N^+ = z_n$. Each system is solved independently to give the population of the energy levels of hydrogen (beginning with the third). Knowing the number z_n it is easy to find the relative intensities of the emission lines.

The energy emitted by the nebula in the Balmer lines, which originate from transition $n \to 2$ is equal to

$$E_{n2} = h\nu_{n2}A_{n2} \quad N_n dV,$$

where the integral extends over the entire volume of the nebula. However, $N_n = z_n N_e N^+$, and if we assume that the electron temperature does not vary inside the nebula, the quantity z_n can be taken outside the integral. Thus we get:

$$E_{n2} = z_n A_{n2} h\nu_{n2} \int N_e N^+ dV. \tag{40}$$

This formula gives the intensity of the Balmer lines up to a constant factor. Dividing (40) by its value for the line $H_\beta (n = 4)$, we get for the relative intensities of the Balmer lines (with respect to the intensity of H_β), i.e. for the *Balmer decrement*, the following expression:

$$\frac{E_{n2}}{E_{42}} = \frac{z_n}{z_4} \frac{A_{n2}\nu_{n2}}{A_{42}\nu_{42}}, (n = 3, 4, \dots). \tag{41}$$

The system of equations (39) was first solved approximately by Cillié [57], who used the first 12 levels ($n = 3, 4, \dots, 14$) and neglected the rest. Subsequent solutions of the systems (38) and (39) were obtained by a number of authors in order to improve upon the original results.* The first in this series of papers is that of Baker and Menzel [58], who solved the systems (38) and (39) for a practically infinite number of levels ($n \to \infty$). They determined the value of the dimensionless parameter b_n, given by the relation

$$b_n = z_n \frac{(2\pi\mu kTe)^{3/2}}{h^3} \frac{e^{-\chi_n/kT_e}}{n^2},$$

i.e. a parameter that indicates by what factor the value of $N_n/N_e N^+$ in the nebula differs from the value of $N_n/N_e N^+$ in thermodynamic equilibrium at the temperature T_e. The value of b_n is fairly close to unity and, as one would expect, $b_n \to 1$ as $n \to \infty$. For the levels with small quantum number the results of Baker and Menzel differed little from those of Cillié.

However, as was shown by Burgess [59], in the calculations of Baker and Menzel there was a mistake in the asymptotic value of the Kramers-Gaunt factor. Furthermore, the error was apparently comitted only in the solution of Eq. (39) for case B. This is shown by the fact that in several cases the quantities N_n given by Baker and Menzel for the approximation that uses just the first 14 levels was greater than for the same quantity computed for the case of an infinite number of levels.** Check calculations carried out by Burgess confirm this assumption.

---

* Formally Eqs. (38) and (39) differ from each other only in that in the first case A the sum over the right hand side of the equation starts with $n = 1$, and in the second case B with $n = 2$.

** Obviously, the quantity N_n with $n \to \infty$ should be greater than for $n = 14$ since the increase of the number of levels in the solution increases the population of the given level N_n by cascade transitions from the higher levels.

Burgess determined the correct value of b_n for twelve levels of hydrogen and two values of the electron temperature, 10^4 and 2×10^4. Table 2-4 gives part of these data for optically thin (case A) and optically thick (case B) nebulae.

If b_n is known we can compute the energy emitted per unit volume of nebula per second in a given line of the Balmer series of hydrogen, or of hydrogen-like atoms, by means of the following equation [8];

$$E_n = N_e N^+ \frac{5.398 \times 10^{-33}}{T_e^{3/2}} b_n \frac{g_{2n}}{n^3} R Z^6 e^{\chi_n}, \qquad (41a)$$

TABLE 2-4

Values of b_n for Case A and Case B

$n \backslash T_e$	Case A		Case B	
	10,000°K	20,000°K	10,000°K	20,000°K
2	0.00378	0.0456	—	—
3	0.0362	0.137	0.108	0.401
4	0.0869	0.215	0.183	0.448
5	0.136	0.272	0.245	0.487
6	0.176	0.312	0.290	0.512
7	0.207	0.339	0.321	0.521

where g_{2n} is the Kramers-Gaunt factor, whose magnitude for the first few Balmer lines of hydrogen, according to [58], are shown on Table 2-5. We also have $\chi_n = hRZ^2/n^2kT_e$, where R is Rydberg's constant. In Table 2-6 χ_n is tabulated for several levels of hydrogen ($Z = 1$) as a function of T_e.

TABLE 2-5

Values of the Kramers-Gaunt Factor g_{2n}

n	3	4	5	6	7
g_{2n}	0.757	0.822	0.844	0.855	0.861

The energy emitted at frequency v of the continuous spectrum beyond the limit of a given series (*i.e.* by recombination of an electron directly to level n or by transitions $\infty \to n$) per unit volume, per unit time, and per unit frequency interval is given by the following formula:

$$E_v = N_e N^+ \frac{2.159 \times 10^{-32}}{T_e^{3/2}} \frac{g_{\infty,n} Z^4}{n^3} e^{\chi_n} e^{-hv/kT_e}, \qquad (41b)$$

TABLE 2-6

Values of $\chi_n = hRZ^2/n^2kT_e$ for $Z = 1$

$T_e°K\backslash n$	2	3	4	5	6	7
5000	7.851	3.490	1.963	1.256	0.872	0.641
10,000	3.926	1.745	0.981	0.628	0.436	0.320
20,000	1.963	0.872	0.491	0.314	0.218	0.160
40,000	0.981	0.436	0.245	0.175	0.109	0.080

which is valid only for $v \geq v_n$, where v_n is the frequency of the series limit. For the continuum beyond the Lyman limit in hydrogen ($\lambda \leq 912$ Å) we have $n = 1$, $g_{\infty,1} = = 0.797$ and beyond the Balmer limit ($\lambda \leq 3646$ Å) we have $n = 2$, $g_{\infty,2} = 0.875$.

Finally, for the intensity of the continuous radiation due to free-free transitions of the electrons in the field of the protons, we have:

$$F_v = N_e N^+ \frac{2.249 \times 10^{-22}}{T_e^{1/2}} \frac{\bar{g}Z^2}{R} e^{-hv/kT_e} ; \qquad (41c)$$

this formula is valid for all frequencies; $\bar{g}$ is the mean value of the Kramers-Gaunt factor for the continuum and differs little from unity.

The main shortcoming in the work of Baker and Menzel (and also of Cillié) lies in the fact that they neglected the degeneracy of the energy levels of the hydrogen atom with respect to the azimuthal quantum number l, *i.e.* they considered that through cascade transitions the atoms were distributed over the levels with equal n but different l according to their statistical weights. This assumption can be realized in thermodynamic equilibrium, but when the excited states are populated only through direct electron captures and spontaneous cascade transitions downwards of the atoms, it is not true. In practice the lower levels that give rise to the lines H_α, H_β, and so forth, are the most interesting since for these levels the deviations from thermodynamic equilibrium are greatest. Therefore, it is impossible to calculate accurately the value of the Balmer decrement unless due account is taken of the influence of the azimuthal and magnetic quantum numbers on the population of a given level of the state with principal quantum number n.

The first attempt to consider the influence of the azimuthal quantum number on the Balmer decrement was that of Ryndina [60] and also of Searle [61]. The latter, for example, solved Eqs. (38) and (39) for the first ten levels taking into account that factor.

The general problem of the Balmer decrement was worked out and solved by Seaton [62]. He also started from the system of equations (38) and (39) and solved it for an infinite member of levels ($n \to \infty$), and in calculating the spontaneous transition probabilities that appear in them, he took into account the fact that the population of a given level depends not only on n but also on l [formula (37)]. In addition, for the calculation of $C_n(T_e)$ Seaton used more exact formulae rather than those proposed by Gaunt [63] and widely used in astrophysics. For the hydrogen atoms these formulae are the following [64]:

$$C_n(Z,T_e) = 5.197 \times 10^{-14} Z(\lambda^{1/2}/n) x_n S_n(\lambda), \qquad (42)$$

where

$$\lambda = 157,980 \, Z^2/T_e,$$

$$x_n = \lambda/n^2,$$

$$S_n(\lambda) = \int_0^\infty \frac{g(n,\varepsilon) e^{-x_n u}}{1 + u} \, du, \quad u = n^2 \varepsilon.$$

Auxiliary tables for the calculation of $C_n(Z,T_e)$ are given in Seaton's paper. The basic data are collected in Table 2-7, which gives the numerical value of $C_n(T_e)$ for hydrogen ($Z = 1$) and for several values of n and T_e of astrophysical interest.

TABLE 2-7

Values of the Function $C_n(T_e) \times 10^{14}$ for Hydrogen

n \ T_e °K	5000	10,000	15,000	20,000	40,000	60,000	80,000	100,000
1	22.148	15.446	12.390	10.549	7.007	5.410	4.466	3.820
2	11.639	7.655	5.889	4.832	2.876	2.064	1.604	1.299
3	7.309	4.544	3.352	2.672	1.469	1.005	0.761	0.597
4	5.010	2.955	2.117	1.639	0.755	0.569	0.421	0.333
5	3.641	2.050	1.423	1.098	0.547	0.350	0.260	0.202
6	2.724	1.490	1.011	0.766	0.372	0.238	0.173	0.134
7	2.093	1.112	0.741	0.560	0.264	0.169	0.120	0.092
8	1.655	0.866	0.570	0.431	0.197	0.125	0.088	0.072
9	0.338	0.670	0.446	0.326	0.150	0.092	0.066	0.052
10	1.103	0.550	0.354	0.259	0.116	0.072	0.052	
11	0.909	0.446	0.288	0.208	0.094	0.058		
12	0.767	0.372	0.237	0.172	0.072			
13	0.656	0.310	0.197	0.143	0.062			
14	0.556	0.263	0.168	0.119	0.052			

He also gives the asymptotic behaviour of the function $C(Z,T_e)$, which is the electron recombination coefficient to all levels (from $n = 1$ to $n = \infty$):

$$C(Z,T_e) = \sum_{n=1}^{\infty} C_n(Z,T_e) = 5.197 \cdot 10^{-14} Z \lambda^{1/2} \left[0.4288 + \frac{1}{2} \ln\lambda + 0.469\lambda^{-1/3} \right]. \quad (43)$$

This formula gives the value of $C(Z,T_e)$ with an error smaller than 0.5 per cent when $T_e/Z^2 \leq 10^5$ and of 3 per cent when $T_e/Z^2 = 10^6$.

TABLE 2-8

Vaules of the Function $C(T_e)$ for Hydrogen

$T_e\,°K$	5000	10,000	15,000	20,000	40,000	60,000	80,000	100,000
$C(T_e) \cdot 10^{13}$	6.725	4.118	3.066	2.477	1.456	1.055	0.831	0.692

Table 2-8 gives the numerical value of the function $C(T_e)$ for hydrogen for several values of T_e.

The Balmer decrements computed by Seaton for optically thin (case A) and optically thick (case B) nebulae, and for electron temperatures of 10,000°K and 20,000°K are shown on Table 2-9. The relative intensity of $H_\beta(n = 4)$ is taken as 100.

From the data on Table 2-9 we see that the Balmer decrement has only a weak dependence upon electron temperature and can be taken as practically constant. However, it does depend on the optical thickness of the nebula at the frequencies of the Lyman series of hydrogen. In particular, the Balmer decrement is steeper for optically thick nebulae (case B) than for optically thin nebulae (case A).

TABLE 2-9

Theoretical Balmer Decrements

n	Case A		Case B		Observed values (corrected)
	$T_e = 10,000°K$	$T_e = 20,000°K$	$T_e = 10,000°K$	$T_e = 20,000°K$	
3	191	199	271	279	277
4	100	100	100	100	100
5	58.9	56.9	50.6	49.9	50
6	37.8	35.6	29.8	28.2	26
7	25.8	23.8	19.2	17.8	18
8	18.4	16.7	13.2	12.0	—

TABLE 2-10

Balmer Decrements of Several Planetary Nebulae

	1	2	3	4	5	6	7	8	9
H_α	—	280	350	—	800	—	—	—	650
H_β	100	100	100	100	100	100	100	100	100
H_γ	47	50	49	55	43	42	37	34	33
H_δ	33	30	24	28	22.5	25	22	20	16.3
H_ϵ	31	26	—	—	11.9	22	19	11	—
H_ζ	12	7.5	17	20	11.6	6.2	—	5	—

1 Anon. 00^h25^m 2 IC 320   3 IC 351   4 IC 5217
5 NGC 40   6 IC 3568   7 NGC 7026   8 Anon. 19^h26^m
9 NGC 7027

By comparison with the data on Table 2-9 it turns out that the error in the value of the Balmer decrement given by Baker and Menzel is about 5% for case A and 20% for case B. Only for $n \geq 30$ both sets of results coincide.

Now let us turn to a comparison of theory with observations. Table 2-10 shows the Balmer decrement of several planetary nebulae according to the measurements of Aller and Minkowski [21, 66]. If we compare with the data given in Table 2-9 we can see several discrepancies. In general the observed Balmer decrement is steeper than the calculated. Furthermore, it varies noticeably from nebula to nebula.

From a correlation that he found between the Balmer decrement and the galactic latitude of the nebula, Shaïn [65] pointed out that the differences between the observed and calculated values of the Balmer decrement could perhaps be explained by the differential absorption of light in the Galaxy, which would cause the observed ratio H_α/H_β to seem larger, and the ratios H_γ/H_β, H_δ/H_β, and so forth to seem smaller than they really are. Berman [67] found a similar correlation between the Balmer decrement and the distance to the nebula. If the absorption of light in the galaxy is taken into account, the observed Balmer decrement shows good agreement with the calculations. In the last column of Table 2-9 we give the observed decrement corrected for differential absorption of light (averaged over 17 nebulae).

We should also point out that the observed Balmer decrement agrees well with the theoretical case B. From this we can conclude that at least the majority, if not all, the nebulae are optically thick at the frequencies of the Lyman series of hydrogen. Only in a few isolated cases may it happen that this is not true (for example, IC 5217).

To conclude this section let us compute the mean time spent by the electrons in a free state. If $N_e N^+ \Sigma_{n=1}^{\infty} C_n(T_e)$ is the number of recombinations per unit time and per unit volume, then, obviously, we can write:

$$t_e = \frac{1}{N_e N^+ \sum_{n=1}^{\infty} C_n(T_e)} = \frac{1}{N_e^2 C(T_e)}. \tag{43a}$$

If $T_e = 10,000°K$ we have, from Table 2-8: $C(T_e) = 4.118 \times 10^{-13}$. Then for T_e we get: in the case of planetary nebulae ($n_e \approx 10^4 - 10^3$ cm^{-3}) from several hours to months; in the case of diffuse nebulae ($n_e \approx 10^2$ cm^{-3}) of the order of tens of years; in the case of the interstellar medium ($n_e \approx 1$ cm^{-3}) of the order of thousands of years.

9. Degree of Excitation

The order of magnitude of the degree of excitation of the hydrogen atoms in the nebula can be obtained from the equations derived for the case of a two-level atom [see, for example Eq. (24)]. In this case we have, approximately:

$$\frac{N_n}{N_1} \approx W \bar{\rho}_{1n} = \frac{W}{e^{h\nu_n/kT*} - 1}. \tag{44}$$

With $T_* = 50,000°K$ and $W = 10^{-13}$ we get, for the lower levels ($n = 2, 3, \ldots$)

$$N_n/N_1 \approx 10^{-14}.$$

The exact solution of the problem of the degree of excitation was obtained in the manner just described. The magnitude of the numbers z_n given by the ratio of the number of atoms in level n to the quantity $N_e N^+$ was determined in the same way. Table 2-11 gives the numerical value of $z_n \times 10^{20}$ for different values of the electron temperature T_e according to Cillié (those sufficient for our purposes). In this case

TABLE 2-11
Values of the Quantity $z_n \times 10^{20}$

n \ T_e	5000°K	10,000°K	20,000°K
3	0.41	0.23	0.13
4	0.59	0.33	0.17
5	0.89	0.48	0.24
6	1.29	0.68	0.34

we have, for the degree of excitation:

$$\frac{N_n}{N_1} = z_n \frac{N^+ N_e}{N_1}. \tag{45}$$

The ratio $N^+ N_e / N_1$ for planetary nebulae, as will be shown in the next section, is of the order of 10^6 - 10^8. Then, using the data on Table 2-11, we find:

$$N_n / N_1 \approx 10^{-15} - 10^{-13} \qquad (n = 3, 4, \ldots).$$

Therefore, the degree of excitation in planetary nebulae is extremely small.* Under special conditions (see Chapter VI) the second state of hydrogen ($n = 2$) may be an exception to this rule. Characteristically the degree of excitation of a nebula increases as we go to higher levels.

10. Ionization in the Nebulae

The formula for the calculation of the degree of ionization in the nebulae can be deduced in the usual way from the equilibrium conditions, that is, the number of atoms that reach the continuum through photoionizations must equal the number of atoms that leave the continuum, *i.e.* that undergo recombinations with free electrons.

From the very small degree of excitation of the hydrogen atoms under nebular conditions we can assume that photoionizations occur only from the ground state. Therefore, if we call n_1, the number of atoms per unit volume in the ground state, the number of ionizations per unit time will be

$$n_1 \int_{\nu_0}^{\infty} k_{1\nu} \frac{\rho_\nu c}{h\nu} \, d\nu, \tag{46}$$

where ν_0 is the ionization frequency.

Recombinations with free electrons can occur to all levels. The total number of recombinations will then be:

$$n_e n^+ \sum_{n=1}^{\infty} C_n(T_e),$$

or

$$\frac{n_e n^+}{p} C_1(T_e), \tag{47}$$

___

* Let us recall that the degree of excitation of the second level of hydrogen under solar photospheric conditions ($T = 5700°K$) is of the order of 10^{-9}, and in the photosphere of a star of spectral class A ($T = 10,000°K$) it is of the order of 10^{-5}.

where p is the fraction of captures to the first level, *i.e.*

$$p = \frac{C_1(T_e)}{\sum_{n=1}^{\infty} C_n(T_e)}. \qquad (48)$$

Application of the equilibrium conditions gives:

$$n_1 W \int_{v_0}^{\infty} k_{1v} e^{\cdots} \frac{\rho_v^* c}{hv} \, dv = \frac{n_e n^+}{p} C_1(T_e), \qquad (49)$$

where we have taken $\rho_v = W \rho_v^* e^{-\tau_v}$, and where ρ_v^* is the radiation density at the surface of the star at the frequency v. Here τ_v is the optical depth of the nebula at the frequency v beyond the Lyman limit.

From (49) we have:

$$\frac{n^+ n_e}{n_1} = Wp \frac{\int_{v_0}^{\infty} k_{1v} e^{-\tau} \frac{\rho_v^* c}{hv} \, dv}{C_1(T_e)}. \qquad (50)$$

In fact this is the ionization equation for a nebula. However, it can be considerably simplified by using the relation that exists between the quantities k_{1v} and $C_1(T_e)$. Detailed calculations can be found, for example, in [68]. The final ionization equation for nebulae takes the following form:

$$\frac{n^+ n_e}{n_1} = W \left(\frac{T_e}{T_*}\right)^{1/2} \frac{2(2\pi \mu k T_*)^{3/2}}{h^3} e^{-hv_0/kT_*} e^{-\tau_c}, \qquad (51)$$

where τ_c is the mean value of the optical distance from the star to a given layer of the nebula at the frequencies of the Lyman series of hydrogen.

We see that Eq. (51) differs from the usual ionization equation basically through the presence of the factor W on the right hand side. Although in nebulae this factor is very small ($\approx 10^{-13}$), nevertheless the degree of ionization n^+/n_1 reaches a very high value, of the order of 10^3. This is explained by the small concentration of free electrons ($\approx 10^3$ cm^{-3}).

In Sec. 3 of Chapter IV we shall give an example of the application of Eq. 51.

11. Ionization and Excitation by Electronic Collisions

At high electron temperature present in the nebulae ($T_e > 10,000°K$) some role must be played in the processes of excitation and ionization of hydrogen atoms by their collisions with free electrons. This, in turn, can have some influence on the Balmer decrement. Several authors have even attempted to explain in this way the differences between the observed and computed Balmer decrement which remain after the differential absorption of light in the Galaxy is taken into account.

The first attempt of this kind was made by Miyamoto [69] who computed the approximate collision cross-sections and redetermined the Balmer decrement including collisions. Later on, similar but more exact calculations were carried out by Chamberlain [70], and also by Kaplan and Gopasyuk [71].

Let us call $\alpha_c(T_e)$ the ionization coefficient of the hydrogen atom from the ground state due to collisions. Then, for the number of ionizations Z_{ion} per unit time and per unit volume we have:

$$Z_{ion} = n_1 n_e \alpha_c(T_e), \tag{52}$$

with

$$\alpha_c(T_e) = \int_{v_0}^{\infty} Q(v) v f(v) dv, \tag{53}$$

where $Q(v)$ is the effective collision cross-section of the electron with a neutral hydrogen atom; $f(v)$ is Maxwell's function for the electron velocities; v_0 is the minimum velocity of the free electron necessary to detach the bound electron.

Chamberlain [70] computed $\alpha_c(T_e)$ for a series of values of electron temperature (see the last column of Table 2-14). To do so he used the expression for the collision cross-sections given by Massey and Mohr [72]. Since then it has become known that this expression, which corresponds to Born's approximation, gives good results when the electron energy considerably exceeds the ionization energy, but gives poor results at threshold, *i.e.* when the electron energy is close to the ionization energy.

Geltman [73] developed a more exact theory (S-wave theory) for the ionization of hydrogen and helium by electron collisions for the case when the electron energy E is close to the threshold E_0. From this theory it follows that, for example, for hydrogen the effective collisional ionization cross-section $Q(v)$ can be represented to a high degree of accuracy by a linear dependence on E/E_0. This dependence can be written in the following way:

$$Q(v) = 0.56\pi a_0^2 [(E/E_0) - 1] = 0.56\pi a_0^2 [(v/v_0)^2 - 1], \tag{54}$$

where $\pi a_0^2 = 0.88 \times 10^{-16}$ cm^2. Substituting (54) in (53) and integrating, we find for the function $\alpha_c(T_e)$:

$$\alpha_c(T_e) = 2.24 \, \pi a_0^2 \left(\frac{kT_e}{2\pi\mu}\right)^{1/2} \left(1 + \frac{2kT_e}{\chi_0}\right) e^{-\chi_0/kT_e}, \tag{55}$$

where χ_0 is the ionization potential from the ground state of hydrogen. Table 2-12 gives the quantity $\alpha_c(T_e)$ computed according to this formula for a number of values of T_e.

We should point out that the values of $\alpha_c(T_e)$ obtained by Chamberlain exceeded several times the values shown in Table 2-12. Miyamoto's values, on the other hand, are several times smaller.

TABLE 2-12

Values of the Function $\alpha_c(T_e)$

(in units of $\pi a_0^2 = 0.88 \times 10^{-16}\,\text{cm}^2$)

T_e	$\alpha_c(T_e)$
10,000°K	5.16
20,000°K	2.23×10^4
40,000°K	1.98×10^6

The number of recombination Z_{rec} to all levels per unit volume and per unit time is given by:

$$Z_{rec} = n^+ n_e \sum_{n=1}^{\infty} C_n(T_e) = n^+ n_e C(T_e). \tag{56}$$

Under equilibrium conditions we obviously must have $Z_{ion} = Z_{rec}$. Then, from (52) and (56) we find:

$$\frac{n^+}{n_1} = \frac{\alpha_c(T_e)}{C(T_e)}. \tag{57}$$

This is the ionization equation valid when ionizations take place through inelastic collisions with electrons. It is interesting to see that in this case the degree of ionization does not depend on the electron concentration.

Table 2-13 gives the quantity n^+/n_1 calculated according to (57) for several values of the electron temperature T_e. The values of $C(T_e)$ were taken from Table 2-8.

The electron temperature in the nebulae is usually less than 20,000°K. Therefore, it follows that the collisional ionization of hydrogen does not play a significant role.

Let us consider now the excitation of hydrogen atoms by electron collisions. The number of collisional excitations per unit volume and per unit time will be,

TABLE 2-13

Degree of Ionization of Hydrogen
Associated wtih Electron Collisions

T_e	n^+/n_1
10,000°K	0.001
20,000°K	8
40,000°K	1200

TABLE 2-14

Values of the Functions α_n and α_c According to
Chamberlain (in units of $\pi a_0^2 = 0.88 \times 10^{-16}\,\text{cm}^2$)

T_e	10,000°K	20,000 K	40,000°K
α_2	264×10^1	102×10^4	26.5×10^6
α_3	7.24	7.6	3.49
α_4	1.72	2.13	1.10
α_5	0.497	0.79	0.462
α_6	0.230	0.41	0.250
α_c	2.94	6.08	7.15

obviously, $n_1 n_e \alpha_n(T_e)$ where $\alpha_n(T_e)$ is defined similarly to (53), except that instead of v_0 we must use v_n—the minimum velocity of the free electron necessary to excite the atom to the n-th level. Table 2-14 gives the quantity $\alpha_n(T_e)$ computed by Chamberlain. For $n > 6$, α_n is approximately determined by $\alpha_n = \alpha_6 (6/n)^{3.20}$.

The problem of the probability coefficients for electron collisions has also been considered by Milford [211] and by Hummer [212]. Table 2-15 shows Milford's values for $\alpha_2(T_e)$ and $\alpha_3(T_e)$ taken from [213].

Hummer gives the following expression for $\alpha_2(T_e)$ for the $1S \rightarrow 2P$ transition, for which the effective cross section is more reliably known:

$$\alpha_2(T_e) = 3.9 \times 10^{-8} e^{-\lambda_2 - 0.20 \ln \lambda_2}, \tag{57a}$$

where $\lambda_2 = 11.9 \times 10^4 / T_e$. For $\alpha_c(T_e)$ Hummer gives:

$$\alpha_c(T_e) = 5.3 \times 10^{-11} T_e^{1/2} e^{-157,890/T_e}. \tag{57b}$$

TABLE 2-15

Values of the Functions $\alpha_n(T_e)$ According to Milford
(in units of $\pi a_0^2 = 0.88 \times 10^{-16}\,\text{cm}^2$)

T_e	$\alpha_2(T_e)$	$\alpha_3(T_e)$
10,000°K	146×10^1	3.25×10^1
15,000°	86×10^3	4.07×10^3
20,000	69×10^4	4.57×10^4
25,000	23×10^5	1.98×10^5
30,000	54×10^5	5.33×10^5
35,000	10×10^6	1.08×10^6
40,000	16×10^6	1.85×10^6

Equation (57a) gives values of $\alpha_2(T_e)$ about 15 to 20% higher than Milford's and Eq. (57b) gives values of $\alpha_c(T_e)$ about 30 to 40% higher than Eq. (55).

For the calculation of the distribution of atoms over the different excited states when the excitation is due entirely to collisions we have the following equilibrium condition:

$$n_n \sum_{k=2}^{n-1} A_{nk} = \sum_{k=n+1}^{\infty} n_k A_{kn} + n_1 n_e \alpha_n(T_e). \tag{58}$$

To elucidate fully the role of electron collisions on the character of the Balmer decrement we solved system (58), using Chamberlain's data for $\alpha_n(T_e)$. It turns out that the degree of excitation due to collisions at $T_e = 10,000°K$ is less $(n_i/n_1 \approx 10^{-15})$ than that due to photoionizations followed by recombinations.

If we compare Table 2-16, which gives the Balmer decrement when excitation is due to collisions only (for a nebula optically thick in the Lyman lines), with Table 2-9, we see that the Balmer decrement becomes steeper than in the case of excitation due to photoionizations and recombinations.

In those cases when the excitation and ionization are due simultaneously to photoionizations (recombinations) and collisions, the resulting Balmer decrement can be obtained as a linear combination of the Balmer decrements due separately to recombinations and collisions, that is:

$$\frac{E_{n2}}{E_{42}} = (1 - x) \left(\frac{E_{n2}}{E_{42}}\right)_R + x \left(\frac{E_{n2}}{E_{42}}\right)_C, \tag{59}$$

where χ depends on n_1/n_e and T_e, and the subscripts R and C refer respectively to recombinations and collisions.

Aller and Minkowski [74] made an attempt to show that the collisions of hydrogen atoms with free electrons have an influence on the Balmer decrement of the nebula NGC 7027. By comparing the intensity ratios of the Paschen and Balmer lines with their theoretical values, which are given by $E_{n3}/E_{n2} = A_{n3} \nu_{n3}/A_{n2} \nu_{n2}$,

TABLE 2-16

Values of the Balmer Decrement, Resulting
from Electron Collisions Alone

T_e	10,000°K	20,000°K	40,000°K
H$_\alpha$	5.76	4.79	4.06
H$_\beta$	1.00	1.00	1.00
H$_\gamma$	0.291	0.347	0.383
H$_\delta$	0.136	0.169	0.194
H$_\epsilon$	0.076	0.097	0.112

they determined the differential absorption of light by the Galaxy for this nebula in the region between $\lambda = 3750$ Å and $\lambda = 8750$ Å. The observed Balmer decrement, corrected for interstellar absorption, fell between the theoretical Balmer decrements shown in Tables 2-9 and 2-16. The authors found that the best agreement between theory and observation was obtained for $T_e = 14{,}000°$K and $x = 0.54$. The large value of x would indicate that in NGC 7027 collisions are as important as recombinations in determining the hydrogen emission.

However, as shown by Sobolev [75, 76], collisions cannot have a great influence on the Balmer decrement of planetary nebulae. The reason is that the fraction of energy of the free electrons which goes into exciting the Balmer lines is very small. This is explained, firstly, by the fact that the mean energy of an ejected electron at the stellar temperature of about 60,000°K is approximately two times smaller than the excitation energy of the hydrogen levels. Secondly, a considerable fraction of the energy of the free electrons goes into exciting the "nebulium" lines, and into continuous emission through recombinations and free-free transitions. Thirdly, in the inelastic collisions of the free electrons with hydrogen atoms most of the energy goes into exciting L_α, and not the Balmer lines, as follows from the data in Table 2-14 (by comparing α_2 and α_c).

We can give a quantitative estimate of the influence of collisions on the Balmer decrement by calculating the relative number of Balmer quanta arising from collisions and from recombinations. Sobolev gives the following relation for the computation of this ratio:

$$\frac{n_e n_1 \sum_{3}^{\infty} \alpha_n}{n_e n^+ \sum_{3}^{\infty} C_n} < \frac{\varepsilon \sum_{1}^{\infty} C_n \sum_{3}^{\infty} \alpha_n}{g v_{12} \sum_{3}^{\infty} C_n \sum_{2}^{\infty} \alpha_n}, \tag{60}$$

where $\bar{\varepsilon}$ is the mean energy of an electron ejected through a photoionization process. Computations show that with $T_* = 60{,}000°$K this ratio is less than 0.04 if $T_e = 10{,}000°$K and less than 0.1 if $T_e = 20{,}000°$K. Thus, for each Balmer quantum which originates through a collision we have a few tens of quanta which originate through recombinations.

In summary, we can say that both the observational evidence and the theoretical calculations show the negligible influence that electron collisions have on the Balmer decrement of planetary nebulae. The conclusions of Aller and Minkowski were based, apparently, on incorrect estimates of the quantities that enter in the calculation of x (the magnitude of the interstellar absorption, the degree of ionization, the electron temperature and the electron concentration in NGC 7027, etc.).

Returning to the question of the collisional excitation of atoms we should point out that hydrogen has the metastable state $2S$, whose excitation energy is 10.15 e.v. The mean life of the atom in this state is of the order of 0.1 sec. Therefore, an accumulation of atoms in this level can occur through collisions with electrons of sufficient energy, that is, the $2S$ level will have an abnormally high population. The atom can leave the $2S$ state by a process of two-photon emission (see Chapter V).

In a stationary state we have:

$$n_1 n_e \alpha_{2S}(T_e) = n_2 A_{2S,1S}, \tag{61}$$

where the left hand side gives the number of transitions $1 \rightarrow 2$ due to collisions; the function $\alpha_{2S}(T_e)$ is similar to (53); for the $1S \rightarrow 2S$ transition; $A_{2S,1S} = 8.227$ sec^{-1}. Calculations carried out for $T_e = 20{,}000°$K give (a graph of the function $Q(T_e)$ for the $1S \rightarrow 2S$ transition is shown in [77]):

$$n_{2S}/n_1 \approx 10^{-7},$$

which exceeds considerably the degree of excitation of the ordinary levels.

12. Chemical Composition of Planetary Nebulae

In principle the determination of the chemical composition of the nebulae is simple: it can be obtained by comparing the theoretical and observed intensities of the emission lines in their spectra. In general, the relative intensities of the lines are compared, and in this way one determines the relative amount of one or other atom or ion in the nebula. To go to absolute abundances one must know the absolute amount of some element present in the nebula, for example, hydrogen. For the latter, it is sufficient to know the electron concentration in the nebula, since the number of free electrons is practically equal to the number of hydrogen atoms. The theoretical determination of the intensity of a line of some atom or ion which arises as the result of a recombination begins by establishing the corresponding equilibrium equations in a way analogous to that used in the case of hydrogen. For the solution of these equations we need to know the recombination coefficients and the spontaneous transition probabilities. The end result of the solution of the equations is a knowledge of the relative number of atoms or ions in the different excited states and, consequently, of the energy emitted by the nebula in each line by means of relations similar to (40). Writing equation (40) for a given atom, hydrogen for example, we find the theoretical relative concentration of these atoms in the nebula. Now, in order to compute the relative number of these atoms in the nebula, all we need to know are the observed relative intensities of their lines.

In the same way we can find the relative concentrations of the ions that emit the forbidden lines in the spectra of the nebulae. The only difference is that in this case, in addition to the spontaneous transition probabilities, we need to know the collision cross-sections.

In spite of the simplicity of the procedure for the determination of the chemical composition of planetary nebulae, in practice the difficulties are so great that the problem becomes one of the hardest in the physics of nebulae. There are several reasons for this. Firstly, the necessary atomic parameters, in particular the recombination coefficients, spontaneous transition probabilities and effective collision cross-section, are not known with a high enough degree of accuracy for all the relevant atoms and ions. Secondly, the atoms of a given element can exist in different stages of ionization, and we just determine the number of atoms in one stage only. True, there are methods to go from one stage of ionization to another, and also from one stage of excitation to another, by means of the ionization and excitation equations. But this cannot be accomplished with sufficient accuracy since it requires the use of several parameters (stellar temperature, dilution coefficients) which themselves are poorly determined. Finally, in several cases the atoms simply do not produce lines in the observable region of the spectrum and hence we can say nothing about their presence in the nebula. In this case we are comitting even a qualitative error in the determination of the chemical composition of the nebula.

The first serious attempt to determine the chemical composition was made by Bowen and Wyse [78] who, on the basis of their data, reached the conclusion that the chemical composition of gaseous nebulae is basically the same as that of the Sun. A fuller investigation of the chemical composition of planetary nebulae was carried out by Aller and Menzel [79]. They found a more or less constant chemical composition for seven planetary nebulae on the basis of the assumption of the identical chemical composition of the nebulae and their stars.

A particularly careful investigation was carried out for the nebula NGC 7027 [74, 80]. Table 2-17 gives the chemical composition of this nebula (*i.e.* the relative number of atoms) and of a "normal" star.

TABLE 2-17

Chemical Compositions of the Nebula NGC 7027 and "Normal" Stars
(Logarithm of the Relative Number of Atoms)

Element	NGC 7027	Stars	Element	NGC 7027	Star
H	13.04	13.27	S	9.21	8.52
He	12.0	12.0	Cl	8.39	8.3:
N	9.56	9.49	A	8.22	9.0:
O	10.00	10.00	K	6.2:	6.39
F	7.4:	7.7:	C	6.6:	7.72
Ne	9.48	10.04			

Hydrogen is the most abundant element in planetary nebulae. Next comes helium, which often amounts to 10% of the number of hydrogen atoms. (However, in several cases there exist large departures from this value. Thus, for example, according to [79] in the nebula NGC 2440 helium is two to three times overabundant, and in the nebula IC 418 it is almost four times underabundant, compared to the "mean" nebula.). Next come carbon, nitrogen, oxygen and so forth. The total abundance of all elements in the nebulae, beginning with carbon, is of the order of one per cent (relative to hydrogen).

TABLE 2-18

Comparison of the Mean Chemical Abundances of
Planetary Nebulae, the Sun and B Stars

Element	Relative Number of Atoms		
	Planetary Nebulae	*Sun*	*B Stars*
Hydrogen	1000	1000	1000
Helium	190	220	160
Oxygen	0.6	0.91	0.6
Carbon	0.4	0.52	0.2
Nitrogen	0.24	0.096	0.15
Neon	0.11	—	0.52
Silver	0.053	0.054	0.03
Argon	0.0076	—	0.008
Chlorine	0.002	—	0.0016
Fluorine	0.00023	—	0.0035

Table 2-18 gives a very approximate "mean composition" of planetary nebulae and a comparison with the results obtained for the atmospheres of the Sun and of τ Scorpii [79]. Although very different methods were used to determine the chemical composition of the various objects, the relative contents of the elements, in particular helium, nitrogen, oxygen and sulphur, seems to be almost identical. In some cases, in particula as regards carbon, significant differences have been observed, but the data are so uncertain that no final conclusion can be made about the reality of these differences. One must keep in mind that the data shown on Table 2-18 are to a certain extent obsolete, and we present them only for lack of new results on the chemical composition of planetary nebulae.

TABLE 2-19
Values of C for Various Transitions

q	SL	$S'L'$	C	q	SL	$S'L'$	C
1	^2P	^1S	1/3	4	^3P	^4S	4/9
2	^3P	^2P	2/3	4	^3P	^2D	5/9
2	^1D	^2P	2/3	4	P	^2P	1/3
2	^1S	^2P	2/3	4	^1D	^4S	0
3	^4S	^3P	1	4	^1D	^2D	1
3	^4S	^1D	0	4	^1D	^2P	1/3
3	^4S	^1S	0	4	^1S	^4S	0
3	^2D	^3P	1/2	4	^1S	^2D	0
3	^2D	^1D	1/2	4	^1S	^2P	4/3
3	^2D	^1S	0	5	^2P	^3P	1
3	^2P	^3P	1/2	5	^2P	^1D	5/9
3	^2P	^1D	5/18	5	^2P	^1S	1/9
3	^2P	^1S	2/9	6	^1S	^2P	2

TABLE 2-20
Values of the Constants B, s and α for Various Ions

Ion	B	s	α	Ion	B	s	α
Ne	2.5	1	4.3	F^{++}	4.5	2	1.7
Fe	3.7	1	4.1	O^{++}	5.23	2	1.30
O	5.7	1	4.0	N^{++}	6.0	2	0.9
N	8.9	1	3.1	Al^{+++}	1.9	2	1
C	15.8	1	1.9	Mg^{+++}	2.3	2	1
B	31	—	—	Na^{+++}	2.7	2	1
Na^+	4.0	2	4.2	Ne^{+++}	2.95	2	1.20
Ne^+	5.0	2	3.8	F^{+++}	3.1	2	1
N^+	6.4	2	3.1	O^{+++}	3.33	3	1.82
O^+	8.1	2	2.45	Si^{++++}	1.6	2.3	1
N^+	10.2	2	2.06	Al^{++++}	1.8	2.3	1
C^+	12.9	2	1.67	Mg^{++++}	1.9	2.3	1
Mg^{++}	2.6	2	2.65	Na^{++++}	2.0	2.3	1
Na^{++}	3.2	2	2.4	Ne^{++++}	2.16	2.31	1
Ne^{++}	3.82	2	2.07	F^{++++}	2.4	2.3	1

13. Continuous Absorbtion Coefficients of Several Elements

To conclude this chapter we present Tables with the numbers needed to determine the continuous absorption coefficients of several atoms and ions of astrophysical interest (Tables 2-19 and 2-20). The calculations are due mainly to Seaton, and also to Burgess, Bates and others.

The continuous absorption coefficient for an atom or ion with an electron in the configuration $2p^q$ can be written [162]:

$$\alpha_\nu = 10^{-18} C(p^q SL \to p^{q-1} S'L') \cdot B\{\alpha(\nu/\nu_0)^{-s} + (1-\alpha)(\nu/\nu_0)^{-s-1}\}\ cm^2. \quad (62)$$

Here ν is the frequency and ν_0 is the ionization frequency of a given atom or ion.
The probable error in the computation of a continuous absorption coefficient for an ion is of the order of $\pm 20\%$.

The value of the coefficient C is given in Table 2-19, while Table 2-20 gives the constants B, α and s.

Chapter III

The Problem of Distances and Dimensions of Planetary Nebulae

1. The State of the Problem

One of the most difficult problems associated with planetary nebulae is that of their distances and dimensions. It seems that in no other problem concerning planetary nebulae have the expended efforts been so nearly fruitless. In fact, at present we do not have a single method allowing us to determine the distances of planetary nebulae from us with an accuracy comparable to the accuracy with which we determine distances, say, of ordinary stars and galaxies. Even in the case of diffuse nebulae the situation is much better, for these objects, as a rule, are associated with ordinary hot stars, open star clusters and stellar associations, whose distances from us may be determined accurately and without particular difficulty.

The reason for such a situation in determining the distances of planetary nebulae lies in the fact that not one of the basic physical or geometrical parameters has a constant value for all planetary nebulae. It is possible to mention at least five such parameters which various authors have attempted to utilize in some way for this purpose:

1) the absolute stellar magnitude M_n of the nebula (plus the nucleus);
2) the absolute stellar magnitude M_* of the nucleus;
3) the difference δ between the apparent stellar magnitude m_* of the nucleus and the integrated magnitude m_n of the nebula; $\delta = m_* - m_n$;
4) the linear diameter D of the nebula;
5) the mass $\mathfrak{M}$ of the nebula.

The values of certain of these parameters ($\mathfrak{M}$, M_*) are different in different nebulae and others (D, δ, M_n), as well as the first, vary with the course of time in one and the same nebula. Nevertheless, the existing methods for determining the distances of planetary nebulae are based on assumptions of constancy of these or other parameters in all nebulae and in all stages of ionization.

Even if one proceeds by assuming that certain parameters are constant in some group of nebulae, he is faced with a new question: how does one determine the zero-point distance? In other words, how does one establish the distance of one or more planetary nebulae to fix the distance scale? The existing methods of astrometry are insufficiently accurate for the solution of this problem. The determination of trigonometric parallaxes for even the nearest of the planetary nebulae is impossible because of their large distances from us [81]. Both the determination of statistical parallaxes based on the analysis of proper motions of planetary nebula, and the analysis of radial velocities on the basis of the theory of galactic rotation have been employed in trying to solve this problem. Generally speaking, statistical parallaxes are able to give values, correct on the average, for the orders of magnitudes of the distances to the groups of planetary nebulae under study, but they are unable to guarantee accurate determinations of the distances of individual objects.

An exception at present is provided by planetary nebulae in other galaxies. In this case the distances of the nebulae are obtained accurately since the distances of the galaxies may be reliably measured. However, in this way one observes and determines the absolute magnitude M_n (see Section 4) of *gigantic* planetary nebulae. The fact of this selection alone is evidence that in reality M_n, for example, is not constant for all planetary nebulae.

At present there exist five catalogues of distances and dimensions of planetary nebulae, compiled by: B.A. Vorontsov-Vel'yaminov [16], L. Berman [18], G. Camm [36], P.P. Parenago [82], and I.S. Shklovskii [17]. The discrepancies in the distance given by the catalogues for any particular nebula are very large in the majority of cases—from a few to ten and even twenty (!) times.

In addition systematic discrepancies appear among the data of their catalogues. The greatest disagreements are found between the catalogues of Berman and I.S. Shklovskii: in the first the distances are, on the average, very large while in the second they are very small.

Under these conditions it is pointless to dwell in detail on a description and analysis of the methods underlying the construction of these catalogues. We shall only enumerate these procedures (in the following section) and indicate their fundamental defects.

The importance of the values of the distances and dimensions of planetary nebulae is very great. The distances are necessary, in particular, to obtain a correct idea of the spatial distribution of planetary nebulae and thereby of the form and structure of the system consisting of these objects. Knowledge of the linear dimensions of planetary nebulae is in itself interesting and moreover gives us the possibility of determining some important physical quantities; the electron concentration, total mass, lifetime of the nebula and so forth, and also the dispersion of these quantities. All of this will eventually lead to an understanding of the nature of planetary nebulae and of their role in the evolution of stars and stellar systems.

Having this in mind, it is particularly important to search for new methods of determining the distances of planetary nebulae. In the future, most probably, we will come to abandon the existing principle of using one *general* method for all planetary nebulae. One must look for the correct solution of the problem, apparently, in the development of *individual* methods for determining distances, applicable only with respect to definite types or categories of planetary nebulae. It may also turn out that a fruitful method is the one (first used by Berman) of dividing the nebulae into groups with more or less similar values of one or more parameters.

2. Methods of Determining Distances of Planetary Nebulae

Zanstra's method [83]. Zanstra discovered the existence of the relation

$$m_* = const. + a(m_* - m_n),\qquad(1)$$

where $a = 0.7$. Provided that the difference $\delta = m_* - m_n$ does not depend on the distance of the nebula, the relation (1) may be interpreted as expressing a connection between δ and the absolute photographic stellar magnitude M_* of the nucleus.

Knowing M_* and m_*, one may establish the distance of the nebula if the value of the constant in (1) is known. This constant may be evaluated from the formula of galactic rotation using the known radial velocities of the nebulae.

However, as was later shown by B.A. Vorontsov-Velyaminov on the basis of a large amount of material, the coefficients a is equal to 0.85, which is sufficiently close to unity that the dependence in question vanishes.

Vorontsov-Vel'yaminov's method [85, 16, 50]. The author of this procedure discovered a statistical dependence between the mean photographic surface brightness H of a nebula and its apparent diameter D'' of the form

$$H = 2.65 + 4 \log D''.\qquad(2)$$

The existence of such a relation may be interpreted as an indication of the constancy for all nebulae of the absolute integrated magnitude M_n. Then one determines the distances of individual planetary nebulae by the usual procedure from the known values of M_n, allowing for interstellar absorption. Later, in order to account for the temperature of the nucleus, a small correction was introduced into relation $M_n = const.$, which then takes the form [16]

$$M_n = 0.04 - 0.22\delta.\qquad(3)$$

The numerical values in (3) are found by various procedures using the statistical parallaxes derived from an analysis of both the components of proper motion and also of the radial velocities in combination with the formulas of galactic rotation. On the average for all nebulae examined (nearly one hundred fifty) a value $M_n = -0^m.64 \pm 0^m.45$ was obtained.

It should be emphasized that the dependence between H and $\log D$ is far from sharp—the scatter in the value of H for a fixed value of $\log D$ amounts to four or more stellar magnitudes. The assumption $M_n \approx const.$ is therefore an extremely rough approximation. Furthermore this assumption is in principle not acceptable. The point is that the total luminosity L of the nebula and consequently also the absolute magnitude M_n can be constant in time only while the nebula has comparatively small dimensions, that is, when its optical thickness τ_c at the limit of the Lyman series is larger than unity. However as the dimensions of the nebula increase, τ_c becomes less than unity and L then decreases with the expansion of the nebula (see Chapter VI, Sec. 9). It is not difficult to show that for a five-fold increase of the diameter of the nebula, L and consequently M_n decrease by $3^m.5$ in the case when $n_e \sim r^{-2}$ (n_e is the electron density) and by $5^m.2$ when $n_e \sim r^{-3}$.

There is reason to suppose that for the majority of planetary nebulae (not stellar) τ_c is less than unity, *i.e.*, they are optically thin; consequently for them the assumption $M_n \approx const.$ cannot be regarded as correct.

The existence of a large dispersion in absolute magnitude M_n is indicated also by observations of planetary nebulae located in other galaxies (see the following section).

I.S. Shklovskii objected to the method of B.A. Vorontsov-Vel'yaminov on the grounds that the statistical dependence expressed by (2) is not real and is completely a result of observational selection in the sense that one does not observe objects simultaneously combining low surface brightness and small angular dimensions. This objection is invalid, for the relation (2) is derived from an analysis of the data pertaining to *real* observed nebulae (brighter than the fourteenth stellar magnitude) and thus the conclusion $M_n \approx const.$ drawn on the basis of this relation pertains as well to *real* existing objects. The relation (2) may be accepted or rejected depending on whether or not we are able to accept a spread in the values of H (*i.e.* M_n) amounting to 4^m or larger. If another method is found for determining distances in which the scatter of the parameter being utilized is smaller than in the present case, then the method of B.V. Vorontsov-Vel'yaminov will have to be abandoned. We note that the values of the distances of individual nebulae shown in this catalogue may differ from their true value by a factor of five or more.

Berman's method [18]. The problem of establishing distances of planetary nebulae was investigated by Berman in connection with the determination of the constants in the formula of galactic rotation. He utilized the following quantities: a) mean statistical parallaxes, derived from an analysis of proper motions; b) angular diameters; c) radial velocities; d) the apparent stellar magnitudes m_n and m_* of the nebula and nucleus respectively.

Berman employs the method of successive approximations. In the first approximation he proceeds from the apparent diameters of the nebulae, considering them as indications of the distance, and then dividing the nebulae into groups according to distance. He also takes account (first) of the effect of interstellar absorption. Finally the true distance r is calculated according to the formula

$$\log r = 0.2[(m_n + 5) - kr - M_n],\qquad(4)$$

where $M_n = 0^m.88 - 0.32\delta$ and $k = 0^m.55$ is the amount of interstellar absorption per kiloparsec. The mean value is $M_n = -1^m.6$.

In criticizing Berman's method, we should note that if, for nebulae with larger apparent dimensions, their angular diameters may be in some way considered as indications of distance, then for nebulae with smaller apparent dimensions (stellar) this assumption is quite inapplicable. A nebula with small angular dimensions may be small in absolute dimensions and still exist near to us, as was mentioned, for example, by Oort [84]. Berman also assumed interstellar absorption to be uniform in all directions and at all distances, which is a highly idealized approximation in no way corresponding to reality.

The distances of planetary nebulae obtained by Berman appear very large, in disagreement with the data of other catalogues. Generally Berman's results inspire little confidence, although the values of the constants in the formulae of galactic rotation derived from these results agree well with the values obtained by other methods.*

Camm's method [36]. Camm postulates the existence of a linear relation between D and M_n, i.e., of the relation

$$D = aM_n + b,\qquad(5)$$

for all planetary nebulae. Here D is the linear diameter of the nebula and a and b are constants which Camm determines from the assumption that the dispersion in linear diameter and absolute magnitude of the nebulae be minimized. In addition, the condition was imposed that, on the average, $\overline{M}_n = +0^m.2$.

The distance of the nebula is determined by the usual procedure from M_n and m_n without allowing for interstellar absorption.

The distances derived by Camm are somewhat smaller than those by Berman but larger than those by B.A. Vorontsov-Vel'yaminov. Nevertheless Camm's catalogue also does not inspire confidence, if only because the relation (5), on which it is based, is not justified and has no physical significance. In a more accurate state-

---

* B.A. Vorontsov-Velyaminov [50] has commented on the error committed by Berman in assuming m_n for the stellar magnitude m of the nucleus is Hubble's well-known relation ($m = 17.64 + 5 \log A$). This remark, however, is not of great significance, the point being that Berman used Hubble's formula as an interpolation formula, after introducing the appropriate correction for changing to the m_n system. There, however, is no direct reference to this in his work.

ment of the problem, in place of (5) one should write, if only for optically thin nebulae $(\tau_c < 1)$:

$$\log D = aM_n + b, \tag{6}$$

in which the values of the constants a and b are to be determined from two independent conditions, without requiring the existence of a minimum dispersion in the quantities D and M_n.

Parenago's method [82]. This does not differ significantly from the procedure of B.A. Vorontsov-Vel'yaminov. Parenago took account of the effect of interstellar absorption more accurately, having kept in mind its considerable nonuniformity in different directions. Moreover, new data on the proper motions of planetary nebulae was included in the determination of mean statistical parallaxes. For the mean absolute photographic magnitude of the nebulae Parenago assumed $M_n = -0^m.8$.

Shklovskiĭ's method [17]. The starting point for this procedure is the well-known formula of V. A. Ambartsumyan, which gives a relation among the mass $\mathfrak{M}$ of the nebula, its total luminosity L and volume V (formula (57), Chapter IV):

$$\mathfrak{M} = C\sqrt{LV} \tag{7}$$

On the other hand, we have $L \sim r^2 I$, where I is the surface brightness and r is the linear radius of the nebula. Comparing this with (7) and remembering that $V \sim r^3$ and $r = R\phi''$, where R is the distance of the nebula from us and $2\phi''$ is the angular diameter, we find

$$R \sim \frac{\mathfrak{M}^{2/5}}{\phi'' I^{1/5}}. \tag{8}$$

This relation is valid, however, only for optically thin nebulae $(\tau_c < 1)$, since for $\tau_c > 1$ the formula (7) does not determine the total mass of the nebula.

The dependence of R on $\mathfrak{M}$ is comparatively weak. I.S. Shklovskiĭ assumed that the dispersion in the masses of planetary nebulae is not great and consequently that one may take approximately $\mathfrak{M} = const$. Then from (8) one may determine the relative distances of the nebulae primarily from their apparent dimensions, but also partially from their surface brightnesses (the dependence of R on I is very weak.) The zero-point was determined by the procedure already applied by other authors.

There is reason to believe, however, that the dispersion in the masses of planetary nebulae is not so small (see below). The extreme values of the masses of planetary nebulae differ from each other by more than two orders of magnitude. This leads to errors as large as a factor of ten in determining the distances of individual nebulae, *i.e.* to errors of the same order or somewhat larger than in the procedure of B.A. Vorontsov-Vel'yaminov. However, the discovery by Osterbrock [88] of a correlation between the surface brightnesses of nebulae and their electron density speaks in favor of Shklovskiĭ's catalogue.

Subsequently O'Dell [214] has somewhat refined Shklovskiĭ's method by utilizing the mean surface brightness in H_β instead of in the photographic region of the spectrum. O'Dell's catalogue of distances of planetary nebulae differs somewhat from Shklovskiĭ's but on the average the distances appear to be of the same order. Apparently at present O'Dell's catalogue should be regarded as the most acceptable.

In conclusion, it should be emphasized that the uncertainty in the zero-point distance of planetary nebulae is the basic defect of all catalogues.

3. The Dispersion of Luminosity and Mass in Planetary Nebulae

A clear idea of the dispersion in the absolute stellar magnitudes M_n may be obtained from observations of planetary nebulae in other galaxies. Of particular interest in this respect are the observations of Lindsay [43] and Koelbloed [44] in the Small Magellanic Cloud. In Figure 1-4a (page 22) is shown the distribution observed by Lindsay of the absolute magnitude of the brightest planetary nebulae over the whole Small Magellanic Cloud and in Figure 1-4b, the distribution observed by Koelbloed in seven areas in the Small Magellanic Cloud. From these results the absolute stellar magnitude M_n of planetary nebulae in the Small Magellanic Cloud are found in the interval from $-5^m.5$ to $-1^m.0$. This estimate of the dispersion of M_n is a lower bound to its actual value. Undoubtedly, in the future, as the limiting stellar magnitude of photographic processes increases, more nebulae with small luminosity will be found. Parallel with this the number of observed planetary nebulae will increase. Obviously the subsequent increase in the number of planetary nebulae will end at some limiting stellar magnitude of photographic processes. This as well establishes the actual value of the dispersion in M_n.

Thus, if one assumes that there is no significant difference in the nature of planetary nebulae in the Small Magellanic Cloud and those in our Galaxy, one may assert that the dispersion in absolute stellar magnitudes of the nebulae in our Galaxy exceeds four or five stellar magnitudes.

It is somewhat more difficult to make a realistic estimate of the dispersion in the masses of planetary nebulae. The difficulty lies in the fact that we at present are unable to determine the maximum value of the difference $\delta = m_* - m_n$ for those nebulae whose nuclei are invisible. Nevertheless the following analysis allows us to reach some conclusions concerning this problem.

For optically thin nebulae we can write, from (7):

$$R \sim \frac{\mathfrak{M}^{2/3}}{\phi'' L^{1/3}}. \tag{9}$$

Let us express the total luminosity L of the nebula (without the nucleus) in terms of the photographic stellar magnitudes m_n and m_* of the nebula and the nucleus alone, respectively. We have

$$L = 10^{-0.4m_n}(1 - 10^{0.4\delta}),\qquad (10)$$

where $\delta = m_* - m_n$. Substituting (10) into (9), we obtain

$$R \sim \frac{\mathfrak{M}^{2/3}}{\phi'' 10^{-0.133m_n}(1 - 10^{-0.4\delta})^{1/3}} \cdot\qquad (11)$$

The quantity in parentheses in the denominator of this expression is close to unity for small values of δ. For example, this factor is of the order of 0.5 for $\delta = 0.1$. Consequently when $\delta > 0.1$ we may write

$$R \sim \frac{\mathfrak{M}^{2/3}}{\phi'' 10^{-0.133m_n}} \cdot\qquad (12)$$

Since the condition $\delta > 0.1$ is satisfied for the majority of planetary nebulae one may conclude from (12) that the distance of a nebula of given mass is a function of its angular diameter ϕ'' and the total (nebula plus nucleus) photographic magnitude m_n, but does not depend on the apparent photographic magnitude m_* of the nucleus.

Let us assume that the masses of all planetary nebulae are equal. Then for the determination of the distance of some planetary nebula, it is sufficient to have ϕ'' and m_n, i.e., occurring at *equal distances* from us but having *different* values of δ. To different values of δ will correspond different values of the ratio of the luminosities L and L_* of the nebula and nucleus, respectively, given by

$$L/L_* = 10^{0.4\delta} - 1.\qquad (13)$$

We observe that the ratio L/L_* does not depend on the distance of the nebula from us.

It appears that such groups of nebulae exist. Without discussing the details, we show only that for the first of these groups δ is in the range $0^m.1 - 0^m.2$ and for the second, $\delta \sim 7^m$. In the second case the cited magnitude disregards the maximum value of δ for all combinations of planetary nebulae in general. It is evident that among them, of the remote and relatively bright specimens, a number of which are *entirely* invisible, there will be nebulae with still larger values of δ. Leaving aside for the present the possibility that a significant dispersion may exist in the physical characteristics of the nuclei, we may assert in general that for larger values of δ one must have giant nebulae. It is then possible to give a more-or-less reasonable estimate for the upper limit of δ, by turning to the planetary nebulae, for example, in the Small Magellanic Cloud. The absolute magnitude of these nebulae is of order $- 5^m$. There is reason to assume that the luminosity of the giant planetary nebulae in the Small Magellanic Cloud exceeds the mean luminosity of the nebulae in our Galaxy by at least 2–3 stellar magnitudes. For the maximum value of δ we then obtain magnitudes of order $9-10^m$. Thus we have $\delta_{min} \approx 0^m.2$ and $\delta_{max} \approx 10^m$. With these values we have from (13)

$$(L/L_*)_{min} \approx 0.2 , \quad (L/L_*)_{max} \approx 10^4, \tag{14}$$

i.e., the extreme values of the ratio (L/L_*) differ by a factor of 5×10^4.

This dispersion in the ratio of the luminosity of a nebula and its nucleus may be a consequence of dispersion in three quantities: a) the luminosity of the nucleus, i.e. of its Planck temperature ($T_* \sim L_*^{1/4}$); b) the radius of the nucleus ($r_0 \sim L_*^{1/2}$); c) the mass of the nebula. In the first case the limiting values of the temperature of the nucleus must differ by a factor of fifteen, in the second case the radius of the star must change by a factor of 200 and in the third case the mass of the nebula must vary by 200 times. In reality, however, for a change in L/L_* the quantities T_*, r_0, and M may decrease simultaneously, so that their dispersions will be less than the above values. But the upper estimate obtained above for the dispersion of the ratio L/L_* cannot be regarded as an upper bound. In some way the conclusion that a large dispersion exists either in the characteristics of the nucleus or in the amount of mass in the nebula should be obvious. If the dispersion in the physical characteristics of the nucleus exceed considerably their probable values, it is also possible that the limiting values of the masses of the nebulae differ by at least two orders of magnitude.

4. Astrophysical Methods of Determining the Distances of Nebulae

Let us dwell now on some purely astrophysical methods of determining the distances of planetary nebulae which, although they are not capable of wide application, are able to give good results in particular cases. In the past these methods have been applied only rarely if at all.

The method of the Balmer decrement. The relative intensity of the hydrogen emission lines, i.e. the Balmer decrement, depends only on the electron temperature of the nebula. But this dependence is very weak and may be assumed to be the same for all nebulae. However, because of selective interstellar absorption, this theoretical decrement is distorted and, as a rule, becomes significantly steeper. The observed decrement will differ more strongly from the theoretical with increases both in the distance of the nebula and in the value of the interstellar absorption per unit length. Since the deviation is strongest for the early lines of the series and decreases toward the late lines, the greatest variation occurs in the quantity H_α/H_β. The

TABLE 3-1

The Observed Values of H_α/H_β for Several
Nebulae

Nebula	H_α/H_β	Reference
NGC 7027	6.50	1
NGC 7662	3.90	2
NGC 2392	3.80	2
NGC 40	8.00	2
NGC 2022	3.06	2
NGC 1535	3.61	2
IC 351	3.50	2
J 320	2.80	2
NGC 3918	3.24	3
NGC 6302	12.	3
NGC 6326	2.32	3
IC 4776	7.67	3

(1) L.H. Aller, I. Bowen, and R. Minkowski,
Ap. J., **122**, 62, 1955
(2) R. Minkowski and L.H. Aller, *Ap. J.,*
124, 93, 1956
(3) D. Evans, *M.N.,* **150**, 1959

theoretical value of H_α/H_β equals 2.70 for nebulae optically thick in the Lyman lines and 1.90 for optically thin nebulae. In Table 3-1 appear the observed values of H_α/H_β for several planetary nebulae. From this data we see significant variations from the theoretical values in the observed ratio H_α/H_β in several cases.

These circumstances may be utilized, if only in certain cases, to determine the distances of nebulae, if one knows the law of interstellar absorption in a given direction and the amount of absorption per kiloparsec (specific absorption coefficient). Designating by a_α and a_β the specific absorption coefficients in the lines H_α and H_β, we have for the determination of the distance R of the nebula

$$R = \frac{a_\alpha}{a_\beta} \frac{2.30}{(a_\alpha/a_\beta) - 1} \log \frac{(H_\alpha/H_\beta)_{obs}}{(H_\alpha/H_\beta)_{theor}} , \tag{15}$$

where a_α and a_β are determined either from the very general relation

$$a_\lambda = a_{pg} \left(\frac{4400}{\lambda} \right)^n \tag{16}$$

or from the data of Whitford [86].

It should be noted that the quantity R is not especially sensitive to changes in the observed values of H_α/H_β, but is sensitive to the exponent n in the law of absorption. Also, in using (15) it is sufficient to have the value of the ratio a_α/a_β without knowing the absolute values of a_α and a_β.

The difficulty in using this method lies in the fact that we do not always have available information on the interstellar absorption processes in the direction of the nebula under study. Also the observational determination of the quantities H_α/H_β with sufficient accuracy for our purposes is difficult, although both lines are among the brightest in the spectra of nebulae.

Sobolev's method [76]. The electron concentration in planetary nebulae, as we shall see in Chapter IV, may be determined by various methods. In one of these procedures, based on the utilization of the intensities of forbidden lines, it is not necessary to know the distance to the nebula, and in another, using the intensities of the Balmer lines or Balmer continuum, the distance must be known. Consequently, by finding the value of n_e from the forbidden lines (equation (47), Chapter IV) and substituting it into the formula (46) or (47) (page 108), we are able to determine the distance to the nebula.

This is a purely astrophysical method of determining the distances to planetary nebulae. But because of the strong dependence of R on n_e (according to formula (46), the distance is proportional to the square of n_e) inaccuracies in the value of n_e may lead to large errors in the determination of R.

One may, however, attempt to obtain reliable results using this method by applying it to a particular part of the nebula. In the majority of cases it is actually possible to establish the equatorial plane of the nebula. In bipolar nebulae, for example, the equatorial plane passes under both "caps". Now, if one takes a sufficiently narrow band of width d near this plane, one may assume with sufficient accuracy that the distribution of electron concentration $n_e(r)$ in this plane is cylindrically symmetrical. The total energy E_n^0 radiated by this band per second in any line H_n of the Balmer series may be written as

$$E_n^0 = i_n \int_V n_e^2(r)dV = i_n 2\pi r_0^3 d \int_0^1 n_e^2(r)rdr, \tag{17}$$

where d and r are measured in units of the external linear radius r_0 of the nebula and i_n equals (see formula (41a), Chapter II):

$$i_n = 1.778 \cdot 10^{-17} \frac{b_n}{T_e^{3/2}} \frac{g_{2n}}{n^3} e^{x_n}. \tag{18}$$

If R is the distance of the nebula from us, then one may write for observed energy flux E_n emitted from the band under discussion in the line H_n (in absolute energy units)

$$E_n^0 = E_n 4\pi R^2. \tag{19}$$

Substituting from here the value E_n^0 into (17) and using $r_0 = R\phi/2$, where ϕ is the angular diameter of the nebula, we find

$$R = \frac{16}{\phi^3 Nd} \frac{E_n}{i_n}, \tag{20}$$

where we have defined

$$N = \int_0^1 n_e^2(r) r \, dr. \tag{21}$$

With the aid of formula (20) one may now determine the distance R of the nebula, if the values of N, ϕ, and E_n are known for it. Usually N must be found by numerical integration, after one has determined, from the solution of Abel's integral, the distribution of electron concentration in the equatorial plane of the nebula (for this it is necessary to have available the isophotes of the nebula). As for E_n, one may write for it in comparison with the sun

$$E_n = E_\odot 10^{-0.4(m_n - m\odot)}, \tag{22}$$

where m_n is the integrated stellar magnitude of the band of the nebula under study and is determined from the isophotes of the nebula.

This method has been applied to the nebula NGC 7293, for which it was found that $R \approx 80$ parsecs (208).

The method of He III. The dimensions of the monochromatic image of some planetary nebulae in the line $\lambda 4686$ Å He II are noticably smaller than in the hydrogen lines. This indicates that the optical thickness of helium beyond the outer boundary of the He^{++} zone in these nebulae is of order of magnitude unity. This fact may be used for the determination of the distance to these nebulae.

Let us write the ionization formula (51) (page 000) for ionized helium

$$\frac{n^{+2}}{n^+} n_e = \left(\frac{R_0}{r}\right)^2 f(T_*) e^{-r}, \tag{23}$$

where n^{+2} and n^+ are the concentrations of doubly and singly ionized helium, τ is the optical depth in frequencies ionizing He^+ and R_0 is the radius of the nucleus.

We have

$$d\tau = k_v(\text{He II}) n^+ dr, \tag{24}$$

where k_v (He II) is the mean value of the continuous absorption coefficient per atom of singly ionized helium.

Substituting the value of n^+ from (23) into (24) and integrating, we find for the linear radius r_0 of the zone of ionization of He III:

$$r_0 = n_e^{-1/3}(n^{+2})^{-1/3}R_0^{2/3}[3(1 - e^{-\tau_0})f(T_*)/k_v(\text{He II})]^{1/2}, \qquad (25)$$

where the function $f(T_*)$ has the form

$$f(T_*) = \frac{1}{2}\sqrt{\frac{T_e}{T_*}} \left(\frac{2\pi\mu k T_*}{h^3}\right)^{3/2} e^{-\chi^+/kT_*} ; \qquad (26)$$

here χ^+ is the ionization potential of He$^+$.

Usually, for planetary nebulae, $n^{+2} \sim 0.1\, n_e$. Assuming that the zone of He III has optical thickness $\tau_0 \approx 1$ and using the value k_v (He II) $= 1.6 \times 10^{-18}$ cm^2, we find from (25)

$$r_0 = 0.54 \times 10^6 n_e^{-2/3}R^{2/3}f^{1/3}(T_*). \qquad (27)$$

The distance R to the nebula is then found from r_0 and D'' (He III), the angular diameter of the He III zone, according to the formula

$$R = 2r_0 \frac{206,265''}{D''(\text{He III})} \qquad (28)$$

The formulae (27) and (28) have been applied to certain planetary nebulae for which D'' (He III) has smaller angular dimensions than D'' (H), observed in the hydrogen lines. The results are given in Table 3-2 (in the fifth column).

TABLE 3-2

Distances R of Planetary Nebulae Determined by the Method of He III

Nebula	D''(H)	D''(He III)	n_e(cm^{-3})	R (parsecs)				
				He III	V.–V.*	S.	B.	C.
IC 2165	8″	4.4	6.4×10^3	2900	2500	2500	4610	3150
NGC 2392	20	12.8	1.1	550	930	460	860	670
NGC 2440	37	5.3	2.7	1100	2000	630	2640	840
NGC 3242	20	14.8	5.2	280	600	700	1040	910
NGC 6720	71	38.	3.5	330	760	390	1640	500
NGC 6818	20	14.4	3.	2200	960	1000	2380	1300
NGC 6886	8	3.4	16.	1750	2090	2000	7240	3630
NGC 7026	15	3.7	4.	1700	1200	1300	3020	1880
NGC 7662	16	13.0	2.4	640	600	900	1200	880

* V.–V. = Vorontsov-Vel'yaminov, S. = Shklovskiĭ, B. = Berman, C. = Camm,

The necessary values of T_*, T_e, and n_e were taken from Tables 4-2 and 4-4. The data on the angular dimensions of the He III zone are taken primarily from the work of Wilson [34] and Wright [23]. In the calculations $r_0 \approx 10^{11}$ cm has been assumed. For comparison, we give in Table 3-2 the distances to the same nebulae according to existing catalogues.

The disadvantages of this method are that we know nothing about the dimensions of the nuclei of planetary nebulae and know poorly the relative helium content in each individual case. However, because of the comparatively weak dependence of r_0 on the indicated quantities, we may obtain from (26) a correct order of magnitude determination of R.

Chapter IV

Temperatures of Nuclei, Electron Temperatures and Electron Concentrations

1. Methods of Determining Temperatures of Nuclei

The nuclei of planetary nebulae are very hot stars. This fact explains in the first place the large apparent brightness of the nebulae, which are many times brighter than their nuclei. Since the emission of the nebula is caused by short-wave radiation from the nucleus, it follows that the nucleus radiates very much more energy in this region of the spectrum than in the visible region. But this can occur only in stars with high temperatures. The presence, in the spectra of nebula of lines belonging to highly ionized elements (He II, Ne V and others) also attests to the high temperatures of their nuclei. Finally, the fact that the nebula has, lying at an enormous distance from the nucleus, a region of comparatively dense hydrogen which consists essentially only of ions and electrons, also indicates the high density of ionizing radiation and consequently the high temperature of the nucleus.* It appears that the nuclei of planetary nebulae are among the hottest stars known to us.

In view of this situation, it appears of particular necessity to develop methods for determining the temperatures of nuclei of planetary nebulae. This possibility was contemplated already when the theory of nebular luminescence was in its early stages. At present we have a number of original and ingenious methods for determining the temperature of nuclei of planetary nebulae. Incidentally, we may note that these methods have been applied as well in determining the temperature of some nonstationary stars possessing gaseous shells, such as Wolf-Rayet stars, P Cygni stars and others.

The high temperatures of the nuclei form an interesting contrast to the comparatively low temperatures of the electron gas in the nebulae and, in the final analysis, to the temperatures of the nebulae themselves. The reason for the electron temperatures of the nebulae has already been discussed in Chapter II. Here we shall encounter a method of determining temperatures from observations. Con-

---

* Although the high degree of ionization is also a consequence of the small density of free electrons in the nebula.

trary to what one might expect, this essentially difficult problem can be solved simply and conclusively.

In general the theory for the determination of the temperatures of nebulae and nuclei, due to the efforts of V.A. Ambartsumyan and H. Zanstra, is one of the most exploited and classical parts of theoretical astrophysics.

The electron concentration is the most important parameter of a planetary nebula. Knowing the electron concentration of a nebula, we are able to determine its mass. There exist a number of methods allowing one to determine or estimate the order of magnitude of the electron concentrations in nebulae. In some cases this problem may be solved together with the determination of the electron temperatures of the nebulae.

In the present chapter we shall become acquainted with methods for determining the electron temperatures and electron concentrations of nebulae and the temperatures of nuclei. Together with this, the principal results obtained for individual results are presented.

The method of recombination (Zanstra's first method [87]). This method is based on the theory of the hydrogen line emission from planetary nebulae. As we have seen in Chapter II, the nebula emits a photon in one line of the Balmer series of hydrogen for each L_c-quantum radiated by the central star and absorbed in the nebula. If N_c is the number of L_c-photons emitted per second in all directions by the star and if the nebula is able to absorb completely these quanta, then the total number of Balmer quanta radiated by the nebula will also be N_c. Consequently, if we are somehow able to compute the number of Balmer quanta emitted by the nebula per unit time, we are then in the position to determine the rate at which the central stars emit ultraviolet radiation and thereby its color temperature, on the assumption that it radiates according to Planck's law.

The number of L_c-quanta radiated by the central star per second in the frequency interval v, $v + dv$ will be

$$dN_c = 4\pi r_*^2 \frac{2\pi}{c^2} \frac{v^2 dv}{e^{hv/kT_*} - 1},$$ (1)

where T_* is the temperature of the star and r_* is its radius.

Let us denote by τ_v the optical thickness of the nebula in some frequency v beyond the Lyman limit. Obviously $dN_c (1 - e^{-\tau_v})$ will be the fraction of the stellar L_c-quanta to be absorbed by the nebula and subsequently radiated in the Balmer series. Therefore, for the total number of Balmer quanta emitted by the nebula per unit time we will have:

$$N_{Ba} = 4\pi r_*^2 \frac{2\pi}{c^2} \int_{v_0}^{\infty} \frac{(1 - e^{-\tau_v})v^2 dv}{e^{hv/kT_*} - 1},$$ (2)

where v_0 is the frequency of the Lyman limit.

Now we must determine from observations the same quantity N_{Ba}, that is, the total number of Balmer quanta radiated by the nebula, and equate N_{Ba} to (2), after which one may determine the required temperature T_* of the nucleus.

Let E_i be the total amount of energy radiated in the i-th Balmer line per unit time by the whole nebula. This quantity may be determined in arbitrary units from measurements of spectrograms of the nebula. Further, let $(\partial E_*/\partial v)_i$ be the total amount of energy radiated by the nucleus per unit time per unit frequency just outside the i-th Balmer line. This quantity may also be found in the same units from measurements of the spectrograms of the nucleus of the nebula being studied. We can then form the ratio

$$A_i = \frac{E_i}{v_i(\partial E_*/\partial v)_i}. \tag{3}$$

On the other hand we have

$$\left(\frac{\partial E_*}{\partial v}\right)_i = 4\pi r_*^2 \frac{2\pi h v_i^3}{c^2} \frac{1}{e^{hv/T_*} - 1}. \tag{4}$$

Therefore, for the total number of Balmer quanta emitted by the nebula, we can write the relation

$$N_{Ba} = \sum_i \frac{E_i}{hv_i} = \frac{1}{h} \sum_i A_i \left(\frac{\partial E_*}{\partial v}\right)_i = 4\pi r_*^2 \frac{2\pi}{c^2} \sum_i^i \frac{v_i^3 A_i}{e^{hv_i/kT_*} - 1}, \tag{5}$$

where the summation runs over all lines of the Balmer series and over the Balmer continuum. Equating (2) to (5) we obtain

$$\int_{v_0}^{\infty} \frac{(1 - e^{-\tau})v^2 dv}{e^{hv/kT_*} - 1} = \sum_i \frac{v_i^3 A_i}{e^{hv_i/kT_*} - 1}. \tag{6}$$

Let us define:

$$x = hv/kT_*, \quad x_i = hv_i/kT_*, \quad x_0 = hv_0/kT_*. \tag{7}$$

Then, taking into account that $\tau_v \sim v^{-3}$, in place of (6) we will have

$$\int_{x_0}^{\infty} \frac{[1 - e^{-\tau_0(x_0/x)^3}] x^2 dx}{e^x - 1} = \sum_i \frac{x_i^3 A_i}{e^{x_i} - 1}, \tag{8}$$

where τ_0 is the optical thickness of the nebula in the frequency v_0. This formula may be simplified somewhat by using the optical thickness τ_c averaged over the Lyman continuum frequencies and by taking out the factor

$$1 - e^{-\tau_0(x_0/x)^3} \approx 1 - e^{-\tau_c}$$

from under the integral in (8). Finally we obtain

$$(1 - e^{-\tau_c}) \int_{x_0}^{\infty} \frac{x^2 dx}{e^x - 1} = \sum_i \frac{x_i^3 A_i}{e^{x_i} - 1}. \tag{9}$$

This is the correct formula for determining the temperature T_* for the nucleus of a nebula having a total optical thickness beyond the Lyman limit equal to τ_c.

Assuming in (9) that $\tau_c \gg 1$, we arrive at Zanstra's formula [87]

$$\int_{x_0}^{\infty} \frac{x^2 dx}{e^x - 1} = \sum_i \frac{x_i^3 A_i}{e^{x_i} - 1}. \tag{10}$$

As will be shown in Chapter VII, for the majority of planetary nebulae the condition $\tau_c \gg 1$ is not satisfied. For double-envelope and ring nebulae τ_c is significantly smaller than unity in every case. Consequently formula (10), generally speaking, gives a lower bound to the temperature of the nucleus, which enters into x_0 and x_i.

Equation (9) is solved by the method of trial and error. By taking various values of T_* and calculating x_0 and x_i, we find the value of T_* for which the right and left sides of the equation are equal. The quantity τ_c is taken as known. To facilitate the solution of this and similar problems, we give in Table 4-1 the numerical values of the integrals

$$\int_{x_0}^{\infty} \frac{x^2 dx}{e^x - 1} \quad \text{and} \quad \int_{x_0}^{\infty} \frac{x^3 dx}{e^x - 1}$$

for various values of x_0. This table was prepared by Böhm and Schlender [89].

Zanstra applied the formula (10), which is correct for $\tau_c \gg 1$, to three planetary nebulae—NGC 6543, 6572 and 7009—and found for the temperatures of their nuclei: 39,000°K, 40,000°K and 55,000°K respectively. However, at least two of these three nebula (NGC 6543 and NGC 7009) are double-shell and consequently for them $\tau_c < 1$; as regards NGC 6572, it still has not been shown to be a double-shell. Therefore the values of T_* shown for the two double-shell nebulae cannot be accepted as correct. As has been shown by calculation, the temperatures of the nuclei for $\tau_c < 1$ differ strongly from those obtained when $\tau_c \gg 1$. In Table 4-2 are given the results of these calculations, based on formula (9) for the nebulae NGC 6543 and NGC 7009. The summation on the right side of (9) included the lines H_β, H_γ, H_δ, and H_ε for which the corresponding values of A_i were taken from the work of Zanstra [87].*

As is seen from the data given in Table 4-2, the values obtained for the temperatures of the nuclei are nearly equal only when they are comparatively low; for large

---

* For the nebula NGC 7009 the quantities A_i were reduced approximately according to the known temperature (55,000°K) since, because of their uncertainty, they were not shown in [87].

TABLE 4-1

Numerical Values of the Integrals $\int_{x_0}^{\infty} \dfrac{x^2}{e^x - 1}\, dx$ and $\int_{x_0}^{\infty} \dfrac{x^3}{e^x - 1}\, dx$

x_0	$\int_{x_0}^{\infty} \dfrac{x^2}{e^x - 1}\, dx$	$\int_{x_0}^{\infty} \dfrac{x^3}{e^x - 1}\, dx$	x_0	$\int_{x_0}^{\infty} \dfrac{x^2}{e_x - 1}\, dx$	$\int_{x_0}^{\infty} \dfrac{x^3}{e_x - 1}\, dx$
0	2.4041	6.4939	3.8	0.5425	2.8620
0.1	2.3993	6.4936	3.9	0.5103	2.7379
0.2	2.3854	6.4915	4.0	0.4797	2.6169
0.3	2.3634	6.4859	4.1	0.4506	2.4991
0.4	2.3342	6.4756	4.2	0.4230	2.3846
0.5	2.2986	6.4596	4.3	0.3969	2.2735
0.6	2.2574	6.4369	4.4	0.3721	2.1659
0.7	2.2113	6.4069	4.5	0.3487	2.0618
0.8	2.1610	6.3691	4.6	0.3266	1.9612
0.9	2.1071	6.3232	4.7	0.3057	1.8641
1.0	2.0502	6.2691	4.8	0.2860	1.7706
1.1	1.9909	6.2068	4.9	0.2675	1.6806
1.2	1.9296	6.1364	5.0	0.2500	1.5941
1.3	1.8669	6.0580	5.1	0.2336	1.5110
1.4	1.8031	5.9719	5.2	0.2181	1.4313
1.5	1.7387	5.8784	5.3	0.2036	1.3550
1.6	1.6740	5.7781	5.4	0.1899	1.2819
1.7	1.6093	5.6714	5.5	0.1771	1.2120
1.8	1.5449	5.5587	5.6	0.1650	1.1453
1.9	1.4810	5.4406	5.7	0.1537	1.0817
2.0	1.4179	5.3176	5.8	0.1432	1.0210
2.1	1.3558	5.1903	5.9	0.1333	0.9632
2.2	1.2949	5.0593	6.0	0.1242	0.9081
2.3	1.2353	4.9252	6.1	0.1154	0.8557
2.4	1.1771	4.7884	6.2	0.1073	0.8060
2.5	1.1204	4.6495	6.3	0.09976	0.7588
2.6	1.0653	4.5091	6.4	0.09271	0.7140
2.7	1.0119	4.3676	6.5	0.08612	0.6715
2.8	0.9603	4.2256	6.6	0.07998	0.6312
2.9	0.9105	4.0835	6.7	0.07425	0.5931
3.0	0.8624	3.9417	6.8	0.06891	0.5571
3.1	0.8162	3.8007	6.9	0.06393	0.5230
3.2	0.7717	3.6608	7.0	0.05929	0.4908
3.3	0.7291	3.5224	7.1	0.05497	0.4604
3.4	0.6883	3.3856	7.2	0.05096	0.4316
3.5	0.6493	3.2510	7.3	0.04722	0.4045
3.6	0.6120	3.1187	7.4	0.04375	0.3790
3.7	0.5764	2.9890	7.5	0.04052	0.3550

x_0	$\int_{x_0}^{\infty} \dfrac{x^2}{e^x - 1}\, dx$	$\int_{x_0}^{\infty} \dfrac{x^3}{e^x - 1}\, dx$	x_0	$\int_{x_0}^{\infty} \dfrac{x^2}{e^x - 1}\, dx$	$\int_{x_0}^{\infty} \dfrac{x^3}{e^x - 1}\, dx$
7.6	0.03752	0.3323	9.2	0.01061	0.1105
7.7	0.03473	0.3110	9.4	0.00903	0.09580
7.8	0.03214	0.2909	9.6	0.00767	0.08296
7.9	0.02974	0.2720	9.8	0.00652	0.07176
8.0	0.02751	0.2543	10.0	0.00554	0.06202
8.1	0.02544	0.2376	10.2	0.00470	0.05354
8.2	0.02352	0.2220	10.4	0.00399	0.04619
8.3	0.02174	0.2073	10.6	0.00338	0.03981
8.4	0.02009	0.1936	10.8	0.00286	0.03428
8.5	0.01856	0.1807	11.0	0.00242	0.02949
8.6	0.01715	0.1686	11.2	0.00205	0.02536
8.7	0.01584	0.1572	11.4	0.00173	0.02178
8.8	0.01463	0.1466	11.6	0.00146	0.01870
8.9	0.01350	0.1367	11.8	0.00124	0.01604
9.0	0.01246	0.1274	12.0	0.00104	0.01375

TABLE 4-2

Temperatures T_* of the Nuclei of the Planetary Nebulae NGC 6543, NGC 6572 and NGC 7009
for Various Values of τ_c

Nebula	$\tau \gg 1$	$\tau_c = 1$	$\tau_c = 0.1$	$\tau_c = 0.01$
NGC 6543	37,500°K	43,000°K	80,000°K	210,000°K
NGC 6572	39,000°K	45,000°K	83,000°K	214,000°K
NGC 7009	55,000°K	80,000°K	130,000°K	> 300,000°K

values of T_* the difference between them is significant. The cases $\tau_c \gg 1$ correspond to Zanstra's scheme.

This method of determining the temperatures of nuclei may be applied as well to other lines formed by recombination, for example, the lines of neutral and singly ionized helium (He I, He II). Such attempts have been made and they led to still higher values of the temperatures of the nuclei. Thus, for example, for the nucleus of the nebula NGC 7009 Zanstra obtained $T_* = 70,000°K$ using the lines of ionized helium, whereas with hydrogen he found $T_* = 55,000°K$. It is possible that this discrepancy gives evidence that in fact the optical thickness τ_c of the nebula beyond the Lyman limit in hydrogen is smaller than unity, while beyond the limit of the fundamental series in ionized helium it is nearer to unity. If this is true, then the temperatures of the nuclei obtained by the recombination method from the lines of ionized helium will be closer to the true temperatures.

The method of "nebulium" (Zanstra's second method). In distinction to the lines of hydrogen and helium, forbidden lines are not formed by recombination. They are excited by means of collisions of the first kind with free electrons, produced primarily by photo-ionization of hydrogen. In other words the energy in forbidden lines is drawn from the kinetic energy of the free electrons.

On the other hand, the energy of the free electrons comes from the ultra-violet radiation of the star. Part of the energy of the radiation is spent in removing electrons from atoms, while the remaining part is transferred to the electrons as kinetic energy. Ultimately part of the energy of the stellar ultra-violet radiation will be re-emitted by the nebula in the form of forbidden lines. Thus, if we can compute as a function of star temperature the fraction of the ultra-violet energy (stellar radiation) which escapes through forbidden lines and if we determine from observation the energy being radiated by the nebula in forbidden lines (in the "nebulium" lines), then we are able to establish the temperature T_* of the nucleus. This is the essence of the method of "nebulium".

The number of L_c-quanta emitted by the nucleus per unit time in the interval v, $v + dv$ is given by (1). Of this number only the fraction dN_c $(1 - e^{-\tau_v})$ will be absorbed by the nebula Therefore

$$4\pi r_*^2 \, \frac{2\pi v^2}{c^2} \, \frac{1 - e^{-\tau_v}}{e^{hv/kT_*} - 1} \, dv$$

will be the number of electrons removed from atoms by this radiation, each electron acquiring kinetic energy equal to

$$\frac{1}{2} mv^2 = hv - hv_0,$$

where v_0 is the ionization frequency of hydrogen. The total kinetic energy, acquired by free electrons per unit time, will equal

$$4\pi r_*^2 \, \frac{2\pi h}{c^2} \int_{v_0}^{\infty} \frac{(1 - e^{-\tau_v})(v - v_0) v^2 dv}{e^{hv/kT_*} - 1} . \tag{11}$$

A fraction η of this energy escapes in the excitation of forbidden lines. We may therefore write for the total energy escaping in this way:

$$\eta 4\pi r_*^2 \, \frac{2\pi h}{c^2} \int_{v_0}^{\infty} \frac{(1 - e^{-\tau_v})(v - v_0) v^2 dv}{e^{hv/kT_*} - 1}, \tag{12}$$

where $\eta \leq 1$.

On the other hand, the energy being radiated by the nebulae in the "nebulium" lines may, in analogy to (5), be represented in the form:

$$4\pi r_*^2 \, \frac{2\pi h}{c^2} \sum_{neb} \frac{v_i^4 A_i}{e^{hv_i/kT_*} - 1}, \tag{13}$$

where A_i are quantities determined from observation according to formula (3), and the summation is taken over all forbidden lines excited by electron collisions.

Equating (13) to (12), we obtain

$$\eta \int_{v_0}^{\infty} \frac{(1 - e^{-\tau_v})(v - v_0)\, v^2 dv}{e^{hv/kT*} - 1} = \sum_{neb} \frac{v_i^4 A_i}{e^{hv_i/kT_*} - 1} \qquad (14)$$

or, utilizing the definitions (7) and removing the factor $(1 - e^{-\tau_c})$ from the integral, we find finally:

$$\eta(1 - e^{-\tau_c}) \int_{x_0}^{\infty} \frac{(x - x_0)x^2 dx}{e^x - 1} = \sum_{neb} \frac{x_i^4 A_i}{e^{x_i} - 1}. \qquad (15)$$

In this equation the unknown is the temperature T_* of the nucleus which enters in x_0 and x_i. The parameters η and τ_c are regarded as given.

Comparing formula (15) with (9), we see that the method of recombination has some advantage over that of "nebulium", because in the first case only one unknown parameter (τ_c) appeared, while in the second case, two parameters (τ_c and η) are unknown. Ideally if both methods could be applied to the same star without error, the simultaneous solution of equations (9) and (15) would make it possible to establish η, the fraction of free electron kinetic energy escaping in the "nebulium" lines.

In the particular case when $\eta = 1$ and $\tau_c \gg 1$, we obtain from (15) Zanstra's second formula [87]:

$$\int_{x_0}^{\infty} \frac{x^3 dx}{e^x - 1} - x_0 \int_{x_0}^{\infty} \frac{x^2 dx}{e^x - 1} = \sum_{neb} \frac{x_i^4 A_i}{e^{x_i} - 1}. \qquad (16)$$

Obviously, formula (16) also gives only a lower bound to the temperature of the nucleus, because in real nebula $\tau_c < 1$ and $\eta < 1$. The optical thickness τ_c has already been discussed. As regards η, V.V. Sobolev [90] has shown that it is usually smaller than unity and ranges in the interval $0.3 \leq \eta \leq 0.6$, that is, only about half of the kinetic energy of the free electrons is transformed into radiation in the "nebulium" lines.

The effect of η on T_* is small, however; taking account of η leads to temperatures only a few thousand degrees higher than those obtained with $\eta = 1$. The greatest effect on the value of T_* comes from τ_c, taking account of which leads to an increase in the temperature by a factor of two or three over that obtained with $\tau_c \gg 1$. In this respect formulae (15) and (9) are equivalent.

Zanstra applied the formula (16) to the nuclei of the planetary nebulae mentioned earlier and found for the nebulae NGC 6543, 6572 and 7009 values of T_* equal to 37,000°K, 38,000°K and 50,000°K respectively.

The results obtained by this method appear to agree quite well with those obtained from the hydrogen lines. This fact is usually taken as evidence that the results obtained by these two methods are not only lower bounds but are close to the true temperatures as well.

It is easy to demonstrate that such an assertion is incorrect. The temperatures of nuclei, determined by the methods of recombination and "nebulium," will be close to the true temperatures only to the extent to which the requirement $\tau_c > 1$ is satisfied by conditions in the nebulae. Thus the degree of uncertainty in the formulae (10) and (16) with respect to τ_c is much the same and the results obtained in both cases, while agreeing well with one another, will be incorrect to the same extent.

In discussing the method of "nebulium" one should note that it is not an independent method. The difference between the method of recombination and that of "nebulium" is purely formal, since in the first case the relevant balance is that between one part of the ultra-violet energy radiated by the star and the energy radiated by the nebula in the hydrogen lines, while in the second case the balance is between the remaining part of the ultra-violet energy and the energy in the forbidden lines. We should be able with the same success to replace both methods by a single method, investigating simultaneously the balance between the ultra-violet energy emitted by the star beyond the Lyman limit and the total energy radiated by the nebula in the lines of hydrogen and "nebulium."

The methods of Zanstra for determining the temperatures of nuclei suffer, in addition, from a general procedural disadvantage: for their application one must obtain spectrograms of the nebula and of the central star; those of the latter are not always possible on account of the weakness of these stars. In addition, obtaining the quantities A_i from these spectrograms presents significant difficulties, although these methods are insensitive to small errors in the determination of A_i. For this reason, in particular, the methods mentioned here have not found widespread application.

Ambartsumyan's method (the method of He II/H). This method uses the ratio of the intensities of the lines H_β and $\lambda 4686$ Å He II. In this method it is necessary to assume that: a) there are enough He^+ ions in the nebula so that all radiation lying beyond the ionization edge $4\nu_0$ of singly ionized helium is absorbed, *i.e.*, $\tau_{HeII} > 1$; b) radiation lying in the frequency interval from ν_0 to $4\nu_0$ is completely absorbed by hydrogen alone, *i.e.*, $\tau_c > 1$.

Let us designate by $\phi(4686)$ the fraction of quanta ionizing He^+ (from $4\nu_0$ to ∞) that are converted to quanta in the line $\lambda 4686$ Å He II, and by $\phi (H_\beta)$ the fraction of the quanta ionizing hydrogen (ν_0 to $4\nu_0$) that are converted into H_β quanta. Then we have for the flux of radiation escaping from the nebula in these lines:

$$\mathscr{F}_{4686} = h\nu_{4686}\,\phi(4686)4\pi r_*^2\,\frac{2\pi}{c}\int_{4x_0}^\infty \frac{x^2 dx}{e^x - 1},$$

$$\mathscr{F}_{H_\beta} = h\nu_{H_\beta}\phi(H_\beta)4\pi r_*^2 \frac{2\pi}{c} \int_{x_0}^{4x_0} \frac{x^2 dx}{e^x - 1},$$

or

$$\frac{\mathscr{F}_{4686}}{\mathscr{F}_{H_\beta}} \frac{\lambda_{4686}}{\lambda_{H_\beta}} \frac{\phi(H_\beta)}{\phi(4686)} = \frac{\displaystyle\int_{4x_0}^{\infty} \frac{x^2 dx}{e^x - 1}}{\displaystyle\int_{x_0}^{4x_0} \frac{x^2 dx}{e^x - 1}}, \tag{16a}$$

where $x_0 = h\nu_0/kT_*$.

According to the above assumptions, ϕ (4686) is obviously equal to the ratio of the number of spontaneous transitions $N_4 A_{43}$ from level 4 to level 3 in singly ionized helium giving rise to the line $\lambda 4686$ Å, to the total number of recombinations, *i.e.*,

$$\phi(4686) = \frac{N_4 A_{43}(\text{He II})}{N^{++} n_e \Sigma c_i(\text{He II})},$$

where N_4 is the concentration of He$^+$ ions in level 4 and N^{++} is the concentration of He^{++}. Analogously we have for (H$_\beta$)

$$\phi(H_\beta) = \frac{n_4 A_{42}(\text{H})}{n^+ n_e \Sigma c_i(\text{H})}.$$

The ratios $N_4/N^{++}n_e$ and $n_4/n^+ n_e$ appearing in the above equations must be determined by methods already known to us (Cillié's Equation [39], Chapter II). The corresponding calculations give for $T_e = 10,000°$K [164], $\phi(4686) = 0.138$ and $\phi(H_\beta) = 0.118$. Substituting these values into (16a), we find finally

$$\frac{\displaystyle\int_{4x_0}^{\infty} \frac{x^2 dx}{e^x - 1}}{\displaystyle\int_{x_0}^{4x_0} \frac{x^2 dx}{e^x - 1}} = 0.90 \frac{\mathscr{F}_{4686}}{\mathscr{F}_{H_\beta}}. \tag{17}$$

The first assumption ($\tau_{\text{HeII}} > 1$) is justified for those planetary nebulae (NGC 2165, 3242, 6818, 7662 and others) for which the dimensions of the monochromatic images in the line $\lambda 4686$ Å He II are smaller than those of the monochromatic images in the hydrogen lines. However, the second assumption ($\tau_c > 1$) cannot be accepted for the majority of planetary nebulae. Consequently the temperatures of nuclei determined in this way will be upper limits to their actual values.

V.A. Ambartsumyan applied his method to two planetary nebulae—NGC 7009 and 7027—and obtained for the temperatures of their nuclei values equal to 115,000°K and 165,000°K, respectively. At present one could apply the method to many planetary nebulae for which values of the ratio $\mathscr{F}_{4686}/\mathscr{F}_{H_\beta}$ are known [21, 92]. The results are interesting: of the 33 nebulae examined only five appear to have T_* smaller than 100,000°K, while for the remaining nebulae T_* is larger than 100,000°K. The minimum value of T_* was found for the nucleus of the nebula NGC 6891 (a planet-like spiral) and is equal to 70,000°K. For four nebulae—NGC 2440, 6309, 6818 and anon. 21^h31^m (the first three are double-envelope, the last of unknown type), T_* appeared larger than 200,000°K and for one nebula NGC 2022 (double-envelope)—even larger than 300,000°K (!).

The supposition has been expressed that the line $\lambda4686$ Å He II may be strengthened under conditions found in nebulae by means of the following fluorescent mechanism: ions of He$^+$ existing in the metastable $2S$ state absorb photons from the hydrogen L_α-line and undergo the transition $2 \to 4$. The reverse transition proceeds spontaneously along the route $4 \to 3 \to 2$ during which are radiated, in particular, photons in the line $\lambda4686$ Å. Obviously this leads to an increase of the ratio $\mathscr{F}_{4686}/\mathscr{F}_{H_\beta}$ and, thereby, to an erroneous enhancement of the temperature of the nucleus. However, a quantitative examination shows that the role of this fluorescent mechanism is small [76]. Moreover, the wavelengths of L_α-photons and those of the $4 \to 2$ transition in He II do not coincide exactly, and in order than an ion of He$^+$ be able to absorb an L_α-photon emitted by a hydrogen atom, H and He$^+$ must have a relative velocity on the order of 100 km/sec.

We must conclude therefore that the large values of the ratio $\mathscr{F}_{4686}/\mathscr{F}_{H_\beta}$, which lead to extremely high values of the temperature of the nucleus, result from the small intensity of the H_β line, that is, from the value of the optical thickness τ_c beyond the Lyman limit being small compared to unity.

In the case when $\tau_c < 1$ and $\tau_{HeII} > 1$, the formula may be written in the form

$$\frac{\displaystyle\int_{4x_0}^{\infty} \frac{x^2 dx}{e^x - 1}}{\displaystyle\int_{x_0}^{4x_0} \left[1 - e^{-\tau_c(x_0/x)^3}\right] \frac{x^2 dx}{e^x - 1}} = 0.83 \, \frac{\mathscr{F}_{4686}}{\mathscr{F}_{H_\beta}}, \tag{18}$$

or approximately:

$$\frac{1}{1 - e^{-\tau_c}} \frac{\displaystyle\int_{4x_0}^{\infty} \frac{x^2 dx}{e^x - 1}}{\displaystyle\int_{x_0}^{4x_0} \frac{x^2 dx}{e^x - 1}} = 0.83 \frac{\mathscr{F}_{4686}}{\mathscr{F}_{H_\beta}} \tag{19}$$

The temperatures of the nuclei as determined by this method decrease as τ_c decreases. In the case of Zanstra's method we have the reverse relation: the temperature of the nucleus increases as τ_c decreases. Therefore, one might suppose that there must exist some value of τ_c for which the temperatures of the nucleus, determined according to the methods of Ambartsumyan and of Zanstra, coincide with one another.

Combining the formulae (9) and (19), we arrive at the following accurate formula for the determination of the temperature of the nucleus of optically thin nebulae, that is, in the case when $\tau_c < 1$ but τ (He II) $> .1$:

$$\frac{\displaystyle\int_{x_0}^{\infty} \frac{x^2 dx}{e^x - 1} \int_{4x_0}^{\infty} \frac{x^2 dx}{e^x - 1}}{\displaystyle\int_{x_0}^{4x_0} \frac{x^2 dx}{e^x - 1}} = 0.83 \frac{\mathscr{F}_{4686}}{\mathscr{F}_{H\beta}} \sum_i \frac{x_i^3 A_i}{e^{x_i} - 1}, \qquad (20)$$

or, with a sufficient degree of accuracy:

$$\int_{4x_0}^{\infty} \frac{x^2 dx}{e^x - 1} = 0.83 \frac{\mathscr{F}_{4686}}{\mathscr{F}_{H\beta}} \sum_i \frac{x_i^3 A_i}{e^{x_i} - 1}. \qquad (21)$$

This is, in fact, Zanstra's formula written for ionized helium—it is a kind of analog to that in the hydrogen recombination method, with the difference that the sum on the right side of (21) relates to hydrogen lines, while of the helium lines only the intensity of $\lambda 4686$ Å is allowed for.

Having determined from (21) the correct value of T_*, we may, with the help of formulae (9) and (19), obtain τ_c. By this means, in fact, we obtain a method of determining the optical thickness, if only for those planetary nebulae for which τ (He II) > 1, that is, for which the dimensions of the monochromatic images in the line $\lambda 4686$ Å are smaller than those in hydrogen lines.

Formula (21) was applied for two planetary nebulae, NGC 6572 and 7009, in the spectra of which appears the line $\lambda 4686$ Å. The results of the calculations appear in Table 4-3. There it may be seen that the actual temperatures of the nuclei of planetary nebulae are very high and significantly exceed the values obtained by Zanstra.

In practical respects Ambartsumyan's method is distinguished by the advantage that for its application it is sufficient to have a spectrogram of the nebula. However,

TABLE 4-3

Accurate Temperatures of the Nuclei and Optical
Thicknesses of the Nebulae NGC 6572 and NGC 7009

Nebula	$\mathscr{F}_{4686}/\mathscr{F}_{H\beta}$	T_*	τ_c
NGC 6572	0.13	75,000°K	0.14
NGC 7009	0.14	94,000°K	0.40

this method may be used only for those nebula in whose spectra appears the line $\lambda 4686$ Å He II (high-excitation nebulae).

The [O III]/[O II] *method.* In this method one utilizes the ratio of the intensity of the green doublet of doubly ionized oxygen $N_1 + N_2$ ($\lambda\lambda 5007$ Å $+ 4956$ Å) to that of the violet doublet of singly ionized oxygen $\lambda\lambda 3726$ Å $+ 3729$ Å (well known as $\lambda 3727$ Å) in the spectrum of the nebula. Obviously, the ratio $E_{N_1+N_2}/E_{3727}$ will increase with the temperature of the nucleus, because for a higher temperature the number of O^{++} ions will be larger in comparison with the number of O^+ ions. However, in a more detailed formulation of the problem the ratio in question will depend also on the dimensions and electron density of the nebula.

The theoretical relations for the determination of the temperature T_* of the nuclei of planetary nebulae corresponding to the known ratios $E_{N_1+N_2}/E_{3727}$ have been derived in [93, 94]. However, no account has been taken of collisions of the second kind, which have practically no effect on the intensities of the lines N_1 and N_2, but which influence the intensity of the line $\lambda 3727$ Å [O II] when $n_e > 10^3$ cm^{-3}. Taking account of this effect, the relations mentioned above for the determinations of T_* take the form:

$$\frac{163{,}000}{T_*} - \log T_* = K, \tag{22}$$

where

$$K = 10.7 - 0.4 m_* - 2 \log D'' - \log n_e - \log \frac{E_{N_1+N_2}}{E_{2737}} + \log \frac{1 + 2.8 \cdot 10^{-4} n_e}{1 + 10^{-6} n_e} : \tag{23}$$

Here m_* is the photographic magnitude of the nucleus, D'' is the angular diameter of the nebula in seconds of arc and n_e is the number of electrons per cm^3.

In deriving formula (22) it was assumed that the ratio W/n_e is constant inside the nebula (W is the dilution coefficient) and that the electron temperatures for all nebulae are the same and equal to 10,000°K.

To determine n_e, which enters into (23), one may employ, in particular, formula (64) (page 120).

In order to use the method of [O III]/[O II] to determine the temperature of a nucleus it is necessary to know the numerical values of a number of auxiliary quantities. Many of them, m_x, $E_{N_1+N_2}/E_{3727}$, D'', are easily found from observations (errors in the determination of the distances and consequently in the linear dimensions D of the nebulae influence slightly the value of the temperatures obtained). The value of H_n—the surface brightness beyond the Balmer limit—entering into the formula (64) presents a considerably more difficult situation; as yet this quantity has been determined for far from all planetary nebulae. One way out is as follows. For eight planetary nebulae Menzel and Aller [95] have measured the quantity H_n. On the other hand, for these and in general for a large number of planetary nebulae the values of H—the photographic surface brightness—are known; they appear in Vorontsov-Vel'yaminov's catalogue [50]. By comparing H and H_n for the eight objects mentioned above, one finds that the ratio of energy radiated by the nebulae in the photographic region to the energy radiated at the Balmer limit (in a bandwidth of 20 Å) is almost constant and equal to 60 (nearly $4^m.5$). By assuming this ratio to be constant as well for the remaining nebulae we may find H_n from the known values of H. The necessity for this procedure vanishes as one succeeds in obtaining the quantities H_n directly by observation of each individual nebula.

The lines N_1, N_2 and $\lambda 3727$ Å are bright and are the most readily available lines in the spectra of all planetary nebulae. Therefore, the ratio $E_{N_1+N_2}/E_{3727}$ may be found for all planetary nebulae without exception independently of their spectral type, that is, of their degree of excitation. The values of $E_{N_1+N_2}/E_{3727}$ obtained from observations, before being substituted into formula (23), are corrected for the effects of interstellar selective absorption. To facilitate calculations by means of formula (22), a curve of the dependence of T_* on K appears in Figure 4-1.

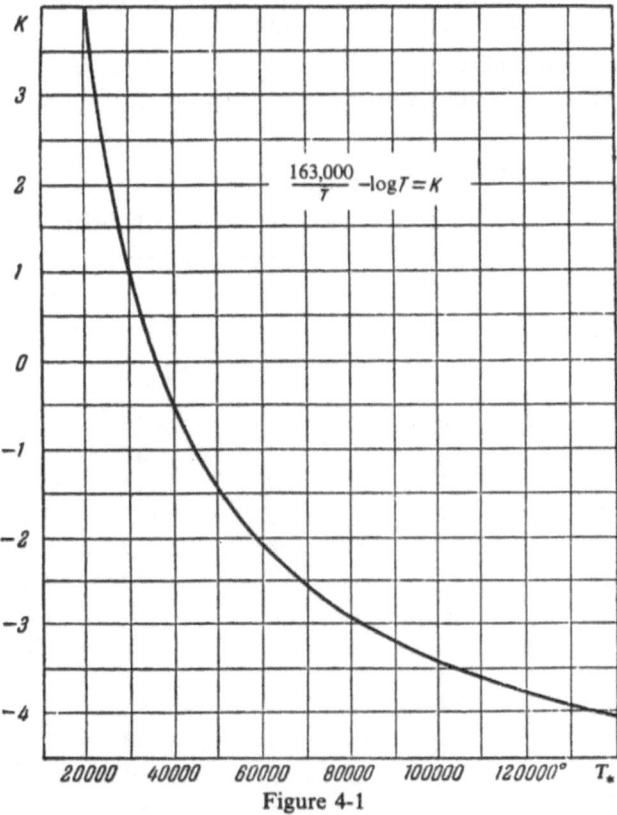

Figure 4-1

The graph of K *versus* T_* for the determination of the temperature of a nucleus by the [O III]/[O II] method.

In Table 4-4 are shown the results of the determination of the nuclear temperatures of more than fifty planetary nebulae by the [O III]/[O II] method (details
may be found in [94]). In these cases, when possible, the most reliable values of
n_e, as determined by various methods, were used; these appear in Table 4-15. In
the remaining cases n_e is determined by the procedure described above with the aid
of formula (64).

The results appearing in Table 4-4 show that the temperatures of nuclei of planetary nebulae obtained by the method of [O III]/[O II] are quite high. The minimum value is of order 40,000°K, the maximum, 150,000°K. The mean value of
the temperature is of order 65,000°K, which is larger than the temperatures obtained
by Zanstra's method and smaller than those obtained by Ambartsumyan's method.

The largest error in the method of [O III]/[O II] may be caused by the fact that
the dimensions of the region of O II emission are larger than those of the region
of O III emission. Moreover, the ionization formula, written for the lines of O III
and appearing at the beginning of the derivation of relation (22), is not valid in
the O II region. However, since the available data shows that the dimensions of
the O II region do not differ very strongly from those of the O III region, one may
think that the effect of this fact will not be very important.

TABLE 4-4

Temperatures of the Nuclei of Planetary Nebulae
Determined by the [O III]/[O II] Method

NGC	$\dfrac{E_{N_1 + N_2}}{E_{3727}}$	Reference	$n_e(10^3 \text{ cm}^{-3})$	$T_*(10^{3\circ}\text{K})$
40	0.2	4	2.3	32
650−1	8	1	0.7	95
I 351	252	2	6.7	71
I 2003	27	2	15	69 (?)
1535	79	4	2.2	52
J 320	57	4	7.7	67
I 418	1.3	4	16	38
2022	49	5	0.8	59
2149	5	2	2.2	41
2165	58	2	6.4	78 (?)
2371−2	8	1	0.3	48
2392	200	5	1.1	50
2440	36	2	2.7	108
3242	114	2	5.2	53
3587	4	1	0.16	66

TABLE 4-4 (continued)

NGC	$\dfrac{E_{N_1+N_2}}{E_{3727}}$	Reference	$n_e(10^3 \text{ cm}^{-3})$	$T_*(10^3 \text{ K})$
II 3568	200	2	4.2	60
II 4593	13	2	2.2	35
II 4634	34	2	6.2	78 (?)
6210	63	4	8.7	46
6445	22	2	6.3	> 120
6543	28	4	15.4	46
6572	36	4	10.2	46
6720	2.1		3.5	68
18^h13^m	0.12	2	8.7	27
6741	24	2	15	63 (?)
6751	26	1	2.8	55
6772	13	2	0.2	96
4846	33	2	55	48
6778	22	1	2.3	60
6781	3	1	0.9	74
6790	156	2	95 (?)	78 (?)
6803	36	2	4.2	48
6807	55	2	33	77 (?)
6818	100	2	3.0	84
6826	41	4	2.2	46
6833	70	2	36	90 (?)
6853	4	1	0.2	77
6884	210	2	11	> 110
6886	27	2	16	63
6891	8	2	18	38
6894	9	1	0.3	40
4997	33	4	110 (?)	42 (?)
6905	360	2	1.2	114
7008	130	1	4.3	103
7009	78	4	6.3	57
7026	31	2	4	58
7027	152	3	4.9	119 (?)
5117	110	2	36	70
21^h39^m	23	2	58	51
7139	4	1	0.3	82
5217	102	2	12	61
7293	3	1	0.16	87
7662	120	3	2.4	56

1. Minkowski, R., *Ap. J.*, **95**, 243, 1942.   2. Aller, L., *Ap. J.*, **133**, 125, 1951.
3. Bowen, I., Wyse, A., *Lick Observ. Bull.*, **19**, 93, 1956.
4. Minkowski, R., Aller, L., *Ap. J.*, **124**, 93, 1956.

Thus, in spite of a number of uncertainties, the results obtained in practice by the application of the methods studied above enable one to conclude that the temperatures of the nuclei of planetary nebulae are very high and that probable values lie within the limits 60,000°K–100,000°K. In addition, however, one must admit that the temperatures of the nuclei as found by different methods differ strongly. Besides the reasons enumerated above for such differences, one might also consider the possibility that the emission of the nucleus deviates from the Planck law. This difference must be particularly strong in those cases when the nucleus has a spectrum of the Wolf-Rayet type. These stars, having extended gaseous envelopes, represent in themselves a kind of planetary nebulae in miniature, in which occur the same processes of photo-ionization and recombination. It is clear that the distribution of the energy emerging from such a shell and falling on the nebula can no longer be represented by the Planck law. If the optical thickness of the envelope of a Wolf-Rayet star beyond the fundamental series limit of some atom is larger than unity, then the envelope will absorb all of the energy of the star beyond the series limit and this energy will not reach the nebula. As confirmation of this may serve, in particular, the discovery by Zanstra and Weenen [96] of a peculiarity in the spectra of planetary nebulae having stars of the Wolf-Rayet type as nuclei. It involves the following. When the lines of He II are present in the spectrum of the nucleus, they are, as a rule, absent from the spectrum of the nebula, although the lines of He I are observed. In other cases, when lines of He I and He II appear in the spectrum of the nucleus, then in general helium lines are absent from the nebular spectrum; at most lines of He I appear. This indicates that in one case the atmosphere of the star absorbs the radiation beyond the limit of the fundamental series of ionized helium partially, and in the other case, completely. Consequently, either only part of this radiation reaches the nebula, causing its emission in the lines of He I, or else none of it reaches the nebula.

In discussing the differences among the values of the temperature found by various methods for one and the same star, one must bear in mind the following. In the methods of Zanstra and Ambartsumyan one compared the radiation of the star in two different regions of the spectrum (in the visible and beyond the Lyman limit): in this one had to deal with a "radiation" temperature. In distinction to these, the method of [O III]/[O II] gives a temperature characterizing the energy being radiated by the star in one region of the spectrum (the energy which ionizes O^+): in this case we deal with an "ionization" temperature. On account of the non-Planckian energy distribution in the spectrum of the nucleus, the temperatures found by these methods lose their uniqueness. In particular it will be impossible to characterize the radiation in a given region of the spectrum by a temperature determined by another part of the spectrum. In all cases, therefore, when speaking about the temperature of the nucleus of one or another planetary nebula, one must also mention the method by means of which the temperature was obtained.

2. Electron Temperatures

Following the photoionization of a hydrogen atom the remaining energy of the photon is imparted to the ejected electron in the form of kinetic energy according to the relation $mv^2/2 = hv - hv_0$, where hv is the energy of the quantum and hv_0 is the ionization energy of hydrogen. This quantity $mv^2/2$ varies for different acts of photoionization. However, the collisions of electrons with one another lead to the situation in which, after some time, the most probable Maxwellian velocity distribution is established. In this case one may speak about the mean energy of the electrons and, therefore, about the mean kinetic temperature of the electron gas in a given volume. We also call this the electron temperature of the medium. Obviously, as the temperature of the nucleus emitting ultra-violet photons increases the average remaining energy of the electrons and, consequently, the electron temperature will also increase.

The dependence of the electron temperature T_e of the nebula on the temperature of the nucleus T_* may easily be derived, proceeding from the following two conditions: a) the condition of stationarity—the number of atoms per unit time entering the continuum by photoionization of hydrogen must be equal to the number of atoms leaving the continuum; b) the condition of radiative equilibrium—the amount of energy expended in the photoionization of hydrogen must equal the amount emitted by recombinations. The application of these conditions leads to a relation of the form $f_1(T_e) = f_2(T_*)$ from which T_e is determined, given T_*. From calculations for an optically thin hydrogen nebula [97] the following values were found:

T_*	T_e
500°K	5000°K
10,000	9500
20,000	18,000
40,000	34,000
80,000	57,000
160,000	92,000

However, all of these considerations relate to the case in which the loss of electron energy occurs only through recombination processes in hydrogen. In the conditions of gaseous nebulae, as we have seen in the preceding chapter, there exists yet another more powerful mechanism for the loss of energy by free electrons. This is the excitation of ions of various elements to metastable levels through collisions of the first kind. In this way a significant fraction of the kinetic energy of free electrons escapes from the nebula in the form of radiation in forbidden lines. This leads to a still larger decrease of the electron temperature. Obviously, the value of the steady-state electron temperature in this case will depend on the relative abundance of the ions giving rise to forbidden lines, that is, ultimately on the chemical composition of the nebula, on the temperature of the nucleus and on the electron concentration of the nebula. The occurrence of such a large number of parameters makes almost impossible, in a practical sense, a theoretical* determination of the electron temperature analogous to that done above for a *hydrogen* nebula. There remains only one correct way of solving this problem: to find a procedure for establishing the electron temperature for each nebula separately directly from observation. The temperature of the electron gas found in this way becomes an independent parameter characterizing the physical conditions in the nebula.

Below we shall dwell on the few procedures which are at present known for determining the electron temperatures of nebulae.

Ambartsumyan's method ([98], *page* 165). Ions such as O^{++} and N^{++} have not only a $^1D_2$ metastable level from which nebular lines originate, but also a higher-lying metastable $^1S_0$ level which decays to $^1D_2$, emitting the auroral lines. For example, the levels $^3P_2$, $^3P_1$, and $^3P_0$ are the lowest in the ground configuration $2p^2$ of the O^{++} ion. Above these is found the level $^1D_2$ and still higher, the level $^1S_0$ (Figure 2.1). The excitation potentials from the ground state of these two levels are 2.5 e.v. and 5.3 e.v. respectively. The transition from the $^1S_0$ level to the $^1D_2$ level gives rise to the forbidden line $\lambda4363$ Å and that from the $^1D_2$ level downward to the levels $^3P_{1,2}$ give rise to the N_1 and N_2 lines. Obviously, as the electron temperature of the nebula increases, so will the relative number of ions in the $^1S_0$ state compared to the number in $^1D_2$ and consequently the lines $\lambda4363$ Å will become brighter in comparison with the lines $N_1 + N_2$. Thus it is possible to judge the electron temperature of the nebula from the observed intensities of the lines $\lambda4363$ Å and $N_1 + N_2$. The problem therefore is to derive a theoretical relation giving the dependence of the electron temperature T_e on the ratio of the intensities of the lines $N_1 + N_2$ and $\lambda\,4363$ Å, that is, on $E_{N_1+N_2}/E_{4363}$.

Let us consider the problem in an extremely general form. Let n_1, n_2 and n_3 denote the concentrations of O^{++} ions in the ^3P, $^1D_2$ and $^1S_0$ levels, respectively. For brevity these levels will be designated by the numerals 1, 2 and 3. Let us write the conditions of stationarity for levels 1 and 2, taking into account electron collisions of the first kind (b_{ij}), collisions of the second kind (a_{ij}), and spontaneous transitions (A_{ij}):

$$n_1(b_{12} + b_{13}) = n_2(A_{21} + a_{21}) + n_3(A_{31} + a_{31}),$$

$$n_2(A_{21} + a_{21} + b_{23}) = n_1 b_{12} + n_3(A_{32} + a_{32}),$$

(24)

---

* Such an attempt, however, bearing a more qualitative character, has been made by Aller ([87], page 136).

where b_{ij} is the probability of a transition induced by a collision of the first kind, a_{ij} is the probability of a transition to a lower state in a collision of the second kind, and A_{ij} is the Einstein coefficient for a spontaneous radiative transition. From (24) we find:

$$\frac{n_2}{n_3} = \frac{b_{12}(A_{31} + a_{31}) + (b_{12} + b_{13})(A_{32} + a_{32})}{b_{12}(b_{12} + b_{13}) + b_{13}(A_{21} + a_{21})}. \tag{25}$$

The coefficients b_{ij} and a_{ij} are related by the expression

$$b_{ij} = \frac{\omega_i}{\omega_j} a_{ij} e^{-(\varepsilon_i - \varepsilon_j)/kT_e} \tag{26}$$

and a_{ij} is given by the formula

$$a_{ij} = 8.54 \times 10^{-6} \frac{\Omega(j, i)}{\omega_i T_e^{1/2}} n_e, \tag{27}$$

where $\Omega(ji)$ is the so-called "collision strength" and ω_i is the statistical weight of the i-th level. For the levels $^1S_0$, $^1D_2$ and ^3P ω_i equals 1, 5 and 9 respectively.

For the intensities of the lines $N_1 + N_2$ and $\lambda 4363$ Å, we have

$$E_{N_1 + N_2} = n_2 h[A(^1D_2 \rightarrow {}^3P_2)\nu_{N_1} + A(^1D_2 \rightarrow {}^3P_1)\nu_{N_2}] = n_2 A_{21} h\nu_{12}, \tag{28}$$

$$E_{4363} = n_3 A_{32} h\nu_{23},$$

where ν_{N_1}, ν_{N_2} and ν_{23} are the frequencies of the lines N_1, N_2 and $\lambda 4363$ Å respectively.

It should be noted that the quantities A_{ij} and $\Omega(i, j)$ are far from being known for all ions and transitions of interest to us; those we do know are of varying degrees

TABLE 4-5

The Quantities A_{ij}, $\Omega(i, f)$ and ε_{ij} for Several Ions

	$A_{21}(\sec^{-1})$	$A_{31}(\sec^{-1})$	$A_{32}(\sec^{-1})$	$\Omega(1, 2)$	$\Omega(1, 3)$	$\Omega(2, 3)$	ε_{21}(e.v.)	ε_{31}(e.v.)
O I	$9.1 \cdot 10^{-3}$	0.078	1.28	—.	—	—	1.96	2.22
O II	$9.72 \cdot 10^{-5}$	0.048	1.102	1.28	0.58	2.12	3.32	1.69
O III	$2.80 \cdot 10^{-2}$	0.230	1.60	1.59	0.22	0.64	2.49	2.84
N I	$1.04 \cdot 10^{-5}$	$5.4 \cdot 10^{-3}$	0.079	—	—	—	2.38	1.19
N II	$4.0 \cdot 10^{-3}$	0.034	1.08	2.39	0.223	0.46	1.89	2.15
Ne III	0.026	2.21	2.80	0.76	0.077	0.27	3.19	3.71
Ne IV	$2.59 \cdot 10^{-3}$	1.06	0.727	0.68	0.23	3.51	5.12	2.62
Ne V	0.521	4.12	2.60	0.84	0.16	0.53	3.64	4.17
S II	$1.07 \cdot 10^{-3}$	0.26	0.44	2.02	0.383	12.7	1.84	1.20

of accuracy. These quantities are generally found by means of quantum-mechanical calculations and since the methods of calculation are constantly being refined and the accuracy improved, it is impossible to say what limits have been reached. In Table 4-5 the latest data for A_{ij} and $\Omega(i,j)$ are shown [53, 94, 100, 215, 216]. Although the relations cited below utilize partially the earlier data of Seaton, this does not have a significant effect on the calculated value of T_e.

From (25)–(28) we find

$$\frac{E_{N_1+N_2}}{E_{4363}} = F(n_e,T_e)e^{\,33,000/T_e} \,, \tag{29}$$

where

$$F(n_e,T_e) = 0.0753\frac{1 + 2.67 \cdot 10^5 \, T_e^{1/2}/n_e + (0.04 + 1.27 \cdot 10^4 \, T_e^{1/2}/n_e)e^{-33,000/T_e}}{1 + 2.30 \cdot 10^3 \, T_e^{1/2}/n_e + 0.09e^{-33,000/T_e}}. \tag{30}$$

For planetary nebulae one usually has $n_e \sim 10^4 \, \text{cm}^{-3}$, $T_e \sim 10^4 \,^\circ\text{K}$. Therefore, with sufficient accuracy we can obtain from (30) and (25):

$$\frac{E_{N_1+N_2}}{E_{4363}} = 0.0753\,\frac{1 + 2.67 \cdot 10^5 \, T_e^{1/2}/n_e}{1 + 2.30 \cdot 10^3 \, T_e^{1/2}/n_e}\,e^{33,000/T_e}. \tag{31}$$

This formula may be simplified still further by setting $n_e = 0$, which is permissible for small values of the density of the nebula ($n_e < 10^3 \, \text{cm}^{-3}$). Then from (30) we find $F(0,T_e) = 8.74$ and formula (29) takes the form

$$\frac{E_{N_1+N_2}}{E_{4363}} = 8.74e^{33,000/T_e}. \tag{32}$$

Thus, for small values of the electron density we have in formula (32) the possibility of uniquely determining the electron temperature T_e by means of the ratio $E_{N_1+N_2}/E_{4363}$ obtained from observations. In the remaining cases one must employ the formulae (29) and (31), for which it is necessary to determine n_e beforehand by some kind of procedure.

The electron temperature may also be found from the relative intensities of the forbidden lines of ionized nitrogen, $E_{6548} + E_{6584}/E_{5755}$. The relation corresponding to (31) may be derived in exactly the same way as in the previous case. Then we find ([8], page 193):

$$\frac{E_{6548} + E_{6584}}{E_{5755}} = 1.625 \cdot 10^{-2}\,\frac{1 + 1.94 \cdot 10^5 \, T_e^{1/2}}{1.03 + 3.32 \cdot 10^2 \, T_e^{1/2}/n_e}\,e^{25,000/T_e}. \tag{33}$$

Formula (32) and in some cases formula (34) were applied to a number of planetary nebulae. Usually the ratio $E_{N_1+N_2}/E_{4363}$ was very high on the average—of order 100. Therefore one must be particularly careful in determining the intensities of these lines (especially the weak line $\lambda 4363$ Å), in order to obtain correctly the electron temperature of the nebulae. In this way, for example, Andrillat [101], employing his measurements, determined the electron temperatures of 24 planetary nebulae according to formula (32). They appeared to be within the interval $9000°K-25,000°K$. Apparently the dispersion in the values of the electron temperature is real.

Sobolev's method [90]. Here the energy balance of the electron gas is considered. The free electrons are assumed to receive energy from the photoionization of hydrogen and to lose it in three ways: 1) by emission in the continuum (recombination and free-free transitions); 2) by collisional excitation of "nebulium" lines; and 3) by inelastic collisions with hydrogen atoms. Obviously in the mathematical representation of such an energy balance we have an important relation between the temperature of the central star and the electron temperature of the nebula. By having another relation of the same type (for example, the modified formula of Zanstra, see further) we obtain a system of equations from which we may find uniquely T_* and T_e.

With the above assumptions the law of conservation of energy of the electron gas reduces to the equation

$$AT_* = BT_e + C\,\frac{N_2}{H_\beta} + D\,\frac{\bar{n}_1}{n^+},\tag{35}$$

where N_2/H_β is the ratio of the intensities of the lines N_2 and H_β in the spectrum of the nebula (determined from observations). $\bar{n}_1/n^+$ is the ratio of the neutral hydrogen and proton concentrations averaged over the nebula, as given by the relation

$$\bar{n}_1/n^+ = (n_1/n^+)\ln(n^+/n_1)_0,\tag{36}$$

where $(n^+/n_1)_0$ is the degree of ionization for $\tau_c = 0$ determined from the usual ionization formula [(51), Chapter II]. To obtain $(n^+/n_1)_0$ one must know the dilution coefficient W and the electron concentration n_e for $\tau_c = 0$. The first quantity is determined from the relation

$$\log W = 7.00 - 0.4m_* - 2\log D'' + 14{,}700/T_*\tag{37}$$

and the second from the formula

$$\log n_e = 8.88 - \frac{3}{2}\log D'' - 0.2m_n - \frac{1}{2}\log r.\tag{38}$$

In these expressions m_* and m_n are the apparent photographic magnitudes of the nucleus and the nebula, respectively; D'' is the angular diameter of the nebula in seconds of arc and r is the distance to the nebula in parsecs.

The coefficient A in relation (35) depends on both the temperature of the star T_* and the optical thickness τ_c of the nebula beyond the Lyman limit. Numerical values of A are shown in Table 4-6. The coefficients B, C and D depend only on the electron temperature and are given in Table 4-7.

The equation (35) has been found also by Aller [203], but without the last term on the right side.

From the relation (35) we may find the electron temperature of the nebula if the temperature of the nucleus is known. However, we can also solve equation (31) together with the partially corrected form of the Zanstra equation (15) (for the case $\tau_c \gg 1$), where η, the fraction of the energy of free electrons expended in the excitation of "nebulium" lines, is given by the expression

$$\eta = \frac{C}{A T_*} \frac{N_2}{H_\beta}. \tag{39}$$

In this way we obtain T_* and T_e simultaneously.

Sobolev's procedure of determining the electron temperatures of nebulae—laborious as it is in practice—is correct in principle. Therefore the application of this procedure, even if to only a few nebulae, is of considerable interest. This was carried out by Sobolev for the case $\tau_c \gg 1$. From these results one can reach the

TABLE 4-6

T_*	A	
	$\tau_c \ll 1$	$\tau_c \gg 1$
20,000°K	0.90	1.24
40,000°K	0.83	1.46
60,000°K	0.77	1.63
80,000°K	0.71	1.76

TABLE 4-7

T_e	B	C	D
5000°K	1.02	12,000	1
7500°K	1.04	12,000	$3 \cdot 10^3$
10,000°K	1.06	12,000	$2.5 \cdot 10^5$
12,500°K	1.08	12,000	$2.5 \cdot 10^6$
15,000°K	1.10	12,000	$1.6 \cdot 10^7$

following conclusions:

1. The coefficient η in formula (15) is smaller than unity and is of order 0.5, so the assumption that free electrons lose all of their energy through "nebulium" emission is incorrect.

2. The electron temperatures of planetary nebulae lie in the interval $9000°K - 14,000°K$.

The above procedure for determining the electron temperatures of gaseous nebulae may be refined, in particular, by using more accurate values of the effective cross section for the electron-hydrogen collisions. It is also necessary to take into account the fact that for the majority of planetary nebulae $\tau_c < 1$. However, it seems that the results obtained after making these improvements will not significantly alter the above conclusions.

The method of the "Balmer continuum". If the continuous spectrum (Balmer continuum) beyond the Balmer limit is assumed to arise from recombinations, the intensity of this spectrum will be determined by the expression [see Formula (41b), Chapter III],

$$I_\lambda = Ce^{-hc/\lambda kT_e}.\tag{40}$$

By comparing the intensities at two wavelengths λ_1 and λ_2, we obtain

$$\log \frac{I_{\lambda_1}}{I_{\lambda_2}} = 0.434 \frac{hc}{kT_e}\left(\frac{1}{\lambda_1} - \frac{1}{\lambda_2}\right)$$

or

$$T_e = 0.629 \frac{1/\lambda_1 - 1/\lambda_2}{\log(I_{\lambda_1}/I_{\lambda_2})}.\tag{41}$$

By means of these formulae one may determine the electron temperature T_e of a nebula from the ratio of intensities in two regions of the continuous spectrum.

This method, despite its simplicity, gives poor results for a number of reasons. In the first place, one cannot always obtain spectrograms of the Balmer continuum with a sufficient density of darkening for the measurements. Secondly, the observed energy distribution in the Balmer continuum must be corrected for interstellar selective absorption and this cannot always be done with sufficient accuracy; secondary difficulties also arise. But the principal reason lies in the fact that the continuous spectrum of planetary nebulae consists of several components (see Chapter V) and in each specific case it is difficult to say to what extent recombination processes in hydrogen are predominant.

The method of the Balmer continuum has been applied to a number of planetary nebulae by Page (see Chapter V). He obtained values of T_e less than $10,000°K$ for almost all of the nebulae he investigated. The most recent application of formula (41) was to the planetary nebula NGC 7027, for which it was found that $T_e = 12,300°K$ [66].

3. Cooling of the Inner Regions of Nebulae

As we have seen, the cooling of a nebula involves such ions as O II, O III, N II, S II, and S III. The monochromatic images of the nebulae in the lines of these ions have the largest dimensions. Moreover, the lowest metastable state in each ion has a very low excitation potential. Therefore, the electron temperatures determined from forbidden lines will be characteristic of the outer regions of the nebulae, where the radiating ions are found and consequently will represent the minimum temperatures in these nebulae.

In the inner regions of the nebulae oxygen, nitrogen, carbon and the remaining elements are no longer present in singly and doubly charged ions. Here these elements are three, four or more times ionized and consequently only these ions can be effective in cooling the gas. The whole question of the temperature in the inner regions depends upon the relation of the excitation potentials ε_{II} of the lowest metastable levels of successive, high stages of ionization. If the metastable states in successive stages have approximately the same excitation potentials, measured from their respective ground states, then the cooling effect will be similar at all points and no temperature gradient will be produced in the nebula.

In order to find the answer to our question, it is evidently necessary to carry out a careful analysis of the energy states of the various ions most frequently encountered in nebulae. Such an analysis has been performed [217], the results of which are presented in Figure 4-2 in the form of a graphical relation between the ionization potential χ and the excitation potential ε of the lowest metastable state, and also in Table 4-8, which shows the dependence of ε on the degree of ionization. The necessary data for this work was taken from the catalogue of atomic energy levels compiled by Moore [218].

In the outer parts of the nebula many elements exist in singly and doubly ionized states, that is, those with ionization potentials less than 60 e.v. (to the left of the broken

TABLE 4-8

Excitation Potential in e.v. of the Lowest Metastable Level

| Element | Stage of Ionization | | | | | |
	II	III	IV	V	VI	VII
C	5.3	6.4	8.0	> 300	—	—
N	1.9	7.0	8.3	9.7	>300	—
O	3.3	2.5	8.8	10.2	12.5	—
Ne	27.0	3.2	5.0	3.7	12.3	—
S	1.8	1.3	8.8	10.0	13.0	—
F	2.6	4.1	3.1	11.0	12.0	—
Fe	2.4	2.4	3.9	3.2	2.3	2.2
Ce	1.4	2.2	1.7	10.7	12.4	15.5

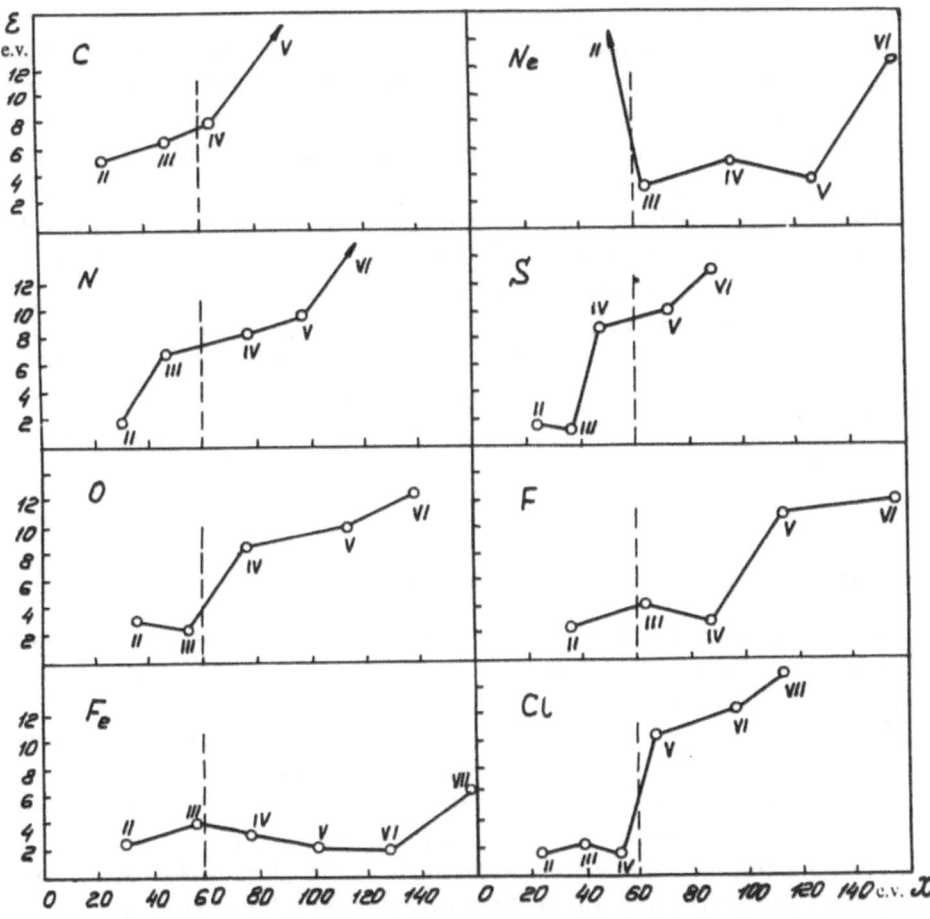

Figure 4-2

The dependence of the excitation potential ε of the lowest metastable level on the ionization potential χ.

line in Figure 4-2). For these ions the excitation potentials of the lowest metastable levels are comparatively small; on the average they reach $\varepsilon_{II} \approx 2.5$. Further in almost all cases one can see a quite rapid increase in ε as the degree of ionization increases. In other words, the more highly ionized the element, the larger is the excitation potential of its lowest metastable state. In almost all of the cases shown in Figure 4-2 ε increases more or less monotically with the degree of ionization. Only in the case of iron does ε remain constant up to Fe VI, after which it increases. But on account of the extreme rarity of iron in nebulae, it appears that cooling by iron will be insignificant.

Thus, ions with ionization potentials greater than 60 e.v. and present in the central regions of nebulae (to the right of the broken line in Figure 4-2) will characteristically have higher excitation potentials to their lowest metastable states. These excitation potentials equal, on the average, $\varepsilon_{II} \sim 8 - 10$ e.v. that is approximately 3–4 times larger than the excitation potentials ε_I of the forbidden lines of the group of ions occurring to the left of the broken line in Figure 4-2 and constituting the mass of the outer region of the nebula. Hence, without any additional calculation we may conclude that the electron temperature in the central region of a planetary nebula must exceed by a noticeable amount the electron temperature in its outer region, which, as we have seen, is easy to determine and is of order 10,000°K.

We conclude, thus, that a planetary nebula is not isothermal throughout.

In connection with this conclusion the following problem arises: is it possible, knowing the electron temperature T_e in the outer regions of a nebula, the temperature T_* of the nucleus and the mean values of the excitation potentials ε_I and ε_{II} for the above-mentioned two groups of ions, to estimate the electron temperature T_{e1} in the inner regions of the nebula. The following section is devoted to this question.

4. The Temperature Gradients in Nebulae

One may solve the problem defined above by writing the law of conservation of energy for the free electrons in the inner and outer regions of the nebula and then comparing the relations for each region.

We will assume that the free electrons obtain their energy from the photoionization of hydrogen and lose it in four ways:

1. By recombination of electrons with protons and by free-free transitions. The energy lost by these processes per unit volume and per unit time is denoted by E_1.

2. By recombination of electrons with singly ionized helium and by free-free transitions of the electrons in the field of singly-charged helium ions (E_2).

3. By inelastic collisions of electrons with neutral hydrogen (E_3).

4. By excitation of all forbidden lines of various atoms and ions through inelastic collisions with electrons (E_4).

If E_e is the amount of energy received by free electrons per unit volume per unit time, then the law of conservation of energy takes the form

$$E_e = E_1 + E_2 + E_3 + E_4. \tag{42}$$

Obviously in writing this expression for various parts of the nebula we find the relation of the electron temperature to the physical parameters of the nebula and nucleus. In this way we are able to estimate the magnitude of the gradient of electron temperature.

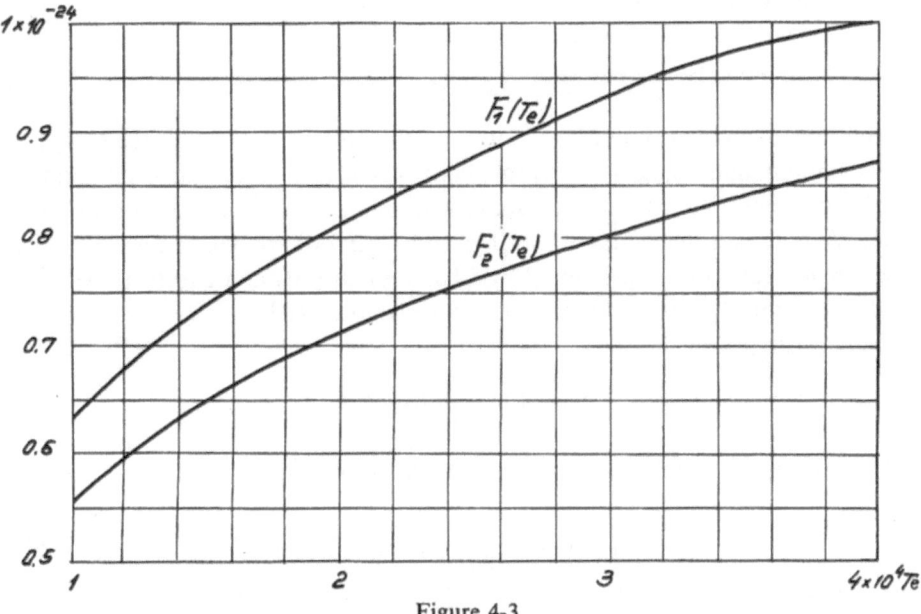

Figure 4-3

The functions $F_1(T_e)$ and $F_2(T_e)$ *versus* the electron temperature T_e.

The energy required by the recombination and free-free collisions with protons may be represented in the form

$$E_1 = n_e n^+ [F_r(T_e) + F_{ff}(T_e)] = n_e n^+ F_1(T_e), \tag{43}$$

where n_e and n^+ are the concentrations of free electrons and protons. The explicit form of $F_1(T_e)$ is given in [62, 213] and is plotted against T_e in Figure 4-3.

In just the same way we may write for the quantity E_2, relating to singly ionized helium:

$$E_2 = n_e n^+ F_2(T_e). \tag{44}$$

The dependence of the function $F_2(T_e)$ on electron temperature is shown in Figure 4-3. In constructing this curve it was assumed that the relative abundance of helium atoms in $n(\text{He})/n(\text{H}) = 0.2$ and that the fraction of ionized helium (degree of ionization) is $n(\text{He}^+)/n(\text{He}) = 0.9$. For other values of the relative abundance of helium and of the degree of ionization one must scale the numerical values of $F_2(T_e)$ in Figure 4-3 by $n(\text{He})/n(\text{H})$ and $n(\text{He}^+)/n(\text{He})$ respectively.

Let us turn to the determination of E_3—the energy expended in inelastic collisions of electrons with ground-state hydrogen. We will limit our investigation to collisional excitation of the second energy level, since the probability of collisional transitions to the third and higher states (including the continuum) are smaller than the probability of transitions to the second level (see Table 2-14) by at least two orders of magnitude. Then we may write

$$E_3 = n_1 n_e \alpha_2(T_e) h\nu_\alpha, \tag{45}$$

where n_1 is the number of hydrogen atoms in the ground state, $h\nu_\alpha$ is the energy of the L_α-photon and $\alpha_2(T_e)$ is the probability coefficient for collisionally-induced $1 \to 2$ transitions. We will use the data of Milford [211] for $\alpha_2(T_e)$ shown in Table 2-14a. We note that the values of $\alpha_2(T_e)$ given by Milford are one and a half to two times smaller than the values found by Chamberlain [70] and two to three times smaller than those given by the formula of Miyamoto [60].

Let us write (45) in the form

$$E_3 = n_e n^+ \frac{n_1}{n^+} \alpha_2(T_e) h\nu_\alpha \tag{46}$$

and replace n_1/n^+ from the ionization formula. We obtain

$$E_3 = n_e n^+ \frac{n_e}{W T_e^{1/2} \phi(T_*)} F_3(T_e), \tag{47}$$

where W is the dilution coefficient and $\phi(T_*)$ and $F_3(T_e)$ are defined by:

$$\phi(T_*) = T_*^{1/2} \frac{2(2\pi\mu k T_*)^{3/2}}{h^3} e^{-h\nu_0/kT_*}; \tag{48}$$

$$F_3(T_e) = \alpha_2(T_e) h\nu_u, \tag{49}$$

where ν_0 is the ionization frequency of hydrogen. Numerical values of the function $F_3(T_e)$ are shown in Table 4-9 and the function $\phi(T_*)$ is graphed in Figure 4-4.

Finally, for the free electron energy expended in the excitation of forbidden lines, we have

$$E_4 = n_e \sum_i n_i b_{ij} h\nu_{ij} = n_e n^+ \sum \frac{n_i}{n^+} b_{ij} h\nu_{ij}, \tag{50}$$

where n_i is the concentration of ions exciting a given forbidden line, b_{ij} is the probability coefficient for collisionally induced transitions and ν_{ij} is the frequency of the given line.

TABLE 4-9

$T_e(°K)$	$F_3(T_e)$
10,000	$2.12 \cdot 10^{-24}$
15,000	$1.24 \cdot 10^{-22}$
20,000	$0.98 \cdot 10^{-21}$
30,000	$0.78 \cdot 10^{-20}$

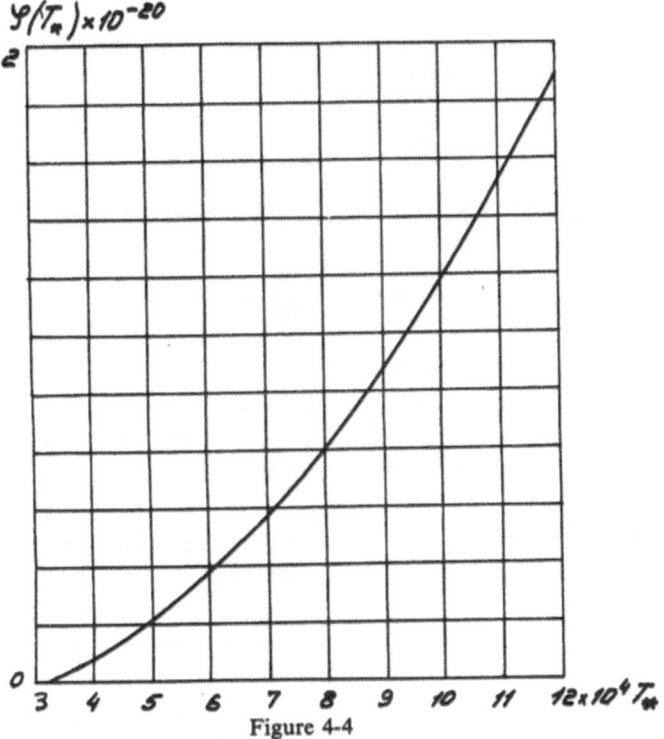

Figure 4-4

The function $\phi(T_*)$ *versus* the temperature of the nucleus T_*.

The following remark is essential for what is to come. Consider, for example, the O^{++} ions, which give rise to the lines N_1 and N_2. In the outer parts of the nebula almost all of the oxygen is doubly ionized. Therefore one may write approximately

$$\frac{n_i}{n^+} = \frac{n(O^{++})}{n^+} \approx \frac{n(O)}{n(H)} .$$

But in the inner parts of the nebula the oxygen will be triply ionized and practically no doubly ionized oxygen will be found. If the chemical composition of the nebula is everywhere the same, one may write, for the inner parts of the nebula:

$$\frac{n_i}{n^+} = \frac{n(O^{+++})}{n^+} \approx \frac{n(O)}{n(H)} .$$

Other ions and atoms behave in a similar manner.

Thus, in spite of significant changes of the physical conditions within the nebula, the ratio n_i/n^+ entering the relation (50) may be assumed to be everywhere constant to a sufficient degree of accuracy.

However, the coefficients b_{ij} are no longer constant throughout the nebula, since they depend on the excitation energy of the metastable level, which, as we have seen in the preceding section, is larger in the inner region than on the outside. Using (26) and (27) we write b_{ij} in the form

$$b_{ij} = C_{ij} \frac{1}{T_e^{1/2}} e^{-\varepsilon_{ij}kT_e}. \tag{51}$$

By substituting b_{ij} into (50) and introducing the expressions for n_i, ε_{ij} and v_{ij}, we obtain

$$E_4 = n_e n^+ C \frac{1}{T_e^{1/2}} e^{-\varepsilon/kT_e}, \tag{52}$$

where C is some quantity which is constant for a given nebula and which depends, in particular, on its chemical composition, and ε is the mean value of the excitation energy of the metastable levels origination in the part of the nebula under consideration. For example, in the outer parts of the nebula $\varepsilon = \varepsilon_1 \approx 2.5$ e.v. and in the inner parts $\varepsilon = 8$ e.v.

Following V.V. Sobolev [90], we write for the total energy E_e received by the electrons per cm^3 per second from photoionization

$$E_e = \bar{\varepsilon} n_e n^+ \sum_1^\infty C_i(T_e) = \bar{\varepsilon} n_e n^+ C(T_e), \tag{53}$$

where $\bar{\varepsilon}$ is the mean electron energy; it depends on the temperature of the nucleus and is given in the following form:

$$\bar{\varepsilon} = K T_* A(T_*), \tag{54}$$

where the numerical values of $A(T_*)$ appear in Table 4-6 for the cases $\tau_c \gg 1$ and $\tau_c \ll 1$.

Substituting (54) into (53) we find

$$E_e = n_e n^+ K(T_*, T_e), \tag{55}$$

where

$$K(T_*, T_e) = \kappa T_* A(T_*) C(T_e) \tag{56}$$

and the numerical values of $C(T_e)$ are to be taken from Table 2-8.

Curves showing the dependence of $K(T_*, T_e)$ on T_e for several values of T_* are given in Figure 4-5 for the case $\tau_c \ll 1$ and in Figure 4-6 for $\tau_c \gg 1$.

Substituting (44), (47), (52) and (55) into (42), we obtain finally

$$K(T_*, T_e) = F_1(T_e) + F_2(T_e) + \frac{n_e}{W T_e^{1/2} \phi(T_*)} F_3(T_e) + C T_e^{-1/2} e^{-\varepsilon/kT_e}. \tag{57}$$

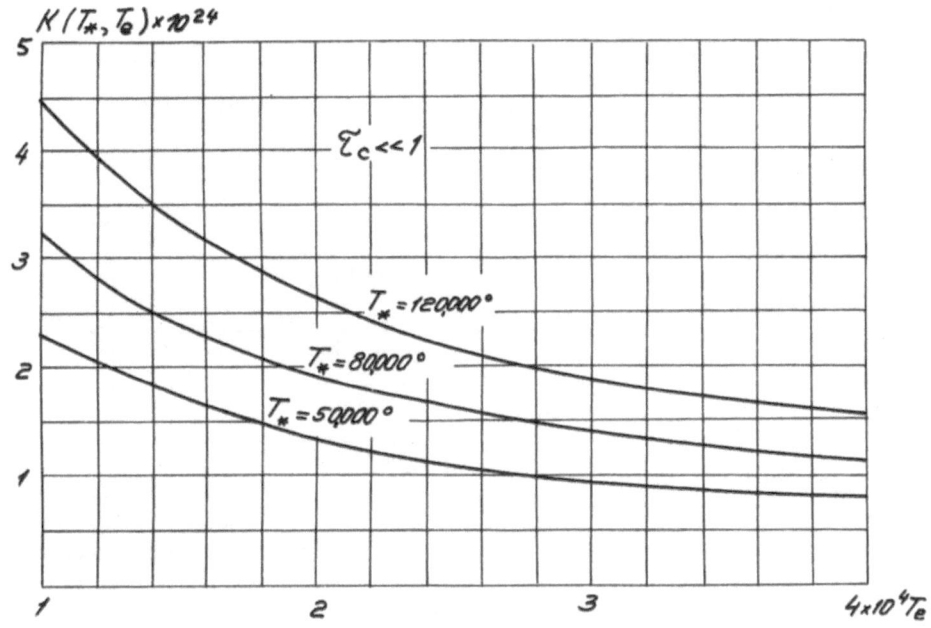

Figure 4-5

The function $K(T_*, T_e)$ *versus* T_e for $\tau_c \ll 1$ and various values of the temperature T_* of the nucleus.

This relation gives the dependence of the electron temperature T_e on the temperature of the nucleus, the electron concentration n_e, the dilution coefficient W and the chemical composition C of the nebula. But primarily it expresses the dependence of T_e on ε. Only T_* and C are constant for a given nebula. Therefore the electron temperature T_e will change noticeably within the nebula (variations in n_e and W have a smaller effect).

Let us apply the relation (57) for some representative models of planetary nebulae ($T_* = 50,000°\mathrm{K}$, $n_e = 10^4\,\mathrm{cm}^{-3}$, $W = 10^{-14}$) with the aim of determining the difference between the electron temperatures in the inner and outer regions.

To do this, one must write the relation (57) first for the outer parts of the nebula, where $T_e = 10,000°\mathrm{K}$ and $\varepsilon_1 = 2.5$ e.v. and then for the inner parts, where $\varepsilon_{II} > \varepsilon_1$ and equals for example, $\varepsilon_{II} \approx 8$ e.v. The electron temperature T_{e1} in the inner region is unknown, but can now be obtained by eliminating C from the two equations. In this way one finds $T_{e1} = 16,400°\mathrm{K}$ for $\tau_c \ll 1$ and $T_{e1} = 21,000°\mathrm{K}$ for $\tau_c \gg 1$. In these calculations it was assumed that the inner radius is half the outer radius, that is, that the dilution coefficient for the inner region is $4W$.

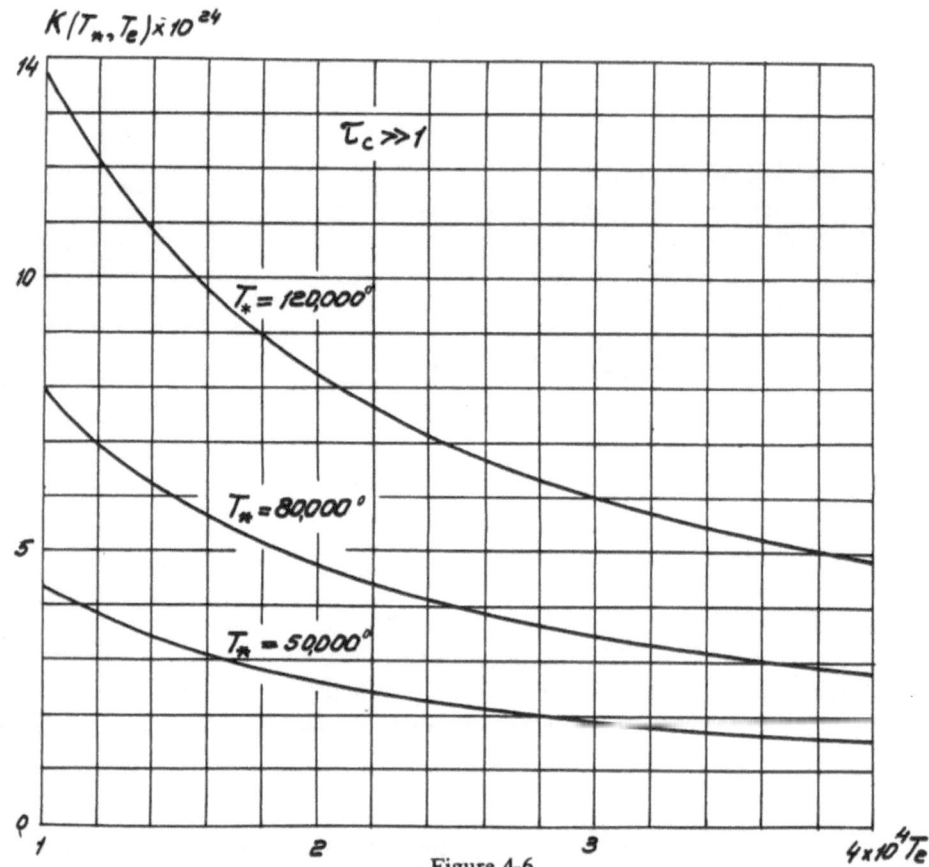

Figure 4-6

The function $K(T_*, T_e)$ *versus* T_e for $\tau_c \gg 1$ and various values of the temperature T_* of the nucleus.

In this way one can find the value of T_{e_1} for other values of ε_{II} and T_*. The results of these calculations appear in Tables 4-10 and 4-11.

Provided that ε increases inward, then T_{e_1} also increases inward. From the results given in Tables 4-10 and 4-11, the electron temperatures in the central regions of nebulae are significantly, that is, two or three times, higher than the electron temperatures in the outer regions. V.I. Pronik [219] and Ōsaki [213] have also shown, by other arguments, the existence of high temperatures in the central regions of planetary nebulae.

This difference in electron temperatures becomes still larger—by a factor of two or three—for nebulae which are optically thick for L_c-radiation. Finally this difference in temperatures also increases with the temperature of the nucleus.

TABLE 4-10

The Dependence of the Electron Temperature
T_{e_1}(°K) on ε for $\tau_c \ll 1$

T_*(°K)	ε(e.v.) 2.5	6	8	10
50,000	10,000	15,500	16,600	17,000
80,000	10,000	17,000	20,000	24,500
120,000	10,000	17,800	22,200	27,500

TABLE 4-11

The Dependence of the Electron Temperature
T_{e_1}(°K) on ε for $\tau_c \gg 1$

T_*(°K)	ε(e.v.) 2.5	6	8	10
50,000	10,000	18,000	21,000	22,500
80,000	10,000	19,800	25,000	30,500
120,000	10,000	21,500	27,000	32,400

TABLE 4-12

Electron Temperature T_{e_1} in the Central Regions of
Planetary Nebulae (for $\varepsilon_{II} = 8$ e.v.)

NGC	T(°K)	W	n_e(cm^{-3})	T_e(°K)	T_{e_1}(°K)
6543	46,000	$1 \cdot 10^{-14}$	$1.7 \cdot 10^4$	9500	17,300
7009	57,000	$3 \cdot 10^{-15}$	$1.5 \cdot 10^4$	13,900	18,400
7293	70,000	10^{-16}	150	10,000	19,000
7662	56,000	$3 \cdot 10^{-15}$	10^4	15,000	20,000

The relation (57) may be applied to the determination of the electron temperature T_{e_1} in the central parts of any planetary nebula for which T_e is known. Values of the temperatures T_{e_1} for several planetary nebula, found in this way for $\varepsilon_{II} = 8$ e.v., are given in Table 4-12. The values of n_e and T_e for these nebulae were taken from Table 4-15, W was calculated according to formula (37) (page 101), and the values of T_* were taken from Table 4-4, that is, determined by the method of [O III]/[O II], and refer just to the energy distribution of that part of the spectrum of the nuclear radiation, which is absorbed by the ions in the central regions of the nebula.

As follows from the data in Table 4-12, the electron temperature in the central regions are from one and a half to two times larger than those in the outer parts of the nebula. In still deeper parts of the nebula, where $\varepsilon > 8$ e.v, the electron temperatures, obviously, must be still higher, of order 30,000°K.

In Table 4-13 we give the percentage of the free electron energy going into various processes for the planetary nebulae considered above in Table 4-12. Calculations were made for both the outer (I) and inner (II) regions of the nebulae. From the results of these calculations one may draw the following conclusions:

1. The largest fraction of free electron energy goes into recombination processes involving hydrogen and helium, that is, into radiation in the continuous spectrum.

2. Inelastic collisions with hydrogen atoms play a role only in the central regions of the nebula; in the outer part they have almost no effect.

3. The fraction of the energy expended in the excitation of forbidden lines is several times larger in the outer parts of the nebula than in its inner parts.

TABLE 4-13

Distribution of Energy of Free Electrons (in %) Among Various Processes

Process	NGC 6543		NGC 7009		NGC 7293		NGC 7662	
	I	II	I	II	I	II	I	II
Radiation in the continuous spectrum of hydrogen	24	48	33	47	22	45	38	50
Radiation in the continuous spectrum of helium	21	42	30	41	19	39	34	44
Excitation of forbidden lines	55	10	36	6	59	14	23	3
Collisions with hydrogen atoms	0	0	1	6	0	2	5	3

One should point out the large fraction of the energy going into recombination processes in helium. This is explained by the fact that the probability of recombination for ionized helium is twice that for hydrogen. In general helium causes an appreciable amount of cooling in the central regions of planetary nebulae—the larger the relative abundance of helium in the nebula, the smaller the electron temperature in its central region. The calculated values of T_{e1} for several values of $n(\text{He})/n(\text{H})$ are given in Table 4-14. From this data it follows that a pure hydrogen nebula, containing no helium, will have a very high electron temperature.

TABLE 4-14

Dependence of T_{e1} on $n(\text{He})/n(\text{H})$

for $T_* = 50,000°\text{K}$ and $\varepsilon_\text{H} = 8$ e.v.

	$T_{e1}(°\text{K})$	
$n(\text{He})/n(\text{H})$	$\tau_c \ll 1$	$\tau_c \gg 1$
0	21,400	23,200
0.1	18,800	22,700
0.2	16,400	18,000

5. The Possibility of Determining the Electron Temperature of the Central Regions of Nebulae from Observations

The most convincing demonstration of the existence of high electron temperatures in the central regions of planetary nebulae would, of course, be that obtained from observations. For this one must use the relative intensities of a pair of lines belongint to a *highly ionized* element, the ionization potential of which is of order 60 e.v. or larger. Unfortunately in the majority of cases one, and sometimes both, components of the line one would employ for this purpose, lie in the far or vacuum ultra-violet. Moreover, for the time being, the atomic parameters for all ionic and atomic configurations are not known. Therefore the determinations of T_{e1} directly from observations is a thing of the future. Nevertheless, it is feasible to carry out some analysis, if only to clarify the prospective possibilities in this direction.

In passing we shall give a survey of the expected structure of the spectra of planetary nebulae in the far and vacuum ultra-violet.

The requirement that the ionization potential be larger than 60 e.v. is satisfied by the following ions (the lowest degree of ionization is given): Ne III, C IV, O IV, N IV, F III, Fe V, S V. The remaining elements are not of interest because of their extreme rarity.

Neon. In the spectra of planetary nebulae the lines of Ne III, Ne IV and Ne V are observed, the ionization potentials of which are 64, 97 and 126 e.v. respectively. The lines of these ions are produced in the central regions of the nebula. From among them one should look for a pair of lines, with the help of which the electron temperatures in the central regions may be established.

The energy levels diagrams of Ne III, Ne IV and Ne V are shown in Figure 4-7. The lines $\lambda 3967$ Å and $\lambda 3868$ Å of Ne III are observed in the spectra of many planetary nebulae. The line $\lambda\,4014$ Å must be four orders of magnitude weaker than $\lambda 3967$ Å and therefore is not observable. The line $\lambda\,3344$ Å, arising in the transition $^1S_0 - {}^1D_2$, has not been detected in the spectra of nebulae. It may be possibly be blended with

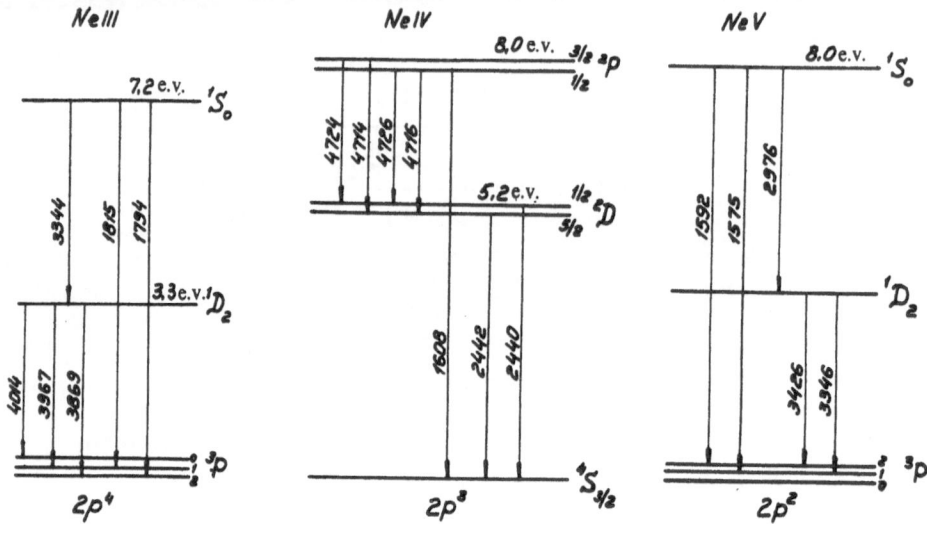

Figure 4-7
Energy levels diagrams for Ne III, Ne IV, Ne V.

the comparatively strong line $\lambda 3346$ Å [Ne V], although it may simply have a small intensity. Employing the known data for A_{ij} and $\Omega(i,j)$ for Ne III, one may easily find the ratio E_{3869}/E_{3344}. It is found to be equal to 133 and 65 for $T_{e1} = 20,000°K$ and $T_{e1} = 30,000°K$ respectively, that is, the line $\lambda 3344$ Å [Ne III] is fairly weak and may be easily lost in the neighboring line $\lambda 3346$ Å [Ne V].

The situation is somewhat more satisfactory for the line $\lambda 1815$ Å [Ne III] in the vacuum ultra-violet. For the relative intensities of these lines we have

$$E_{3869}/E_{1815} = 10.5 e^{43,500/T_{e1}}. \tag{58}$$

For $T_{e1} = 20,000°K$ and $T_{e1} = 30,000°K$ this ratio has the values 92 and 45, respectively, that is, the line $\lambda 1815$ Å [Ne III] must be only somewhat brighter than $\lambda 3344$ Å [Ne V]. But because no other forbidden lines (belonging to the elements listed above) are expected in the immediate neighborhood of $\lambda 1815$ Å [Ne III], its detection does not appear to be a desperate matter. The relation (58) may then be used to determine the electron temperature in the inner regions of the nebula.

In the far ultra-violet Ne IV ($\lambda 1608$ Å, $\lambda 2440$ Å) is also able to provide suitable lines for finding T_{e1}. From the ratio of the intensities of the lines $\lambda 2440$ Å [Ne IV] and $\lambda 2442$ Å [Ne IV] one may obtain the *electron concentration* in the central regions of the nebula, which is also of great interest. For the same purpose apparently, one may also employ the ratio of the intensities of the lines $\lambda 4724$ Å [Ne IV] and $\lambda 4714$ Å [Ne IV].

The most promising lines for determining the electron temperature in the inner regions appear to be the short-wave forbidden lines of Ne V. The most intense line among them is expected at $\lambda 1575$ Å [Ne V]. For the ratio of the intensities of the lines $\lambda 3426$ Å + $\lambda 3346$ Å [Ne V] and $\lambda 1575$ Å [Ne V] we have:

$$\frac{E_{3486+3346}}{E_{1575}} = 3.80 e^{50,000/T_{e1}}.$$

For $T_{e1} = 20,000°K$ and $T_{e1} = 30,000°K$ this ratio is found to be 46 and 20, respectively. The line $\lambda 2976$ Å [Ne V], originating from the transition $^1S - ^1D$, must be weaker than the line $\lambda 1573$ Å [Ne V] by a factor of three.

Carbon. All forbidden lines of triply ionized carbon exist in a very remote short-wave region of the spectrum (see Figure 4-8). Moreover, since the lines $\lambda 420$ Å [C IV] and $\lambda 330$ Å [C IV] must be very weak, one should not hope to use them to determine the electron temperature. The lines $\lambda 1551$ Å [C IV] and $\lambda 1548$ Å [C IV] may be of some interest; their ratio depends on the electron density.

As regards C V, all of its forbidden lines occur in the X-ray region of the spectrum (shorter than $\lambda 40$ Å).

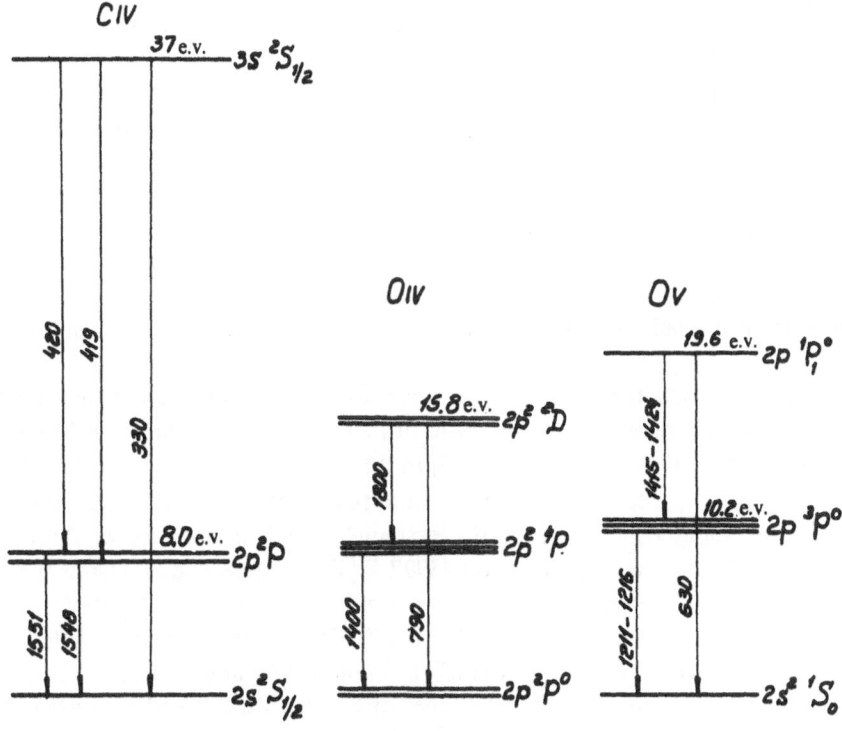

Figure 4-8
Energy level diagrams for C IV, O IV, O V.

It is interesting to note that carbon, in every stage of ionization, has no forbidden lines in the visible region of the spectrum. Because of the high excitation potential of the lowest metastable level (larger than 5 e.v.) all forbidden lines of both neutral and ionized carbon appear with wavelengths shorter than 3000 Å.

Oxygen. Forbidden lines, belonging to O IV and O V, exist in the vacuum ultraviolet (see Figure 4-8).

The line $\lambda790$ Å [O IV] is evidently very weak. Of some interest may be the group of lines with wavelengths near $\lambda1800$ Å, arising in the transitions $^2D-^4P$ ($2p^2$ configuration). Another group of lines ($\lambda\lambda1400-1412$ [O IV]) arises in the transitions $2p^2\ ^4P-2p^2P^0$, although they are probably all very weak. This group is adjoined immediately by a number of forbidden lines ($\lambda\lambda1415-1424$ Å) in the O V spectrum, arising in the transitions $^1P^0-^3P^0$ ($2p$ configuration).

It is interesting that transitions from the $^3P^0$ level ($2p$ configuration) to the ^1S level ($2s^2$ configuration) give rise to forbidden lines ($\lambda1215.7$ Å) in almost exact coincidence with the L_α-line of hydrogen ($\lambda1215.67$ Å). The possibility is not excluded, therefore, that the $^3P^0$ level of four times ionized oxygen is anomalously excited by the influence of L_α-radiation of hydrogen.

It appears that both O IV and O V are unable to furnish a convenient combination of lines which could be used to establish the electron temperature in the inner regions, even if the relevant atomic parameters were known.

Nitrogen. The structure of the energy levels of N IV is analogous to that of O V, and the structure of N V, with that of C IV. All forbidden lines arising in transitions with the ^2S level lie in the region of wavelengths shorter than $\lambda300$ Å. Of some interest (for the determination of electron density) are the pair $\lambda1243$ Å and $\lambda1239$ Å [N IV], emitted in the transitions $^2P^0 - {}^2S$. Triply ionized nitrogen has groups of lines with wavelengths of order $\lambda1580$ Å [N IV] ($^1P^0 - {}^3P^0$), 765 Å [N IV] ($^1P^0 - {}^1S$) and the lines $\lambda\lambda1485-1488$ Å [N IV] ($^3P^0-{}^1S$). Since the necessary atomic parameters are not available, it is difficult to say anything of their use in finding T_{e1} and n_{e1}.

Fluorine. The structure of the energy levels of F IV is analogous to that of O III. Hence one may foresee the possibility of utilizing the ratio of the intensities of the lines $\lambda\lambda4060$ Å $+ 3997$ Å [F IV] ($^1D-^3P$) to the intensity of $\lambda3533$ Å [F IV] ($^1S-^1D$) for determining T_{e1}, especially as the atomic parameters for F IV have been computed. Unfortunately, because of the extremely small abundance of fluorine in nebulae, the lines $\lambda4060$ Å [F IV] and $\lambda3997$ Å [F IV] appear to be just at the limit of detection, that is, some hundred times weaker than the line H_β (for example, in NGC 7027). In its turn, the line $\lambda3533$ Å [F IV] must be at least some ten times weaker than $\lambda4060$ Å [F IV] and therefore is completely unobservable.

For the same reasons the forbidden lines of F V, Fe VI, S V and S VI are of little interest, although some of them may give convenient combinations of lines in the visible region of the spectrum.

6. The Electron Concentration

One may obtain a general estimate of the electron concentration in planetary nebuale with the help of the ionization formula [(51), Chapter II] by assuming that the optical thickness τ_c of the nebula beyond the Lyman limit is of order unity. By assuming $n_e \approx n^+$ and $e^{-\tau_c} \approx 1$ we find from the ionization formula for the case $T_* = 50,000°K$, $T_e = 10,000°K$ and $W = 10^{-14}$:

$$n_e^2/n_1 \approx 10^6.$$

On the other hand, according to our assumptions, we have $\tau_c = \chi_c n_1 r = 1$, where $\chi_c = 0.65 \cdot 10^{-17}\,\text{cm}^2$ is the absorption coefficient per atom of ground-state hydrogen at the Lyman limit and r is the radius of the nebula. Taking $r = 10,000$ a.u. $= 1.5 \times 10^{17}\,\text{cm}$, we find $n_1 \sim 1\,\text{cm}^{-3}$ and consequently $n_e = n^+ \approx 10^3\,\text{cm}^{-3}$. These results attest to the high degree of ionization in planetary nebulae.

For the determination of the electron concentration of planetary nebulae from observational data there are a number of methods, which we shall now consider.

Method of surface brightness (H *method*). This method was proposed by Menzel and Aller [95] and is based on the idea that the surface brightness of a nebula in the hydrogen lines depends on its electron concentration.

The amount of energy emitted by the nebula in some line of the Balmer series of hydrogen, is determined by the expression

$$E_{ki} = z_k A_{ki} h\nu_{ik} n_e^2 V, \tag{59}$$

where $z_k = n_k/n^+ n_e$ and V is the volume of the nebula, which may be expressed in the form

$$V = 4\pi A^2 D, \tag{60}$$

A being the radius of the nebula and

$$D = d(1 - d/A + d^2/3A^2).$$

Here d is the thickness of the shell. For a nebula which extends into its star, $d = A$ and $D = A/3$.

From (59) we have

$$n_e = \left(\frac{E_{ki}}{z_k A_{ki} h\nu_{ik} V}\right)^{1/2}. \tag{61}$$

Going from the luminosity E_{ki} of the nebula in a given line to the surface brightness S_{ki} in the same line by means of the relation $S_{ki} = E_{ki}/4\pi A^2$, and using (60), we have in place of (61),

$$n_e = \left(\frac{S_{ki}}{D z_k A_{ki} h\nu_{ik}}\right)^{1/2}. \tag{62}$$

Let us denote by H_{ki} the surface brightness of the nebula in the given line, expressed in stellar magnitudes per square minute. The relation between S_{ki} and H_{ki} is of the form

$$S_{ki} = 0.84 \cdot 10^3 (2.512)^{-H_{ki}},$$

Representing z_k in the form $z_k = b_k z_k^0$, where z_k^0 is the value of z_k for thermodynamic equilibrium, and applying the formula (62) to the line H_β, we find

$$n_e = 1.84 \cdot 10^{11} \frac{T_e^{3/4} e^{-\chi_4/kT_e}}{(Db_4)^{1/2}} (2.512)^{-\frac{1}{4}H_\beta} \tag{63}$$

From this expression we may determine the electron concentration in the nebula if its surface brightness H_β in the line H_β is known.

One may derive in analogy to (63) a formula where H_β is replaced by H_n—the surface brightness of the nebula in the continuous spectrum beyond the Balmer limit (in a bandwidth of 20 Å), again expressed in stellar magnitudes per square minute. This formula has the form

$$n_e = 8.54 \cdot 10^{11} \frac{T_e^{3/4} (2.512)^{-\frac{1}{4}H_n}}{A^{1/2}} . \tag{64}$$

Finally, assuming that the luminosity of the nebula in the line H_β amounts to $\frac{1}{15}$ the total luminosity in the visible spectrum and converting by the usual procedure from luminosity to apparent stellar magnitudes m_n, we obtain still another relation for the electron concentration, which was employed above [formula (38)].

Formulae (63), (64) and (38) were applied for many nebulae, with the result that electron concentrations of order $10^3 - 10^4$ cm^{-3} were obtained [95, 102] (see Table 4-15).

The method of E_{3729}/E_{3726}. The metastable levels $^2D_{3/2}$ and $^2D_{5/2}$ of the O$^+$ ion in the ground configuration $2p^3$ are located very near one another and about 3.32 e.v. above the $4S_{3/2}$ level, the lowest level of the configuration. The well-known ultraviolet doublet $\lambda 3729$ Å and $\lambda 3726$ Å is emitted in the transitions $^2D_{5/2} \to {}^4S_{3/2}$ and $^2D_{3/2} \to {}^4S_{3/2}$ (see Figure 2-1). In distinction from the nebulium lines, the ratio of the intensities of which does not depend on the conditions in the nebula, the ratio of the intensities of $\lambda 3729$ Å and $\lambda 3726$ Å depend on n_e and T_e.

For very low matter densities in a nebula the time between two successive collisions will be sufficiently large that an ion, having been raised into a metastable state by collisions, has returned to the initial state through a spontaneous transition, emitting a photon in a forbidden line. In these conditions the components of the ultra-violet doublet will have an intensity ratio equal to the ratio of the effective collision cross sections, hence $E_{3729}/E_{3726} = 1.47$. For very high values of the matter density the time between collisions will be small, in consequence of which each ion will be excited to the metastable state and de-excited again by collisions many times before a spontaneous transition can occur with the emission of a photon. In these conditions the ratio E_{3729}/E_{3726} is determined by the condition of thermodynamic equilibrium and equals 0.45. For intermediate values of the matter density and the electron concentration, the ratio will be between the above values.

Having the theoretical relation between E_{3729}/E_{3726} and n_e and T_e, we can then estimate the electron concentration n_e if we know the value of the ratio E_{3729}/E_{3726}, obtained from observation, and the electron temperature.

The method of E_{3729}/E_{3726} was developed by Seaton, who made the necessary calculations to determine the effective cross sections for the atomic collisions [105]. In his final version, this ratio has the form [66]:

$$\frac{E_{3729}}{E_{3726}} = 1.5 \frac{1 + 3.26x}{(1.02 + 10.4 + 4.0e^{-19,500/T_e})x} \tag{65}$$

or approximately, for $T_e \sim 10,000°K$,

$$\frac{E_{3729}}{E_{3726}} = 1.5 \frac{1 + 3.26x}{1.02 + 10.5x}, \tag{66}$$

where

$$x = 10^{-2} \frac{n_e}{T_e^{1/2}}. \tag{67}$$

For large values of n_e (in practice, for $n_e > 5 \cdot 10^4 \ cm^{-3}$), we have

$$\frac{E_{3729}}{E_{3726}} \to 0.45 \quad (n_e \to \infty),$$

while for small values of $n_e (n_e < 50 \ cm^{-1})$ we find

$$\frac{E_{3729}}{E_{3726}} \to 1.47 \quad (n_e \to 0).$$

Curves showing the dependence of E_{3729}/E_{3726} on the electron concentration are given in Figure 4-9 for $T_e = 10,000°K$ and $T_e = 20,000°K$. From these curves we see that the dependence of E_{3729}/E_{3726} on T_e is very weak. Therefore, for the determination of the electron concentration from formula (66) it is sufficient to know T_e roughly.

Seaton [105] first applied this method to several nebulae. Subsequently Osterbrock [88] measured the ratio E_{3729}/E_{3726} for a large number of planetary nebulae. The electron concentrations of these nebulae were then determined on the basis of this data, using the known electron temperatures. The results appear in the fifth column of Table 4-15. In those cases when no data on the electron temperature is available, it is taken to be $T_e = 10,000°K$.

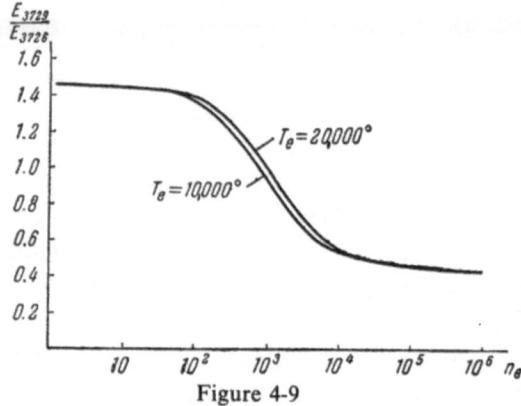

Figure 4-9

The dependence of E_{3729}/E_{3726} on the electron concentration n_e.

TABLE 4-15

T_e and n_e for Several Planetary Nebulae

Nebula	$T_e(10^{3\circ}K)$	$n_e(10^3\ cm^{-3})$			
		H	[O III], [N II]	E_{3729}/E_{3726}	D
NGC 40	—	—	—	2.5	—
NGC 650–1	—	—	—	0.35	—
NGC 1535	14.3	2.2	—	1.4	—
IC 48	18.3	16	8	10	24
IC 2149	14.2	2.2	—	3.2	—
NGC 2392	19.6	1.1	—	24	—
NGC 2440	17	—	40	4.3	—
NGC 3242	17.7	5.2	—	3.0	—
NGC 3587	—	—	—	0.15	—
IC 4593	7.7	2.2	—	3.2	—
NGC 6210	10.8	8.7	—	17	—
NGC 6543	9.5	6.3	—	17	—
NGC 6572	13.5	10.2	50	36	40
NGC 6720	—	—	—	10	—
NGC 6803	12.5	4.2	—	6.9	—
NGC 6818	15.7	3.0	—	—	—
NGC 6826	11.2	2.2	—	80	—
NGC 6853	—	—	—	0.5	—
IC 4997	10	—	—	10	—
NGC 7009	13.9	6.3	—	15	28
NGC 7027	16.8	4.9	77	19	45
NGC 7293	—	—	—	−18	—
NGC 7662	15	2.4	50	10	50

The electron concentrations, as determined by the present method, appear to be significantly larger than those obtained by the H method. Moreover, by this method the electron concentrations of extremely rarified (low-excitation) planetary nebulae were successfully determined, thereby giving some idea of the lower limit of electron concentrations which occur. Thus, for example, in the most dense parts of the giant planetary nebula in Aquarius, NGC 7293, the electron concentration was equal to $140-180$ cm^3 and for the bipolar nebula NGC 3587 the mean electron concentration was found to be 150 cm^{-3}, in places going to 70 cm^{-3}. In the central parts (on the connection) of the bipolar nebula NGC 605-1, the electron concentration appeared to be of order 300 cm^{-3}, while in the shell itself, $n_e = 130$ cm^{-3} [196].

The method E_{3729}/E_{3726} is simple and convenient to apply. However, it possesses the disadvantage that high dispersion spectrographs must be used to separate the lines $\lambda3729$ Å and $\lambda3726$ Å in measuring the intensity ratio of these lines.

The ion S^+ also has metastable levels analogous to those of O^+; the red doublet $\lambda6717$ Å and $\lambda6731$ Å is emitted by transitions from the levels $^2D_{3/2,5/2}$ to $^4S_{3/2}$. The dependence of the ratio E_{6731}/E_{6717} on n_e and T_e has the form

$$\frac{E_{6731}}{E_{6717}} = 1.5 \frac{1 + 0.35x}{1 + 0.96x}. \tag{68}$$

This formula has been applied at present to three or four planetary nebulae for which the intensities of the relevant lines had been measured.

The method of [O III], [N II]. We have already mentioned this method above in connection with the problem of determining the electron temperatures of nebulae. Combining formulae (31) and (33), we obtain a system of equations, the left sides of which are determined by observations, while in the right sides appear the required quantities T_e and n_e. From the simultaneous solution of these equations we can find T_e and n_e.

In spite of its simplicity this method is limited in its application: it gives good results only for comparatively dense nebulae. With intermediate values of the density ($n_e \sim 10^4$ cm^{-3}) the method becomes very sensitive to errors in measurement of the appropriate lines. In this way, apparently, one can explain the fact that the values of the electron concentration obtained by this method appear noticeably higher than the values obtained by the H method.

Seaton's method (*method of the Balmer discontinuity*). The magnitude of the discontinuity in the continuous spectrum at the limit of the Balmer series depends on n_e and T_e. In particular, the higher the electron concentration of the nebula, the smaller the magnitude of the discontinuity $D = \log (I_{3646-}/I_{3646+})$. But the Balmer discontinuity is particularly sensitive to the electron temperature. Therefore, if one can establish T_e beforehand by other methods, then it is possible to determine n_e.

However, D depends on T_e and n_e in a rather complicated way. The appropriate calculations were performed by Seaton [102] for a number of planetary nebulae under the assumption that the continuous spectrum in nebulae originates from: a) recombination of hydrogen; b) recombination of ionized helium; c) two-photon processes (see Chapter V).

This method gives good results when T_e is small and n_e is not very large. However, the necessity of obtaining a sufficiently well exposed spectrogram of the nebular continuum limits the applicability of this method.

A summary of the electron concentrations obtained by the methods of this section for several planetary nebulae appears in Table 4-15. In the last column (D) appear the values of n_e found by Seaton's method. The results obtained by the H and D methods, on the average, agree well between themselves. But in general a large spread appears in results of Table 4-15, especially for the nebulae NGC 7027, 6572, 2392 and so forth. On the other hand, it does not seem possible to find faults of a fundamental character in these methods. To be sure, the values of the effective collision cross-sections are not always known with sufficient accuracy—the errors in some cases amount to 40%. But since these values enter the formulae in ratios, this circumstance is unlikely to influence essentially the final results. Therefore, the spread in the values in Table 4-15 for one and the same nebula must be attributed, firstly, to the sensitivity of the methods to errors in the observed quantities (intensity ratios, surface brightness) and, secondly, to incorrect application of these methods. The latter circumstance plays a significant role. For example, the method E_{3729}/E_{3726} cannot be applied to nebulae with electron concentrations larger than 10^4cm^{-3} and the [O III], [N II] method is applicable only when $n_e > 10^4 \text{ cm}^{-3}$. Hence, these two methods cannot be applied to the same nebula.

The H method is free from such difficulties. However, this method has the disadvantage that for its application the distance to the nebula must be known. The advantage of the methods [O III], [N II] and E_{3729}/E_{3726} depends, in particular, on the fact that for their applications the distance to the nebula is not required.

7. Planetary Nebulae with Very High Densities

For the majority of planetary nebulae the ratio $E_{N_1+N_2}/E_{4363}$ is very large, of order 100–300. But in some stellar planetary nebulae anomalously small values of this ratio have been discovered—ratios of only a few units, that is, smaller than usual by almost two orders of magnitude. The values of $E_{N_1+N_2}/E_{4363}$ for these nebulae are given in Table 4-16 [106, 107].

If we attempt to establish the electron temperature using the usual formula (32), then for the nebula IC 4997 we obtain $T_e = 75,000°\text{K}$ and for the nebula anon. 19^h21^m, $T_e = 350,000°\text{K}(!)$, a clearly inadmissible value. Formula (32) is generally inapplicable to the nebula anon. 01^h52^m, since the minimum value of $E_{N_1+N_2}/E_{4361}$,

TABLE 4-16

Nebula	$E_{N_1+N_2}/E_{4363}$
IC 4997	13.6
anon. 01^h52^m	7.9
anon. 19^h21^m	9.6

obtained for $T_e \to \infty$, equals 8.74, whereas the observed ratio equals 7.9. Hence it follows that formula (32), derived for small values of n_e ($n_e < 10^3$ cm^{-3}) is inapplicable for nebulae with small values of the ratio $E_{N_1+N_2}/E_{4363}$.

For brevity, let us define $\delta = E_{N_1+N_2}/E_{4363}$. Obviously, a small value of δ occurs when collisions of the second kind play an appreciable role, that is, when the electron concentration is high. The effect of collisions of the second kind in the cases under consideration have a selective character: the fraction of the O^{++} ions leaving the $^1D_2$ level on account of collisions of the second kind will be significantly larger than the fraction leaving $^1S_0$ for the same reason, since the lifetimes of the two levels differ by two orders of magnitude. Thus, the effect of collisions of the second kind is to decrease δ. A similar effect, incidentally, is observed in the first period of the nebular phase of the outburst of novae.

The small value of δ indicates, moreover, that the number of spontaneous (radiative) transitions becomes comparable with the number of induced transitions downwards under the influence of collisions of the second kind. Calculations show that for this to occur the density in the nebula must be of the order 10^7 cm^{-3}. However, a more accurate value of n_e may be inferred from formula (29), which is correct for all values of n_e and T_e.

We investigate first two limiting cases.

1. For $n_e \to 0$, (in practice $n_e < 10^3$ cm^{-3}), the function $F(0,T_e)$ given by formula (30) depends weakly on T_e and for $T_e \sim 10^4$ it equals 8.74. In this case formula (29) is equivalent to formula (32).

2. For $n_e \to \infty$ ($n_e > 10^3$ cm^{-3}) we have $F(\infty, T_e) = 0.0753$. Then we obtain from (29) and (30):

$$\delta = \frac{E_{N_1+N_2}}{E_{4303}} = 0.0753e^{33,000/T_e}. \tag{69}$$

For the case $T_e = 10,000°$K, the numerical value of δ lies well within the limits of 237 for $n_e < 10^3$ cm^{-3} [formula (32)] and 2.05 for $n_e \gg 10^3$ cm^{-3} [formula (69)]. It follows that for large values of n_e one must expect very small values of δ.

Formulae (29) and (30) are represented in Figure 4-10 in a graph showing the dependence of $\log \delta$ on $\log n_e$ for various values of T_e. From this figure one can draw a number of interesting conclusions. First, for a given value of δ the electron con-

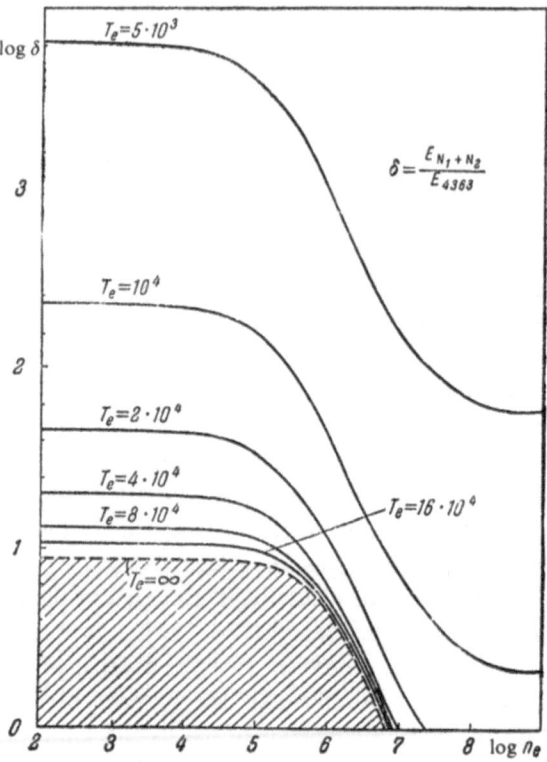

Figure 4-10

The dependence of $\delta = E_{N_1+N_2}/E_{4363}$ on the electron concentration n_e for various values of the electron temperature.

centration decreases as the electron temperature increases. In the second place, for small values of δ (less than 10) the minimum value of n_e occurs for $T_e = \infty$. In other words, not every value of n_e and δ will correspond to a physical value of the electron temperature (the shaded region in Figure 4-10). in the third place, the minimum values of n_e, just mentioned, are higher than those generally observed in nebulae. Thus, for example, with $\delta \sim 5$ (log = 0.7) we have $n_e \sim 10^6 \, \text{cm}^{-3}$ for $T_e = \infty$. But in reality $T_e \ll \infty$, so that we must always have $n_e > 10^6 \, \text{cm}^{-3}$ for $\delta \sim 5$.

The last conclusion is of greatest interest. It indicates that in all those cases when δ is on the order of several units, the electron concentration of the nebula must be larger than $10^6 \, \text{cm}^{-3}$, independently of the electron temperature. In real nebulae $T_e \sim 10^4$, so that we must have $n_e \sim 10^7 \, \text{cm}^{-3}$ when $\delta \sim 5\text{--}10$.

TABLE 4-17

Electron Concentrations in Very Dense Nebulae

Nebula	$n_e(cm^{-3})$	
	$T_e = 10,000°K$	$T_e = 20,000°K$
IC 4997	$4 \cdot 10^6$	$8 \cdot 10^5$
anon. 01^h52^m	10^7	$1.8 \cdot 10^6$
anon. 19^h21^m	$7 \cdot 10^6$	$1.4 \cdot 10^6$

From formulae (29) and (30) or with the help of Figure 4-10, one can find the electron concentration for the three nebulae listed above, for a given electron temperature. The results appear in Table 4-17 for $T_e = 10,000°K$ and $T_e = 20,000°K$.

To uniquely determine n_e and T_e it is obviously necessary to have still another relation between these quantities. For this purpose one may take advantage of, for example, the ratio of intensities of the forbidden lines in the $2p^2$ configuration of ionized nitrogen [formula (34)].

The radii of nebulae for which $n_e \sim 10^6 - 10^7$ cm^{-3} will be on the order of some hundreds of astronomical units and, in all cases, smaller than a thousand astronomical units.

It is interesting to note that, for a given value of δ, T_e has a minimum as $n_e \to \infty$. The asymptotic expression for this value is

$$(T_e)_{min} = 14,300 (1.123 + \log \delta)^{-1}. \tag{70}$$

For the nebula IC 4997, for example, $(T_e)_{min} = 6300°$; for anon. 01^h52^m, $(T_e)_{min} = 7150°$.

8. The Masses of Planetary Nebulae

Knowing the electron concentration and the linear dimensions of a nebula, it is easy to establish its mass $\mathfrak{M}$:

$$\mathfrak{M} = \frac{\pi}{6} D^3 n_H m_H, \tag{71}$$

where D is the diameter of the nebula, n_H is the concentration of hydrogen atoms and m_H is the mass of the hydrogen atom. Assuming $n_H = n_e \approx 10^3 - 10^4$ cm^{-3} and $D = 20,000$ a.u., we find

$$\mathfrak{M} \sim 0.01 - 0.1 \mathfrak{M}_\odot. \tag{72}$$

Another, more accurate, procedure for determining the masses of nebulae is based on use of formula (61), which relates the electron concentration to the energy radiated in the hydrogen lines. From this formula we may write for the ion concentration

$$n^+ = n_e = \left(\frac{E_{ki}}{z_k A_{ki} h\nu_{ik} V}\right)^{1/2},$$

and since

$$n^+ = \frac{\mathfrak{M}}{m_{\mathrm{H}} V},$$

we will have

$$\mathfrak{M} = m_{\mathrm{H}} \left(\frac{E_{ki} V}{z_k A_{ki} h\nu_{ik}}\right)^{1/2} \tag{73}$$

Relating the energy E_{ki} to the total apparent luminosity through the expression $E_{ki} = \lambda_{ik} L$, we have in place of (73):

$$\mathfrak{M} = C\sqrt{LV} \tag{74}$$

This expression was first derived by V.A. Ambartsumyan.

The factor C, given by the formula

$$C = m_{\mathrm{H}} \sqrt{\frac{\lambda_{ik}}{z_k A_{ki} h\nu_{ik}}}, \tag{75}$$

may be assumed to be approximately constant for all planetary nebulae, since λ_{ik} does not vary much from one nebula to another. For example, if the nebula is observed in the H_β line and the ratio N_2/H equals 3, then $\lambda_{24} = 1/12$ (under the assumption that the luminosity of the nebula is determined on the basis of the lines N_1 and N_2). In this case $C \approx 1.5 \cdot 10^{12}$ (in CGS units).

Formula (74) may also be written in the form

$$\log \mathfrak{M} = C_1 + 0.2m_n + 3/2 \log D, \tag{76}$$

where C_1 is a new constant, m_n is the integrated apparent stellar magnitude of the nebula and D is its linear diameter.

Application of the formulae (75) or (76) to individual nebulae is hindered by the fact that our knowledge of their linear diameters is very poor. However, judging from the observed dispersion of luminosity and, in particular, in volume, the masses of individual planetary nebulae deviate from the average value indicated in (72) by at least ten times.

Formula (74) gives the total mass of any nebula—planetary or diffuse—when its optical thickness τ_c beyond the Lyman limit is less than unity. In the case $\tau_c > 1$ formula (74) refers to the mass of only the bright (ionized) part of the nebula. Since τ_c is usually larger than unity for diffuse nebulae, we cannot obtain a correct picture of their total masses. In the dark part of the nebula hydrogen exists almost entirely in the ground state. Nevertheless, it may still be detected by the monochromatic radio emission at 21 cm. From the intensity of this radiation, we are able to determine the mass of the dark part of the nebula (for details, see [141]). Thus, by combining astrophysical methods with observations from radio astronomy, we can establish the total mass of a gaseous nebula, even if $\tau_c > 1$.

It should be noted that formula (74) was derived for a uniform nebula in which $n_e = const.$ throughout. In the case when the electron density is not constant and varies appreciably throughout the nebula, formula (74) gives only an upper bound for the mass of the nebula. Therefore the real mass of the nebula could be expressed as

$$\mathfrak{M}_0 = f\left(\frac{d}{d_0}\right)\mathfrak{M} = f\left(\frac{d}{d_0}\right)C\sqrt{LV}, \tag{77}$$

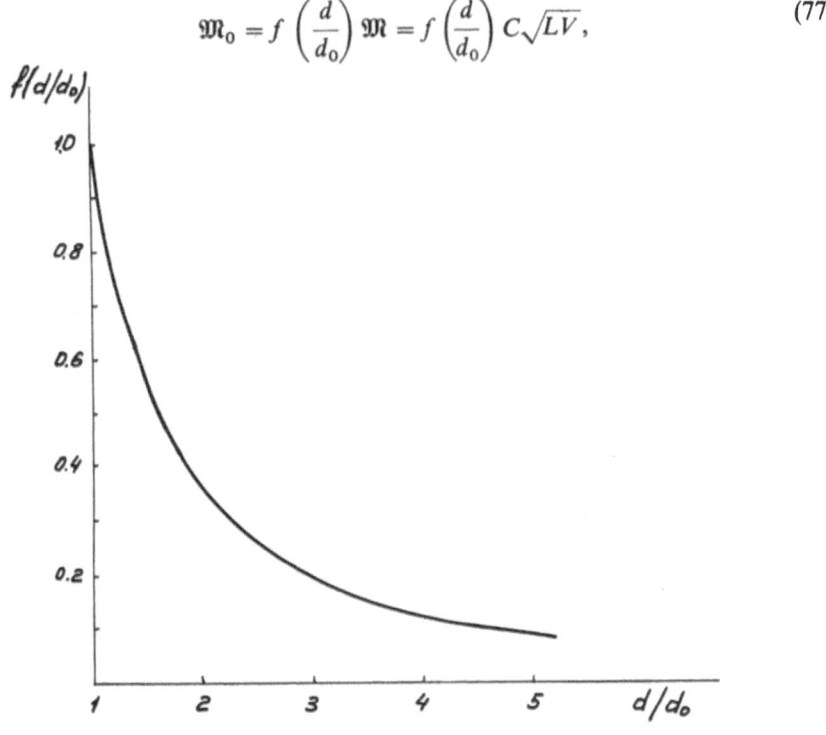

Figure 4-11

The coefficient of nonuniformity versus the ratio of d—the average spacing of the irregularities—to d_0—their average size.

where $f(d/d_0)$ may be called the coefficient of nonuniformity; it depends on the ratio of the mean distance d between the nonuniformities (single spots) on the image of a nebula to the average size d_0 of these nonuniformities. If the mean value of the ratio d/d_0 is known from observations for a given nebula, then the numerical value of $f(d/d_0)$ can be found from the following approximate relation [220]:

$$f(d/d_0) = (d/d_0)^{-3/2}. \tag{78}$$

A graph of the function $f(d/d_0)$ appears in Figure 4-11. It is apparent from this graph that even for small irregularities ($d/d_0 \sim 1.5-2$), the masses of the nebulae calculated from (74) must be reduced two or three times.

Chapter V

The Continuous Spectrum of Planetary Nebulae

1. Observational Data

The continuous spectrum of planetary nebulae beyond the Balmer limit of hydrogen ($\lambda < 3646$ Å) has been observed for a long time. It results from the capture of free electrons by protons into the level $n = 2$ of neutral hydrogen. We shall refer to this radiation as the Balmer continuum and designate its intensity by B_c. A continuous spectrum is also present on the long-wavelength side of the Balmer limit, that is, in the visual region of the spectrum ($\lambda > 3646$ Å), which originates from: a) free-free transitions of electrons in the field of a proton; b) capture of free electrons by protons into the levels $n = 3,4,5, \ldots$ of neutral hydrogen, during which they emit in the Paschen ($\lambda < 8202$ Å) and higher continua, extending through the infra-red into the visible region of the spectrum. We aesignate the intensity of this radiation* by V_c. It is found from observations that V_c is considerably larger than would be predicted on the assumption that this radiation originates purely from recombination processes, with the given electron temperature of the nebula.

The first survey of the continuous spectrum of a large number of planetary nebulae was made by Page for the spectral region from $\lambda 5000$ Å to $\lambda 3000$ Å [107, 108]. He discovered in many planetary nebulae a relatively strong continuous spectrum of approximately constant intensity in the wavelength interval 4800 Å − 3600 Å. For the majority of these nebulae he determined the quantity V_c/H_δ, that is, the ratio of the intensity in this continuous spectrum (within a bandwidth of 1 Å) to the intensity of the H_δ line. The logarithms of these ratios for the group of nebulae under consideration lay in the interval from -1.0 to -2.8. In other words the integrated intensity of the visual continuous spectrum in a region several hundred Ångstroms wide is comparable with the intensity of the H_δ line. There is reason to believe that the integrated intensity of this continuous emission over the whole spectrum in

---

* Sometimes called the visual continuum.

several cases is comparable with the total intensity of all lines in the Balmer series. Page also determined the quantity B_c/H_δ, that is, the ratio of the intensity of the Balmer continuum to the intensity of the H_δ line.

Subsequently, similar surveys were conducted by other workers. Barbier and Andrillat [109], for example, obtained measures of the continuous spectrum for six planetary nebulae and determined for them, in particular, the value D of the Balmer discontinuity, that is, the logarithm of the ratio of the intensity of the continuous spectrum beyond the Balmer limit to the intensity up to the limit. These values of D appeared rather high, in the range 0.60 to 1. The intensity of the continuous spectrum in a narrow region near the head of the Balmer continuum was measured (in stellar magnitudes) for a number of planetary nebulae by Menzel and Aller [95] for the purpose of determining their electron concentration (see Chapter IV).

In all of these cases discrepancies were discovered between the observed and theoretical values of the measured quantities (V_c/H_δ, B_c/H_δ, D). These discrepancies were considerably larger than the errors of measurement and although a number of hypotheses and assumptions were advanced for their explanation, none of these proved tenable. Among these were the existence of dust grains in planetary nebulae, the possible roles of continuous emission by the hydrogen negative ion and the scattering of the continuous radiation from the nucleus by the free electrons in the nebula.

The solution was found finally when A.Ya.Kipper [110] and independently Spitzer and Greenstein [111] advanced the idea of hydrogen two-quantum emission as a possible source of the additional radiation in gaseous nebulae. This idea was subsequently developed in a series of papers by the present author [112, 113, 102] and it is now evident that, to a large extent, this idea explains for the basic features of the continuous spectrum in planetary nebulae, although as usual there remain a number of substantial discrepancies between theory and observations.

2. Theory of Two-Quantum Emission

An atom, finding itself in an excited state, either ordinary or metastable, has a certain non-zero probability of undergoing a transition into a lower level, accompanied by the emission of two photons, and not one as is usual. In the majority of cases this probability is extremely small but for certain metastable levels it is sufficiently large to be comparable, for example, with the transition probability downwards from these levels with emission in some forbidden line. The frequencies of the quanta for so-called two-quantum transitions ($2q$ transitions) may be arbitrary, but the sum of their energies equals the excitation energy of the original level.

The two-quantum emission of hydrogen, originating in transitions from the metastable $2S_{1/2}$ state to the ground state $1S_{1/2}$, is of greatest interest. Atoms arrive in the $n = 2$ states by cascade transitions downwards after the capture of an electron by a proton. When an atom reaches the $2P$ level it must continue on its way to the ground state, emitting a Lyman-photon (Figure 5-1). But a certain fraction of the atoms, say X, in the process of recombination, reach the $2S$ level, which is favorable to two-quantum emission. The quantity X, generally speaking, depends on the physical conditions in the nebula, but in the first approximation (for small electron densities) it is equal to the ratio of the statistical weights of the $2S$ and $2P$ levels. This ratio is of order 0.3. Consequently nearly one third of the atoms participating in the downwards cascade transitions reach the ground state by means of two-quantum emission. It should be noted, however, that a quadrupole transition from the $2S$ to the $1S$ state with the emission of one quantum is possible. But since the probability of such a transition is many thousand times smaller than the probability of a two-quantum transition, one may assume that for practical purposes $2S \rightarrow 1S$ transitions occur only with the emission of two quanta.

In the theory of the luminescence of gaseous nebulae we saw that at least one L_α-photon was produced by each L_c-photon absorbed by the nebula. Now we see that this is not exactly so: of the total number N_c of L_c-photons, on the average, $2/3 \, N_c$ L_c-photons are produced, the remainder $1/3 \, N_c$ going into two-quantum emission. Strictly speaking, this is not a case of each L_α-photon "breaking up" into two photons (see next section), although the total energy in two-quantum emission equals the energy of the "lost" (by number $1/3 \, N_c$) L_α-photons.

From all this it follows that the number of two-quantum transitions per unit time is X times the number of Balmer photons emitted per second by the nebula. In other words the intensity of two-quantum emission depends on the intensity of radiation in the Balmer series of hydrogen. Taking X to be constant, as a first approximation, for all nebulae, we arrive at the interesting conclusion that the ratio of intensity of the two-quantum emission to the intensity in any emission line of the Balmer series is constant and does not depend on the physical conditions in the nebula.

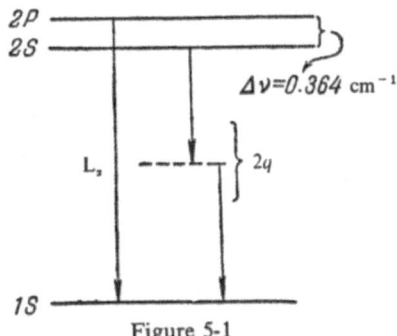

Figure 5-1

On the problem of two-photon ($2q$) emission of hydrogen.

The theory of two-quantum $2S \rightarrow 1S$ transitions in hydrogen was developed by Breit and Teller [114] on the basis of the more general theory of two-quantum processes of Goppert-Mayer [115]. In connection with its application to astrophysical problems, this theory has been subsequently revised and refined [110, 111]. The basic results of this theory are presented below.

Let us denote by $y\nu_{12}$ and $(1 - y)\nu_{12}$ the frequencies of the photons being emitted during a two-quantum transition in hydrogen (ν_{12} is the frequency of a L_α-photon), and by $A(y)dy$ the probability per unit time that a photon be emitted in the frequency interval $y\nu_{12}$ to $(y + dy)\nu_{12}$. The quantity $A(y)$ has the form

$$A(y) = \frac{9\alpha^6\nu_0}{2^{10}} \psi(y), \tag{1}$$

where ν_0 is the ionization frequency of hydrogen and $\alpha = 2\pi e^2/hc$ is the fine-structure constant. The energy radiated per unit frequency is proportional to the quantity $h\nu A(y)$ or $yA(y)$. Values of the function $\psi(y)$ and of the relative intensity $yA(y)$ are given in Table 5-1. Since $\psi(1 - y) = \psi(y)$, the numerical values of this function are not given for $y > 0.5$. The Einstein coefficient for the two-quantum $2S \rightarrow 1S$ transition is

$$A_{2S,1S} = \frac{1}{2} \int_0^1 A(y)dy = 8.227 \text{ sec}^{-1}, \tag{2}$$

where the factor $1/2$ appears because two photons are emitted.

TABLE 5-1

The Relative Probability $\psi(y)$ and the Emissivity $\tau_\lambda = y\psi(y)$ for
Two-Photon Emission of Hydrogen

y	$\lambda(\text{Å})$	$\psi(y)$	$y\psi(y)$
.00	–	0.	0.
.05	24313	1.725	0.0863
.10	12157	2.783	0.2783
.15	8105	3.481	0.5222
.20	6078	3.961	0.7922
.25	4862	4.306	1.077
.30	4052	4.546	1.363
.35	3473	4.711	1.649
.40	3039	4.824	1.929
.45	2702	4.889	2.200
.50	2431	4.907	2.454

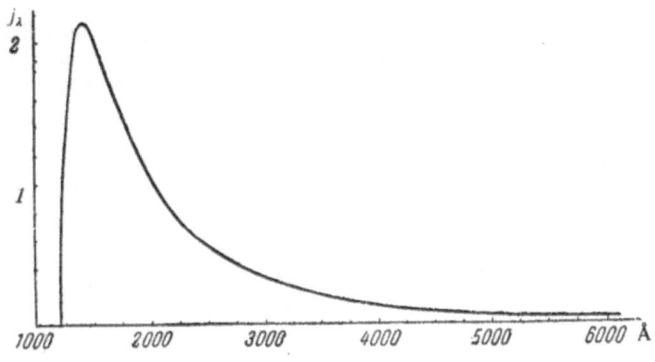

Figure 5-2

The spectrum of two-photon emission of hydrogen (Spitzer and Greenstein).

The data of Table 5-1 are represented in Figure 5-2 in the form of a curve show-ing the dependence of the emissivity $j_\lambda = y\psi(y)$ (in arbitrary units) for the two-quantum process in hydrogen on wavelength (in Ångstroms). Two-quantum emis-sion in hydrogen gives rise to a continuous spectrum which begins at $\lambda = 1216\,\text{Å}$ and extends to $\lambda \to \infty$. The emissivity reaches a maximum near $\lambda = 1400\,\text{Å}$ and quickly drops away on the long wavelength side. Thus, for example, at $\lambda = 4000\,\text{Å}$ it comprises nearly 5% and at $\lambda = 6000\,\text{Å}$, only 1%, of the maximum emissivity.

The two-quantum emissivity of hydrogen is almost constant in the interval $\lambda 4800\,\text{Å} - \lambda 3600\,\text{Å}$, as may be seen from Table 5-1 or Figure 5-2. This fact is con-sistent with observations. However the value of the intensity of the two-quantum emis-sion depends on the physical conditions in the nebula (n_e, n^+ and T_e). The final ex-pression for the intensity of two-quantum continuous emission has the form [111]

$$I_{2q} = \frac{B(n_e, T_e)h\beta^2}{4\pi\rho} F(T_e),$$ (3)

where ρ is the density of matter. The numerical values of the function $F(T_e)$—they are the same as the intensity of two-quantum emission in units $B h\beta^2/4\pi\rho$—are given in Table 5-2 for a series of values of T_e, and $B(n_e, T_e)$ and β are obtained from the relations

$$B(n_e, T_e) = \left(\frac{2^{11}k}{3^3\pi m^5}\right)^{1/2} \frac{he^2}{c^3} T_e^{1/2} n_e n^+$$

and

$$\beta = \frac{h\nu_\alpha}{kT_e} = \frac{15{,}800}{T_e}.$$

TABLE 5-2

Logarithm of the Intensity of Two-Photon Emission ($\log I_{2q}$) in Units $Bh\beta^2/4\pi\rho$

$\lambda(\text{Å})$	$T_e = 14,400°\text{K}$	$T_e = 10,000°\text{K}$	$T_e = 7200°\text{K}$	$T_e = 5050°\text{K}$
8202	− 2.04	− 2.15	− 2.25	− 2.36
6080	− 1.85	− 1.96	− 2.06	− 2.17
4861	− 1.72	− 1.82	− 1.92	− 2.04
4050	− 1.62	− 1.72	− 1.82	− 1.94
3646	− 1.56	− 1.66	− 1.76	− 1.88
3470	− 1.53	− 1.64	− 1.74	− 1.85
3040	− 1.47	− 1.57	− 1.67	− 1.79
2700	− 1.41	− 1.51	− 1.61	− 1.73
2430	− 1.36	− 1.47	− 1.57	− 1.68
2210	− 1.31	− 1.43	− 1.53	− 1.64
2020	− 1.29	− 1.39	− 1.49	− 1.61

The intensity I_{2q} is given by formula (3) and in Table 5-2, as was mentioned, in units of $Bh\beta^2/4\pi\rho$. This unit is convenient since in terms of it one may also obtain the intensity of the recombination emission of hydrogen and ionized helium (see Section 5), so that in forming the ratio I_{2q}/I_{H_δ} (relative intensity) it drops out.

3. The Effects of Collisions

An atom can also leave the $2S$ state through being excited to the higher-lying, nearby $2P$ level in collisions with electrons or protons. Evidently this process leads to a reduction of the intensity of two-quantum emission. Denoting by $W_{2S,2P} \cdot n_e$ the probability per unit time of a collisional $2S \to 2P$ transition, we may write for the fraction of atoms in the $2S$ state which undergo two-quantum transitions to the ground state

$$\frac{A_{2S,1S}}{A_{2S,1S} + W_{2S,2P} \cdot n_e} = \frac{1}{1 + W_{2S,2P}\dfrac{n_e}{A_{2S,1S}}} \tag{4}$$

On the other hand, approximately 0.32 of the atoms formed by recombination to all levels starting with the second enter the $2S$ state. Therefore in the "pure" case, that is when collisions are neglected, we have $X = 0.32$. Taking account of the collisions (de-activation factor) we obviously will have:

$$X = \frac{0.32}{1 + W_{2S,2P}\dfrac{n_e}{A_{2S,1S}}} \tag{5}$$

In de-activation processes it appears that protons play the basic role. In this case $W_{2S,2P} \approx 5 \cdot 10^{-4}$ cm^3 sec^{-1} for $T_e = 10,000°$K [112, 105]. Then instead of (5) we may write:

$$X = \frac{0.32}{1 + 0.6 \cdot 10^{-4} n_e}. \qquad (6)$$

Through the influence of proton, and to a smaller extent, electron, collisions, the intensity of two-quantum emission becomes dependent on the physical conditions of the nebula, in particular, on its electron concentration. The intensity of this emission diminishes as n_e increases and, for example, for $n_e \sim 10^4$cm^{-3}, its value is almost half and, for $n_e \sim 10^5$cm^{-3}, almost one-tenth of its theoretically possible maximum. In other words, two-quantum emission will comprise a larger fraction of the total intensity in comparatively rarified nebulae ($n_e < 10^3$cm^{-3}); in dense nebulae ($n_e > 10^5$cm^{-3}) it will be practically absent.

4. Processes Degrading L$_\alpha$-Photons

As was stated above, of the total number N_c of L$_c$-photons absorbed by the nebula, on the average $2/3\ N_c$ L$_\alpha$-photons are produced and the remainder, $1/3\ N_c$, go into two-quantum emission. The question arises whether it is possible in the physical conditions in nebulae for a L$_\alpha$-photon to be split into two quanta of arbitrary frequencies. In other words, the problem arises of looking for other mechanisms which produce continuous emission in nebulae. At present a number of mechanisms degrading L$_\alpha$-photons are known. None of these, it appears, plays an essential role under ordinary nebular conditions. Nevertheless they are of sufficient interest to consider them at some length, especially since some of them have not yet been completely investigated.

The first mechanism destroying L$_\alpha$-quanta was pointed out by Spitzer and Greenstein [111], and involves the following. A neutral hydrogen atom goes from the state $1S_{1/2}$ into one of the $2P_{1/2,3/2}$ states by absorbing a L$_\alpha$-photon. Because of the large density of L$_\alpha$-radiation in nebulae, the populations of the $2P_{1/2}$ and $2P_{3/2}$ states might not be negligibly small in comparison with the ground state. Therefore, there is a definite probability that an atom in one of the $2P_{1/2,3/2}$ states, in collisions with free electrons, goes over into the $2S_{1/2}$ state and then to the ground state via two-quantum emission. However, a more detailed investigation of this problem shows that collisions with protons are more important than those with electrons in inducing transitions from the $2P_{1/2,3/2}$ states to the $2S_{1/2}$ state [112]. The probability per scattering p_1 of degrading a L$_\alpha$-quantum by this mechanism (the cycle $1S \rightarrow 2P \rightarrow 2S \rightarrow 1S$), will evidently depend on the electron (proton) density $n_e (= n^+)$ of the medium. This is given by the formula [113]:

$$p_1 = \frac{3.765 \cdot 10^{-13} n_e}{1 + 8.58 \cdot 10^{-5} n_e}. \tag{7}$$

The quantity p_1 as determined from this formula, is of order $10^{-9} - 10^{-10}$ for planetary nebulae, $10^{-11} - 10^{-12}$ for diffuse nebulae and for HI regions around planetary nebulae and 10^{-17} for interstellar media.

The actual probability that a L_α-photon is degraded during its stay within the nebula depends on the total number of scatterings and absorptions its experiences. This, in turn, depends on solving the L_α transfer problem in nebulae, taking into account the actual conditions in the medium. This will be investigated in Chapter VI. Here, however, we cite the final formula giving the total number $N(t)$ of acts of scattering and absorption experienced by a photon in a nebula with a total optical thickness t for L_α-radiation:

$$N(t) = 2t\sqrt{\pi \ln t}. \tag{8}$$

This formula is valid for large values of t. In planetary nebulae the quantity t is of order 10^5 if one assumes that the optical thickness of the nebula for L_c-radia tion is of order unity. This gives $N \sim 10^6$. The total probability that a L_α-photon will be degraded during its stay in the nebula by this mechanism will be $p_1 N = - 10^{-3} - 10^{-4}$.

Thus the mechanism of Spitzer and Greenstein can split only an insignificant number of the L_α-photons present in the nebula. In very dense nebula ($n_e \to \infty$, $p_1 = 0.44 \cdot 10^{-8}$) it is possible to split a few percent of these photons.

The second mechanism for degrading L_α-quanta was pointed out by A.Ya. Kipper and V.M. Tiit [116] and involves the following. The transition $2P \to 1S$ in hydrogen with the emission of one L_α-photon has a transition probability of $w_1 = 6.24 \times \times 10^8 \text{ sec}^{-1}$. But this transition may also occur with the emission of two photons, only with a very much smaller transition probability, $w_2 = 1.938 \cdot 10^{-5} \text{ sec}^{-1}$. In other words the probability that for each single-photon transition of a hydrogen atom there will be one two-photon transition equals

$$p_2 = \frac{w_2}{w_1} = 3 \cdot 10^{-14}. \tag{9}$$

This is also the probability per scattering of degrading a L_α-photon by this mechanism (cycle $1S \to 2P \to 1S$). We note that it does not depend on the physical conditions of the medium.

The total probability of degrading a L_α-photon while it is in the nebula in this case is still smaller than in the first, of order 10^{-8}, and so the mechanism cannot play an appreceiable role in planetary nebulae. Only in certain comparatively dense hydrogen clouds or in HI-zones around planetary and diffuse nebulae could this mechanism, we might suppose, play some role because of the extremely large increase in the total number of scatterings.

One may mention still another mechanism*—the third—degrading L_α-photons under conditions found in nebulae [203]. This is connected with the fact that the energy difference between the $2P$ and $2S$ levels in hydrogen is small and comparable with the kinetic energy of atoms caused by their thermal motion. In frequency units, the difference is $1.092 \cdot 10^{10}$ sec^{-1}, which is equivalent to a difference of 1.32 km/sec in the velocities of two hydrogen atoms. Therefore, if the atom absorbing a L_α-quantum has a negative component of radial velocity of 1.32 km/sec with respect to an atom emitting a given L_α-photon, the absorbing atom can ascend into the metastable $2S$ level. In the conditions of the nebulae, this atom has only one way of returning to the ground state—the $2S \rightarrow 1S$ two-photon transition. The effectiveness of this mechanism (the cycle $1S \rightarrow 2S \rightarrow 1S$) obviously depends first of all on the probability per second of the transition $1S \rightarrow 2S$ in hydrogen. This transition is forbidden for electric dipole radiation by a selection rule and an exact calculation gives for the lifetime of the $2S$ state a value of order 10^5 sec for magnetic dipole radiation [114]. However, the results of experiments investigating behavior of positrons in an electric field [204] indicate that positrons and electrons may possess a small electric dipole moment. Feinberg [205] and Salpeter [206], in their investigations on the effect of this moment on atomic energy levels, have pointed out, in particular, the possibility that the $2S$ and $2P$ states of hydrogen will be mixed. The mixing of these levels will result in a significant reduction in the lifetime of atoms in the $2S$ state; it will be between 0.12 sec (with two-photon decay of the $2S$ level) and $1.6 \cdot 10^{-9}$ sec (the lifetime of the $2P_{1/2}$ state). An experiment by Fite *et al.* [207] appeared to show that the lifetime of the $2S$ state was on the order of $2.4 \cdot 10^{-3}$ sec. Hence we have for the Einstein coefficient for a spontaneous, single-quantum, transition $A_{2S,1S} \approx 0.4 \cdot 10^3$ sec^{-1}, *i.e.* approximately 50 times larger than the Einstein coefficient A_{2q} for a two-quantum transition. However, this would imply that the existing estimate for the intensity of two-quantum radiation must be reduced by a factor of 50, so that two-quantum emission in planetary nebulae would be too weak to account for the appearance of the continuous spectrum. It is unlikely that processes splitting L_α-photons could lead to the observed spectrum. It is most likely that the experiments of Fite *et al.* gave a large overestimate for the value of $A_{2S,1S}$; evidently it is of order 1 sec^{-1}. However, for the following this circumstance is not of vital importance.

A velocity difference of 1.32 km/sec corresponds to a kinetic temperature of the medium equal to 100°K. Therefore the present mechanism for degrading L_α-quanta

---

* [*Ed. note*] Stein, Carrico, Lipworth and Weisskopf (*Phys. Rev. Letters* **19**, 741, 1967) have obtained experimentally an upper bound on the hypothetical electron electric dipole many orders of magnitude smaller than the value needed for the mechanism described here to operate.

is able to operate in all cases for which the electron temperature of the medium exceeds 100°K, *i.e.* in clouds of interstellar hydrogen as well as in ordinary gaseous nebulae. The problem lies in knowing the probability per scattering of degrading L_α-quanta in this way.

The number of $1S \rightarrow 2S$ and $1S \rightarrow 2P$ transitions in hydrogen atoms per unit time and unit volume induced by L_α-radiation will be $N_{1S,2S} = n_1 B_{1S,2S} \rho_\alpha$ and $N_{1S,2P} = \eta_1 B_{1S2P} \rho_\infty$ respectively. Here n_1 is the concentration of ground state hydrogen atoms, $B_{1S,2S}$ and $B_{1S,2P}$ are the Einstein coefficients for induced transitions and ρ_α is the density of L_α-radiation. Thus we may write for the probability p_3 of degrading L_α-quanta by means of the $1S \rightarrow 2S \rightarrow 1S$ cycle:

$$p_3 = \frac{N_{1S,2S}}{N_{1S,2P}} \frac{A_{2q}}{A_{2q} + A_{2S,1S}} = \frac{g_{2S} A_{2S,1S}}{g_{2P} A_{2P,1S}} \frac{A_{2q}}{A_{2q} + A_{2S,1S}} \cong \frac{g_{2S} A_{2q}}{g_{2P} A_{2P,1S}}, \quad (10)$$

since $A_{2S,2S} \gtreqless A_{2q}$.

Substituting into (10) the values $g_{2S}/g_{2P} = 1/3$, $A_{2q} = 8.227 \, \text{sec}^{-1}$ and $A_{2P,1S} = 6.24 \cdot 10^8 \, \text{sec}^{-1}$, we find, per scattering, the probability $p_3 = 0.44 \cdot 10^{-8}$. This value is at least one or two orders of magnitude larger than that for Spitzer and Greenstein's mechanism, and five orders of magnitude larger than the probability for Kipper and Tiit's mechanism. Only for very dense planetary nebulae ($n_e > 10^5 \text{cm}^{-3}$) does Spitzer and Greenstein's mechanism become comparable with the present one.

A more precise statement of the problem takes account of the influence of electron and proton collisions on the population of the $2S$ level (formula (4)), in which case we have, instead of (10),

$$p_3 = \frac{0.44 \cdot 10^{-8}}{1 + 0.6 \cdot 10^{-4} n_e}. \quad (10a)$$

The dependence of p_3 on n_e is significant only for large values of the electron concentration ($n_e > 10^4 \, \text{cm}^{-3}$). For comparatively small values of n_e this probability is essentially independent of the physical conditions of the medium. Therefore, other conditions being equal, this degradation mechanism appears more effective than the others in planetary nebulae of average and low density, in diffuse nebulae, in zones of neutral hydrogen around gaseous nebulae and in clouds of interstellar hydrogen. In all of these situations the degradation of a L_α-quantum into two requires that it experience on the average 10^8 scatterings. In planetary nebulae the number of scatterings may amount to 10^6, *i.e.* only about 1% of the L_α-quanta occurring in the nebula can be split into two. However, a layer of neutral hydrogen surrounding a nebula and having a thickness on the order of a thousand astronomical units and a concentration of $10^3 - 10^4 \, \text{cm}^{-3}$ will have an optical thickness in the L_α-line of order 10^7, if its temperature is equal to $1000°K$. According to equation (6), in this case $N \sim 10^8$, i.e. all of the L_α-quanta will be split.

Thus, although the probability per scattering p_3 of degrading L_α-quanta does not depend on the state of the medium, the net probability that L_α-quanta on the whole will be split does depend significantly on the physical conditions of the medium. When these conditions differ strongly between nebulae, the intensities of the continuous spectra relative to the emission lines will also differ. The scatter in the relative intensities of continuous spectra, observed in both planetary and diffuse nebulae, cannot be explained within the framework of existing theories. It is not impossible that in these and other cases the degradation of L_α-quanta plays a certain role.

5. The Intensity of Continuous Radiation

The basic sources of the continuous spectrum in planetary nebulae are considered to be a) recombination processes of hydrogen (capture of electrons into levels $n = 2,3, \ldots$ and free-free transitions); b) recombination processes of ionized helium (capture of electrons on levels $n = 4, 5, \ldots$ and free-free transitions); c) two-quantum emission of hydrogen in $2S \rightarrow 1S$ transitions.

Two-quantum emission by ionized helium is insignificant and may be neglected. This problem has been investigated quantitatively by Seaton [102].

For convenience we introduce the quantity

$$J_\lambda = 10^3 \frac{I_\lambda}{H_\delta}, \tag{11}$$

where I_λ is the energy in one Ångstrom of the continuous spectrum at wavelength λ and H_δ is the energy in the H_δ line of hydrogen.

In its final form the expressions for J_λ is

$$J_\lambda = J_\lambda(H) + 0.55PJ_\lambda(\text{He II}) + \frac{1}{1 + 0.6 \cdot 10^{-4}n_e} J_\lambda(2q), \tag{12}$$

where $J_\lambda(H)$, $J_\lambda(\text{He II})$ and $J_\lambda(2q)$ characterize the various components of the continous spectrum of the nebula and P is the ratio of the intensity of the He II line 4686 Å to the intensity of the line H_β in the spectrum of the nebula.

The quantities $J_\lambda(H)$ and $J_\lambda(\text{He II})$ are calculated either from the formulae given in [58] or those appearing in Section 5 of Chapter IV, and $J_\lambda(2q)$, by means of formula (3). The numerical values of these quantities for optically thick nebulae are given in Table 5-3 for electron temperatures $T_e = 10,000°K$ and $T_e = 20,000°K$ [the data for $J(2q)$ relates to the case $X = 0.32$].

TABLE 5-3

Components of the Continuous Spectrum of Nebulae

$\lambda(\text{Å})$	$J_\lambda(\text{H})$		$J_\lambda(\text{He II})$		$J_\lambda(2q)$	
	$T_e = 10{,}000°\text{K}$	$T_e = 20{,}000°\text{K}$	$T_e = 10{,}000°\text{K}$	$T_e = 20{,}000°\text{K}$	$T_e = 10{,}000°\text{K}$	$T_e = 20{,}000°\text{K}$
5000	1.5	3.0	2.9	4.1	1.0	1.2
4545	1.4	3.2	2.7	4.3	1.5	1.7
4167	1.2	3.3	2.4	4.4	2.0	2.3
3846	1.1	3.3	2.1	4.5	2.6	3.0
3646+	1.0	3.3	1.9	4.5	3.1	3.5
3646−	17.5	15.2	10.6	10.8	3.1	3.5
3571	16.7	15.2	10.2	10.8	3.3	3.8
3333	14.5	15.2	8.8	10.7	4.1	4.7

It may be seen from Table 5-3 and formula (12) that the intensity of the two-quantum emission (for $X = 0.32$ and $n_e \rightarrow 0$) in the visible region of the spectrum ($\lambda > 3646$ Å) is comparable with the intensity of the energy emitted during recombinations and free-free transitions in hydrogen and significantly larger than the energy emitted by ionized helium (for $P \sim 0.4$).

The theoretical energy distribution calculated in this manner was compared by Seaton with the observed distribution in two ways. First, by summing the energy of the continuous spectrum in the interval $\lambda\lambda 3646 - 3333$ Å, he determined the theoretical values of the relative energy of the Balmer continuum as a function of electron temperature T_e and compared these with the values of B_c/H_δ obtained from the observations of Page for a number of nebulae with known electron temperatures. The agreement appeared to be satisfactory, although the observed and theoretical values of B_c/H_δ differed by as much as a factor of five in isolated cases.

TABLE 5-4

Balmer Discontinuity in the Spectra of Planetary Nebulae

Nebula	D_{obs}	D_x	D_0	$D_{0.32}$
NGC 6543	0.98	0.95	1.26	0.76
NGC 6572	0.79	0.84	1.00	0.59
NGC 6826	0.61	0.89	1.15	0.66
NGC 7009	0.82	0.72	0.90	0.59
IC 418	0.48	0.52	0.69	0.45
Average	0.75	0.78	0.98	0.59

In the second method the value of the Balmer discontinuity D_x depending on the parameter X ($D = \log 3646^-/3646^+$) is calculated for a given nebula using the known values of n_e and T_e (which are determined beforehand using the methods discussed in Chapter IV) and is compared with the observed values of the Balmer discontinuity D_{obs} [109]. The results are shown in Table 5-4, where D_0 is the value of the Balmer discontinuity found neglecting two-quantum emission (with $X = 0$) and $D_{0.32}$, including two-quantum emission ($X = 0.32$) but neglecting collisional effects. As follows from this data, the observed and calculated values of the Balmer discontinuity appear in good agreement with each other.

6. Continuous Emission of Unknown Origin

In general, the theory of the continuous spectrum of planetary nebulae gives a satisfactory explanation of the basic observational facts. At the same time there exist cases in which large discrepancies appear between theory and observations. Moreover, for some nebulae a number of anomalous phenomena have been observed in the character and distribution of the continuous emission, which apparently cannot be explained, even qualitatively, within the framework of the theory just discussed.

It, therefore, seems natural to assume that, in addition to the processes already known, there exist in planetary nebulae still other mechanisms, as yet unknown to us, which emanate continuous energy. We shall present some observational facts in confirmation of this point of view.

In Figure 5-3 there appear curves showing the energy distribution in the continuous spectrum of nebulae near the Balmer limit, drawn according to formula (12) and the data of Table 5-3 for $T_e = 10,000°K$, $P \sim 0.4$ and neglecting collisional de-activation ($X = 0.32$). The heavy curve represents the sum of the separate contributions, that is, the sum $J_\lambda(H) + 0.55J_\lambda(He\,II) + J_\lambda(2q)$. From the drawing it appears that the intensity of the continuous spectrum beyond the Balmer limit ($\lambda < 3646$ Å) decreases approximately linearly with decreasing wavelength (in any case up to $\lambda = 3300$ Å).

On the other hand Page has drawn curves showing the observed energy distribution beyond the Balmer limit in the spectra of a number of planetary nebulae [108]. In Figure 5-4 these curves are given for seven planetary nebulae. The general character of these curves agrees well with the theoretical curves appearing in Figure 5-3.

The situation is quite different for the remaining five (of twelve) nebulae, the energy distributions of the spectra (in the same region of the spectrum) of which are represented in Figure 5-5. From this data it appears that the observed energy distributions are significantly *larger* than the theoretical distributions in the ultraviolet region of the spectrum (starting with $\lambda \sim 3600$Å).

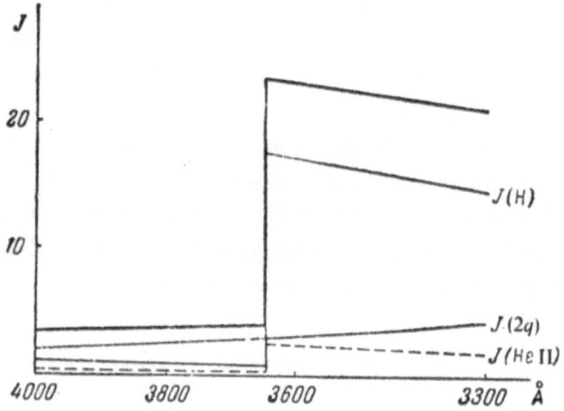

Figure 5-3

Theoretical distribution of the continuous spectrum near the Balmer discontinuity
for recombination processes connected with hydrogen, $J_\lambda(\text{H})$, with helium, $J_\lambda(\text{He II})$
and for two-photon emission of hydrogen $J_\lambda(2q)$. The heavy curve designates the sum
$$J_\lambda(\text{H}) + J_\lambda(\text{He II}) + J_\lambda(2q).$$

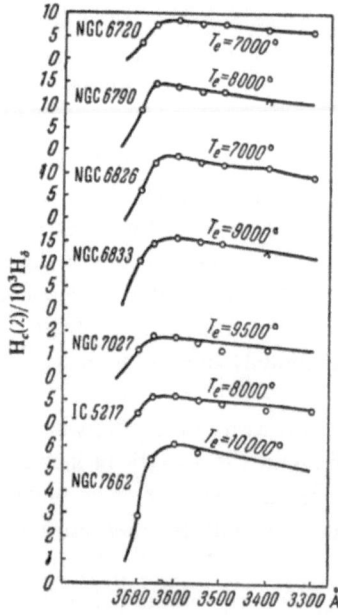

Figure 5-4

The energy distribution in the continuous spectrum beyond the Balmer limit for a
number of planetary nebulae (Page).

Evidently in the cases shown in Figure 5-5 there exists additional continuous emission of quite different and as yet unexplained character. Page's explanation, that the observed excess energy in the ultraviolet region of the spectrum is a consequence of excess emission of the nucleus in the same spectral region, is incorrect, since the energy distribution in the Balmer continuum depends on the electron temperatures of the nebula.

From the absence of this discrepancy in the nebulae listed in Figure 5-4, one may make an inference about the selective character of this unknown source of radiation. Thus in one nebula this radiation plays an observable role, while in others it is either entirely absent or appears very weakly.

Page mentions an interesting peculiarity in the distribution of continuous radiation in the nebula NGC 7662 [107]: the relative intensity of the continuous spectrum (the quantity V_c/H_δ) is larger in the denser parts of the nebula. However, as was stated above, with an increase in the density of matter and therefore of electron density, the amount of energy going into two-quantum emission and hence, the total continuous emission is reduced, so that one would have expected the intensity to decrease in the denser parts of the nebula. It is possible that this observation is evidence of the existence of unknown sources of continuous emission in planetary nebulae.

In this respect the results of Minkowski's [117] spectrophotometric observations of the small circular nebula IC 418 are of special interest. The spectrogram of this nebula was obtained at the coude focus of the 200-inch reflector with exceptionaly good atmospheric conditions. On this spectrogram the Balmer continuum

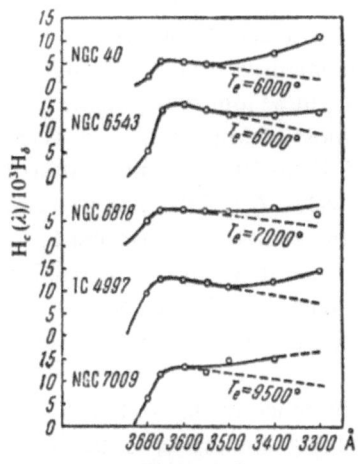

Figure 5-5

Anomalous energy distribution in the continuous spectrum beyond the Balmer limit
in a series of planetary nebulae (Page).

($\lambda < 3646$ Å) clearly shows the circular structure of the nebula; the intensity appears to increase at the edge of the spectrogram and to reach a minimum at the center. In contrast to this, the continuous spectrum in the visual ($\lambda > 3646$ Å) has a totally different intensity distribution. As the microphotometer profiles show, in this case the intensity of the continuum appears in the central portion of the nebula and decreases towards the outer regions. This result implies that the value of the Balmer discontinuity varies throughout the nebula and increases from center to edge. If the continuous emission of this nebula originates entirely from recombinations, the observed behavior of the Balmer decrement must imply that the electron temperature decreases away from the center. The electron temperature, as determined from the Balmer discontinuity, appeared to exceed 15,000°K in the outer regions of the nebula. The electron temperature in the inner regions must be still higher. On the other hand, the electron temperature found by a more reliable method— that of V.A. Ambartsumyan—is of order 7000°K.

All of these facts indicate that, in all probability, unknown sources of continuous radiation exist also in the nebula IC 418.

Above it was mentioned that substantial differences between theory and observations, even in relative quantities, appear in some cases. Of this the nebula NGC 7027 is a noteworthy example. This object had the weakest continuous spectrum among all the planetary nebulae studied by Page [107]. This was in 1939. Sixteen years later, Minkowski and Aller noticed in the spectrogram of this nebula "a comparatively strong continuum which is prominent at the red end of even moderately exposed plates" [66].

For visualization, in Figure 5-6 there are shown curves of the continuous energy distribution of NGC 7027, obtained by two observations of Page in 1936 and 1942 [107, 108] and by two observations of Minkowski and Aller in 1946 and 1956 [66]. In drawing these curves the intensity near $\lambda = 3700$ Å is taken as unity. Large differences are obvious between these four observations and especially between Page's series of observations and those of Minkowski and Aller. For example, the Balmer discontinuity in the period 1936–1942 was not especially large: $D = 0.3$–0.5. By 1946 it had reached a large value, $D \approx 0.9$. Later, in the observations of Minkowski and Aller, the intensity of the continuous spectrum to the red of the Balmer limit increases quite confidently with increasing wavelength. In contrast to this, Page finds an almost constant intensity in this region of the spectrum. He also obtains totally different energy distributions in the Balmer continuum in the observations of 1936 and 1942.

It is difficult to judge how much of this scatter is real and not due to errors of measurement. In all probability this scatter is, to some extent, due to differences in the method of measurement and in the apparatus used. The possibility of systematic errors, which has not yet been ruled out, must be kept in mind particularly in discussing measurements of continuous spectra of planetary nebulae. Therefore

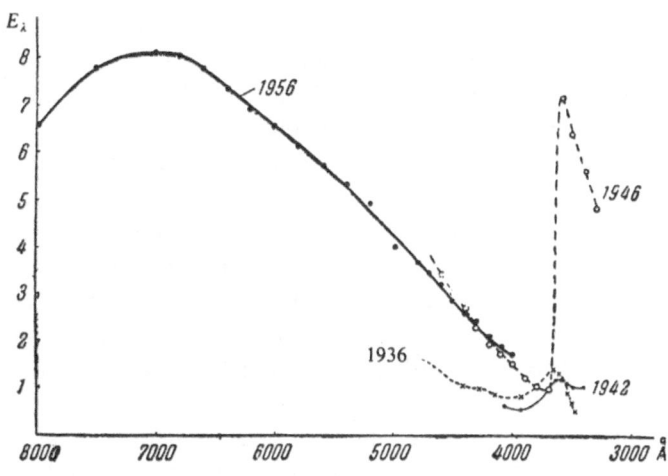

Figure 5-6
The energy distribution in the continuous spectrum of nebula NGC 7027 according
to the observations of Page (1936 and 1942) and Minkowski and Aller (1946 and 1956).
The intensity near $\lambda = 3700$ Å is set to unity.

before drawing any final conclusions, one must carefully check that these effects
are real, and for this it is necessary to have available more homogeneous material.

However, if one assumes that for NGC 7027 the differences in the energy distribution of the continuous spectrum are to some extent real, then we must conclude
that *planetary nebulae exhibit variations in the strength of their continuous emission.*
If this is so, then we are obliged to speak about new, at present, unknown, physical processes operating in planetary nebulae, a particular manifestation of which
is the generation of continuous radiation. Moreover this conclusion may significantly modify our ideas about the nature of the nuclei of planetary nebulae, since
their connection with these processes is apparent.

In the following section we shall return to the analysis of the continuous spectrum
of NGC 7027 in an attempt to show that, to some extent, in the visible region of
the spectrum, the radiation may be of synchrotron origin.

Convincing evidence of the possibility that energy of an unknown type is emitted
may be obtained by photographing the nebulae through filters passing various
spectral regions. This allows one to show simultaneously all regions emitting continuous radiation in a given spectral interval. No such work has been carried out
especially for this purpose, although there is one case in which a positive result
was obtained as though in this way. This concerns the planetary nebula NGC 6826,
studied by Curtis [15]. This is a round, double-shell nebula in the center of which,
near the nucleus, one can see an abnormal condensation. Curtis stated that on the

photograph obtained through a violet filter this condensation emerged much brighter than the rest of the nebula. The possibility can not be excluded that this condensation is a region emitting energy of an unknown nature.

From the point of view of the questions considered here, the planetary nebula NGC 7293 is of particular interest. On the photograph of this nebula, obtained in red light with the 200-inch telescope, there were discovered new, previously unknown features, which apparently are significant for our understanding of the nature of planetary nebulae. These features appeared as brilliant, comet-shaped elongated formations scattered on the inside boundary of the nebula. It is characteristic that all of these formations are directed toward the center where the exciting star is found. In this central region in place of the comet-shaped forms appear several tens of almost circular specks. According to the available data, the nebula NGC 6720 displays the same structure.

Zanstra attempted to explain the existence of these formations by assuming them to be regions of lower temperature and of higher matter density (condensation) [118]. These condensations are formed, according to Zanstra, as a consequence of the loss of energy by the free electrons in exciting forbidden lines, which leads to a decrease in electron temperature. The nuclei of the condensations are small dust clouds which, like reefs, are scattered in space. The last assumption seems rather artificial, since in this case we are left with the unexplained absence of similar condensations in the outer regions of the nebula. In any case, the formation itself of the condensations remains unexplained.

In all probability, the existence of these features also is relevant to the emission of energy of an unknown nature. One may test the assumption by photographing the nebula through various filters and also by comparing two photographs of the nebula obtained over a comparatively large time interval, in order to find changes of the brightness or location of these formations.

V.A. Ambartsumyan has shown [119] that some nonstationary objects which also have a recent origin—such as the Herbig-Haro objects, T Tauri and UV Ceti stars and cometary nebulae produce especially the particular type of radiation referred to as "continuous emission." From its character this radiation appears unable to have thermal origin and it is therefore ascribed to some agent ("intra-stellar substance") which, being ejected from the interiors of the stars, carries with it energy which is released beyond the boundaries of the star.

Planetary nebulae must be assigned to the class of nonstationary objects, although according to the flow speed characterizing nonstationary processes, they occupy an intermediate position between typical nonstationary and stationary objects. Therefore one may conjecture the existence of "continuous emission" also in planetary nebulae which could account for the anomalous continuous radiation. The difficulty in detecting this radiation in planetary nebulae lies in the fact that, in distinction, say, to cometary nebulae, where the nebula and nucleus are roughly of equal brightness, in planetary nebulae the brightness of its envelope (nebula) usually exceeds that of the nucleus by some ten or a hundred times. If one assumes that the emanation of continuous emission occurs in planetary nebulae with the same strength as in the Herbig-Haro objects or in cometary nebulae, then one finds that continuous emission amounts to hardly one percent of the total energy radiated by the nebula through the usual processes (fluorescent and forbidden-line radiation). Only by delicate spectrophotometric and photometric investigations would it be possible to check the assumption of continuous emission.

The above facts are far from sufficient to allow us to reach a final conclusion on this question. It is still necessary to make special observations in this direction. Nevertheless, it is already clear that in the micro-structure of the radiation of planetary nebulae deviations appear from the well-known regular features. In their nature these "deviations" border upon phenomena usually observed in nonstationary and recently formed objects.

7. The Possibility of Synchrotron Radiation

As will be shown later (Chapter IX), magnetic fields of order $10^{-3} - 10^{-4}$ gauss exist in planetary nebulae. On the other hand, according to the ideas developed in the final chapter of this book, planetary nebulae are remnants of star-producing processes and the nuclei are recently formed stars. Under these conditions it is not impossible that relativistic electrons with energies of order $10^{11}-10^{12}$ e.v. exist in nebulae. The motion of these electrons in the magnetic fields of the nebula may generate synchrotron radiation (magnetic bremsstrahlung) which has a continuous spectrum. The question arises whether or not the unidentified continuous radiation discussed in the previous section originates from the synchrotron mechanism.

At present it is possible to mention two procedures by which one could verify the presence of, and even isolate, synchrotron radiation in the continuous spectrum of planetary nebulae. The first method is based on an analysis of the energy distribution of the continuous spectrum of a given nebula. The second method utilizes the measurement of the polarization of the continuous radiation of the nebula. We shall first discuss the second method.

There is one distinctive feature in which synchrotron radiation differs strongly from the usual continuous (thermal) radiation. The point is that both recombination (line and continuous) and two-quantum radiation are not polarized. On the other hand synchrotron radiation must be polarized to a very high degree: the theoretical degree of polarization may reach 70% [120], although for integrated continuous radiation the figure will be very much reduced. Therefore if one succeeds in detecting even a small amount of polarization in the continuous spectrum between the emission lines of some nebula, this is evidence of the possibility that synchrotron radiation is emitted by that nebula.

Unfortunately, because of the weakness of the continuous spectra of planetary nebulae, such measurements are extremely difficult to perform. Nevertheless such a measurement was attempted with NGC 7026, from which the degree of polarization of the continuous radiation was found to equal $p = 5.3 \pm 1.7\%$ in the interval $\lambda\lambda 3740$–4600 Å [121]. This determination is preliminary and has not yet been verified. Therefore, before one makes some decision from such data, it is necessary to undertake similar experiments involving other planetary nebulae, first of all those which have strong continuous spectra or appear bipolar. Among such nebulae are IC 418, NGC 7027, and NGC 7662.

We turn now to the analysis of the energy distribution of the continuous spectrum of the nebula.

More or less reliable data of this nature exists at present for the nebula NGC 7027. We have in mind the results of Minkowski's and Aller's measurements, already mentioned above, which cover a large wavelength interval from 3400 Å to 8000 Å. The results are shown in curve of Figure 5-7 (see also Figure 5-5). The most characteristic feature of this curve, as was pointed out, is that it indicates a rapid increase in the intensity of the continuous radiation in the direction of increasing wavelength. The intensity increases from the Balmer limit up to $\lambda = 7000$Å after which it falls off. However this part of the curve is apparently not very reliable and the existence of the maximum at $\lambda = 7000$ Å is still unsettled.

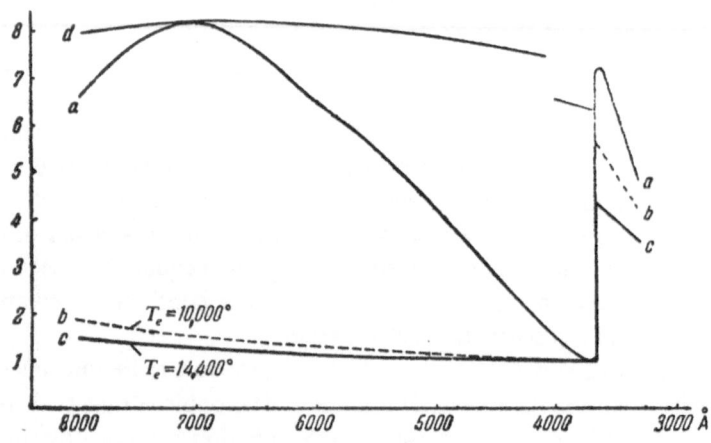

Figure 5-7

The energy distribution in the continuous spectrum of a nebula: (a) the observed distribution for NGC 7027 (Minkowski and Aller); (b) the theoretical distribution (the combination of hydrogen and two-photon emission) for $T = 10{,}000°$K; (c) the same for $T = 14{,}400°$K; (d) the theoretical distribution for synchrotron radiation due to a monochromatic current of electrons.

In Figure 5-7 also appear theoretical curves of the energy distribution in the continuum of a nebula calculated on the assumption that it originates from recombination in hydrogen and from two-quantum processes. Curves are given for the cases b) $T_e = 10,000°K$ and c) $T_e = 14,000°K$ and are quoted from the work of Spitzer and Greenstein [111]. In all cases the intensity of the continuous spectrum near $\lambda = 3700$ Å is set to unity.

From Figure 5-7 the difference between theory (curve b and c) and experiment (curve a) is apparent, particularly in the visible and infra-red regions of the spectrum. This difference is significantly less beyond the limit of the Balmer series and may be neglected in the first approximation. Incidentally the disparity between theory and observation and the difficulty in explaining it forced Minkowski and Aller to question the accuracy of their results; they referred to the difficulty in determining the precise level of the continuum in the visible and infra-red regions of the spectrum. But in the present case the theoretical data is 5 to 8 times smaller than the observed and such a "stretch" hardly saves the situation. In all probability these disparities are real and are due to other causes. In particular the surplus radiation in the nebulae may originate from the synchrotron mechanism.

The energy distribution in the spectrum of synchrotron radiation depends on the ratio of the magnetic field strength to the energy of the electrons, and also on the spectrum of the relativistic electrons. In the particular case when all electrons have the same kinetic energy E, that is, when we have a monochromatic current of electrons, then the intensity of the synchrotron radiation will reach a maximum at some frequency ν_{max}, given by the relation [122, 123]

$$\nu_{max} = \frac{1}{2\pi} \frac{eH}{mc} \left(\frac{E}{mc^2}\right)^2 = 2.8 \cdot 10^6 H \left(\frac{E}{mc^2}\right)^2 \sec^{-1}, \tag{13}$$

where H is the component of magnetic field strength perpendicular to the direction of motion of the electrons. In the high frequency limit $\nu \gg \nu_{max}$ the intensity of radiation diminishes according to the law

$$I_\nu \sim \left(\frac{\nu}{\nu_{max}}\right)^2 e^{-\frac{4}{3}\nu/\nu_{max}}, \nu \gg \nu_{max}. \tag{14}$$

In the low frequency limit ($\nu \ll \nu_{max}$), the intensity falls off according to a quite different law:

$$I_\nu \sim \left(\frac{\nu}{\nu_{max}}\right)^{1/3}, \nu \ll \nu_{max}. \tag{15}$$

At first sight the maximum in the intensity of the continuum of NGC 7027 might be though to correspond to that given by the theory. What is more, in this case one might infer that the synchrotron radiation of the nebula NGC 7027 is caused by a monochromatic current of relativistic electrons. However, the observed curve falls off on both sides of the maximum considerably faster than would be the case for radiation from a monochromatic current of electrons. For visualization in Figure 5-7 the intensity distribution of the synchrotron radiation (curve d) is drawn with the assumptions that the radiation is emitted by a monochromatic current of electrons and that the intensity maximum coincides with the maximum of the observed curve. The data for plotting this curve is taken from the work of V.V. Vladimirskiĭ [122].

Thus the synchrotron radiation in the nebula NGC 7027 (if it really is of this type) cannot have been produced by a monochromatic current of relativistic electrons. The problem lies in finding that spectrum of relativistic electrons for which the distribution over frequency of the radiated energy will coincide with the observed distribution (curve d in Figure 5-7).

If the energy spectrum of the relativistic electrons is continuous and can be described in the form

$$N_e(E) = KE^{-\gamma}, \tag{16}$$

where K and γ are parameters of the spectrum, then one can always find electrons with energy E, which for a given value of the magnetic field strength H, will radiate in both the visible and the infra-red. The spectrum of this radiation will also be continuous, but with an entirely different gradient, depending in particular on γ. The intensity of this radiation will depend on the density of relativistic electrons and also on the extent of the region S where the magnetic field strength has the given value.

For the intensity I_ν of the synchrotron radiation at frequency ν per unit frequency per unit area of nebula perpendicular to the direction of propagation of the radiation (to the observer), we have the following expression (see, for example, [124]):

$$I_\nu = \frac{3}{2}(2\pi)^{(1-\gamma)/2}\,\frac{e^3 H}{mc^2}\left[\frac{2eH}{m^3 c^5}\right]^{(\gamma-1)/2} U(\gamma)K\nu^{(1-\gamma)/2}S. \tag{17}$$

Let us write (17) in the form

$$I_\nu = C(\gamma)KH^{(\gamma+1)/2}\nu^{(1-\gamma)/2}S, \tag{18}$$

where the numerical values of the function $C(\gamma)$ are tabulated in [125].

Thus the dependence of intensity on wavelength has the form (since $I_\nu|d\nu| = I_\lambda|d\lambda|$),

$$I_\lambda = \lambda^{(\gamma-5)/2}. \tag{19}$$

Let us represent the observed intensity distribution in the visible and infra-red regions of the spectrum of the nebula in the form

$$I_\lambda \sim \lambda^n. \tag{20}$$

Having assumed that the continuous radiation in the interval from $\lambda = 6500$ Å to $\lambda = 5000$ Å has a purely synchrotron origin, that is, neglecting the effect of the recombination and two-quantum components, we find from curve a in Figure 5-7, $n = 2.5$. Substituting this into (20) and comparing with (19), we find $\gamma \sim 10$. Then the energy spectrum of the relativistic electrons (10) may be written in form

$$N_e(E) = KE^{-10}. \tag{21}$$

It remains to determine K, the magnitude of which depends on the brightness of the nebula in the continuous radiation.

Let us designate by E_c the energy radiated by the nebula per unit time in the photographic range of the continuous spectrum and by E_n the total energy emitted by the nebula in the continuous spectrum and in the lines. Let σ denote the ratio of these two quantities,

$$\sigma = E_n/E_c. \tag{22}$$

Then for the integrated apparent magnitude m_c of the nebula in the continuum we shall have:

$$m_c = m_n + 2.5 \log \sigma, \tag{23}$$

where m_n is the integrated apparent magnitude of the nebula in the photographic region.

Knowing m_c, the linear dimension S of the nebula, and its apparent area in steradians, we can obtain the magnitude of K from the following relation [125, 126]:

$$K = \frac{F_\odot}{C(\gamma)} \frac{H^{-(\gamma+1)/2}}{\Omega \Delta \nu_{pg} S} \nu^{-(1-\gamma)/2} 10^{-0.4(m_c - m_\odot)}, \tag{24}$$

where $m_\odot$ and $F_\odot$ are the bolometric apparent magnitude and the total radiation flux of the sun, $\Delta \nu_{pg}$ and ν are respectively the width of the photographic region in frequency units and the mean frequency of the photographic region of the spectrum and the function $C(\gamma) = C(10) = 3.09 \cdot 10^{62}$ [the function $U(10)$ in (17) is equal to 0.70].

For the nebula NGC 7027 we have: $m_n = 10.4$, $D'' = 11'' - 18''$ or $15''$ on the average and $S = 15 \cdot 10^3$ a.u. [16]. On the basis of these data with given values of σ and H we can find the number K from (24) and then obtain $N_e(E)$ from (21). The total density of relativistic electrons with energies larger than a given value E_0 is found from the following relation

$$N_e(E > E_0) = \int_{E_0}^{\infty} N_e(E) dE = \frac{1}{\gamma - 1} KE_0^{1-\gamma} = \frac{1}{9} KE_0^{-9}. \tag{25}$$

The quantity E_0 may be determined from (13) and equals (for $v \sim 0.7 \cdot 10^{15}$ sec^{-1}):

$$E_0 = \quad 3 \cdot 10^{11} \text{ e.v. for } H = 10^{-3} \text{ gauss}$$

$$E_0 = 1.1 \cdot 10^{12} \text{ e.v. for } H = 10^{-4} \text{ gauss}.$$

From calculations performed for the values $\sigma = 2$ and $\sigma = 10$ in the case $H = 10^{-3}$ gauss, one is able to estimate the order of magnitude of the total density of relativistic electrons in NGC 7027. From the data of Table 5-5, this quantity was found to be of order $10^{-10} - 10^{-11}$ cm^{-3}.

TABLE 5-5

Total Concentration of Relativistic Electrons in NGC 7027 for $H = 10^{-3}$ gauss

	$\sigma = 2$	$\sigma = 10$
K	$6 \cdot 10^{-13}$	10^{-13}
$N_e(E_0 > 3 \cdot 10^{11})$ cm^{-3}	$3 \cdot 10^{-10}$	10^{-11}
$N_e(E_0 > 10^{12})$ cm^{-3}	10^{-15}	$2 \cdot 10^{-16}$

It is obvious that the problem of the generation of synchrotron radiation in planetary nebulae has been studied here only in outline. When more reliable data on the energy distribution in the continuous spectra of nebulae becomes available, one could set up a more precise problem. In particular, one might try to isolate the synchrotron radiation from the total continuous radiation of the nebula. For this it is necessary to know the precise value of the electron temperatures of the nebula (determined, for example, by the $E_{N_1 + N_2}/E_{4363}$ method), the ratio $\sigma = E_n/E_c$, the apparent and linear dimensions of the nebula and so forth. However, the analysis performed above shows that the assumption that synchrotron radiation is generated in planetary nebulae and, in particular, NGC 7027, is not inconceivable and that it may explain the anomalous phenomena discussed above.

Finally the question of the synchrotron nature of the excess continuous radiation in the nebula NGC 7027 may be solved by polarimetric investigation of the radiation.

8. The Possibility of Radio Emission

Free-free electron transitions in the field of a proton may also emit energy at radio frequencies. The spectrum of this radiation is a continuum. Since planetaries contain electrons and protons in equal numbers, they should act as sources of radio waves. The total radio-frequency energy radiated by a nebula is proportional to $\int n_e^2 dV = n_e^2 V$, where n_e is the concentration of free electrons and V is the volume of the nebula.

Given n_e and V, we can calculate the flux of radio energy reaching the Earth-bound observer in each particular case. Calculations show that the expected radio flux is very small, of the order of 10^{-26} W $\cdot$ m^{-2} Hz^{-1}. The smallness of this figure explains why reliable radio observations of planetaries were made not earlier than the 1960's.

Thermal emission of planetaries and of gaseous nebulae in general is characterized by certain definite properties. This radiation is generated mainly by free-free electron transitions in the field of a proton, so that the intensity of radio emission depends primarily on the electron temperature, T_e, of the medium. Let τ_ν be the optical thickness of a nebula at the frequency (in the radio range); there is an obvious relation between the effective radiation temperature $T(\tau_\nu)$ and T_e:

$$T(\tau_\nu) = T_e(1 - e^{-\tau_\nu}), \tag{26}$$

where $\tau_\nu = \int K_\nu ds$; the radio-frequency absorption coefficient K_ν is given by

$$K_\nu = 0.125 \frac{n^+ n_e}{\nu^2 T_e^{3/2}} \text{ cm}^2. \tag{27}$$

The expression for τ_ν is often written as

$$\tau_\nu = 0.125 \ \frac{1}{\nu^2 T_e^{3/2}} \int n^+ n_e ds = 0.125 \frac{\text{EM}}{\nu^2 T_e^{3/2}}, \tag{28}$$

where EM $= \int n^+ n_e ds$ is called the *emission measure*.

The intensity of radio emission $\mathscr{I}_\nu$ from a nebula is given by the Rayleigh-Jeans formula

$$\mathscr{I}_\nu = \frac{2k\nu^2}{c^2} T(\tau_\nu) = \frac{2k\nu^2}{c^2} T_e(1 - e^{-\tau_\nu}). \tag{29}$$

This is the theoretical spectrum of thermal radio emission. For sufficiently low frequencies (long wavelengths), we see from (28) that τ_ν may be greater than unity. Then from (29)

$$\mathscr{I}_\nu = \frac{2k\nu^2}{c^2} T_e \sim \frac{1}{\lambda^2}, (\tau_\nu \gg 1), \tag{30}$$

i.e., the intensity of thermal radio emission decreases with increasing wavelength.

For $\tau_\nu \ll 1$, which may obtain at sufficiently high frequencies, we have from (29) and (28)

$$\mathscr{I}_\nu = \frac{2k}{c^2} 0.125 \frac{\text{EM}}{T_e^{1/2}} = 0.38 \cdot 10^{-37} \text{EM} T_e^{-1/2}, \tag{31}$$

i.e., the intensity of short-wave thermal radio emission is independent of wave-length.

Figure 5-8 shows schematically the theoretical spectra of thermal radio emission from (30) and (31) (solid lines). Any deviation of the observed radio spectrum from relations (30) and (31) should naturally be regarded as indication of its nonthermal origin (the dashed line in Figure 5-8).

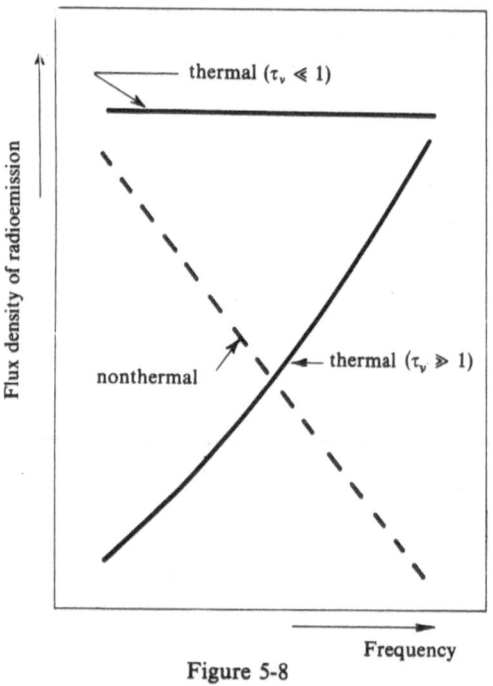

Figure 5-8

Theoretical radio spectra of planetary nebulae. Solid lines = thermal emission; dashed line = nonthermal radiation.

Since the emissivity of gaseous nebulae in the optical region (in hydrogen lines) also depends on the emission measure, the ratio of the intensity of radio emission at centimeter wavelengths to the intensity of one of the hydrogen lines is a function of the electron temperature of the medium only, *i.e.*,

$$\frac{\mathscr{S}_\nu}{\mathscr{S}_{H_i}} = f_i(T_e). \tag{32}$$

For the second Balmer line of hydrogen H_β at $T_e = 10,000°K$ we have (assuming a gaseous nebula which is optically thick in the lines of the Lyman series of hydrogen)

$$\frac{\mathscr{I}_\nu}{\mathscr{I}_{H_i}} = 3.23 \cdot 10^{-14}. \tag{33}$$

Relations (32) and (33) are valid if light and radio waves are not absorbed in the interstellar medium and in the nebula.

The first successful determinations of the radio flux for a number of planetaries were carried out by Lynds at 10-cm wavelength [128]. Similar measurements were later made (with a much higher accuracy) by Menon and Terzian [233, 234], Slee and Orchiston [235], and others [236, 237]. Their observations covered the range of frequencies from 195 MHz (154 cm) to 16,200 MHz (1.85 cm). We thus have now fairly reliable data on the radio fluxes at the high frequencies which were found to range from 0.1 to about 10 (in units of 10^{-26} W/m²Hz). The largest radio flux is that of NGC 7027, being 7.42 ± 0.4 at 16,200 MHz. For a number of planetaries the fluxes are less than 0.1, and we cannot be certain that they emit radio waves at all.

Analysis of the radio observations of planetaries and, in particular, analysis of trons (with energies of the order of 1 Me v.) clearly originate in the central stars.

1. The radio emission of the great majority of planetaries is of thermal origin; the observed radio fluxes are in good agreement with the theoretical fluxes calculated from (33) using the intensity of the H_β line.

2. Some planetaries (NGC 3242, NGC 7652, NGC 7293) are optically thin at 800 to 16,200 MHz. Specimen spectra of these nebulae are given in Figure 5-9.

3. Most planetaries are optically thin at high frequencies and optically thick at low frequencies; the critical frequency f_0 which corresponds to unit optical thickness is different for different nebulae. For NGC 6853 we have $f_0 = 400$ MHz (75 cm), for NGC 6543 $f_0 = 1500$ MHz (20 cm), and for NGC 7027 $f_0 = 3000$ MHz (10 cm). Specimen spectra of such nebulae are given in Figure 5-10.

4. The radio emission of some nebulae is of nonthermal, probably synchrotron origin; the radio spectra of these planetaries are qualitatively different from the spectra of thermal radiation. About a dozen nebulae of this class are known, NGC 6153, NGC 6445, NGC 40, NGC 7008, NGC 7635, probably NGC 6833, NGC 6875, and others). This fact is nevertheless of the greatest significance for understanding the nature of planetary nebulae and of their central stars.

Synchrotron radiation indicates primarily the presence of relativistic electrons with energies of $10^9 - 10^{10}$ e.v. in the planetaries. Energetic, if not relativistic, electrons (with energies of the order of 1 Me.v.) clearly originate in the central stars. This in itself is an undisputed indication of the high activity of the central stars, and in particular, of nuclear reactions in the outermost layers of the stellar atmospheres.

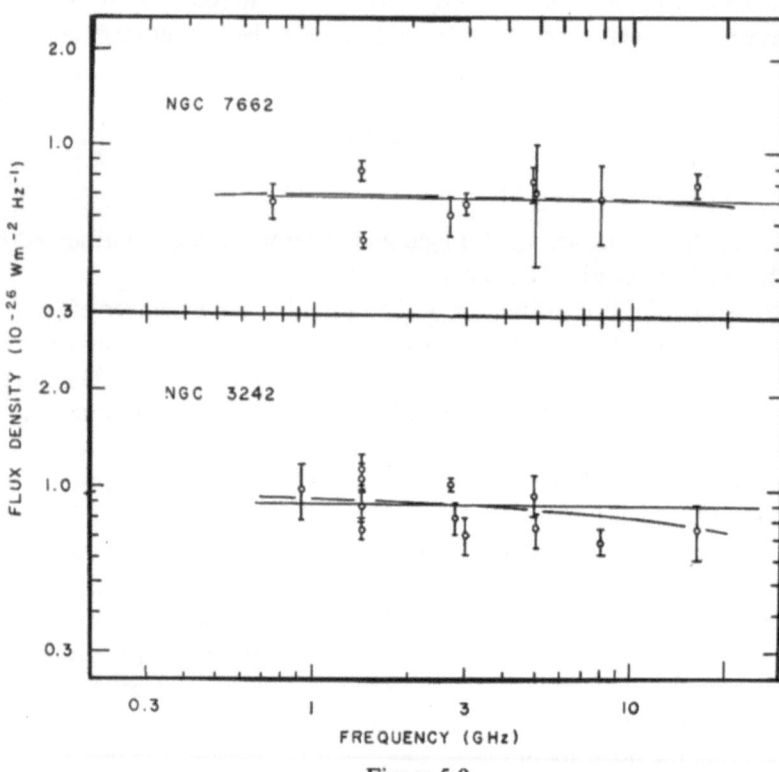

Figure 5-9

Radio spectra of NGC 7662 and NGC 3242. (These nebulae are transparent to short
and long wavelengths.)

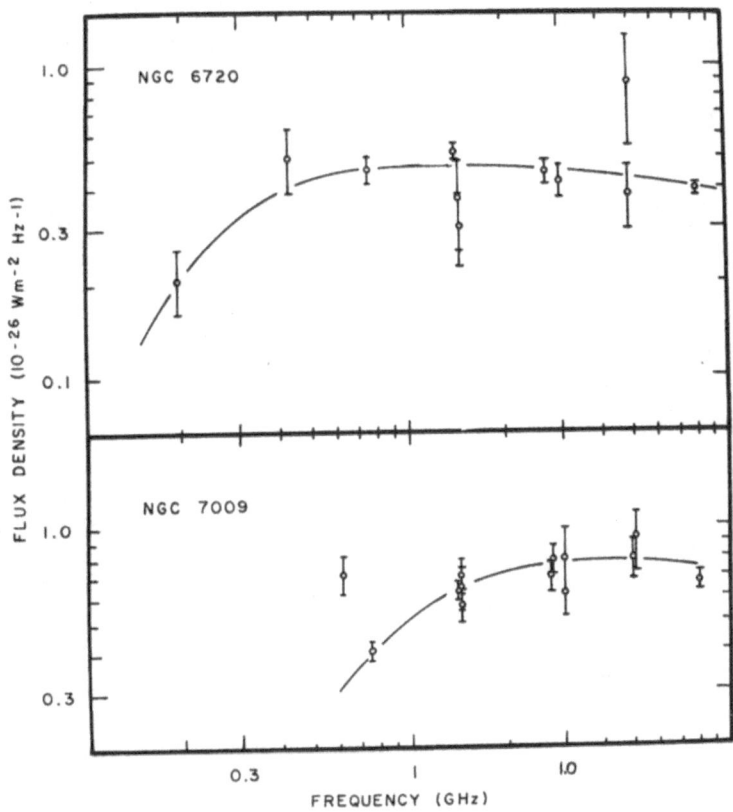

Figure 5-10(*a*)

Radio spectra of (*a*) NGC 6720 and NGC 7009, (*b*) NGC 7072 and NGC 6853, (*c*) IC 418 and NGC 6572. (These nebulae are transparent to short waves and opaque to long waves.)

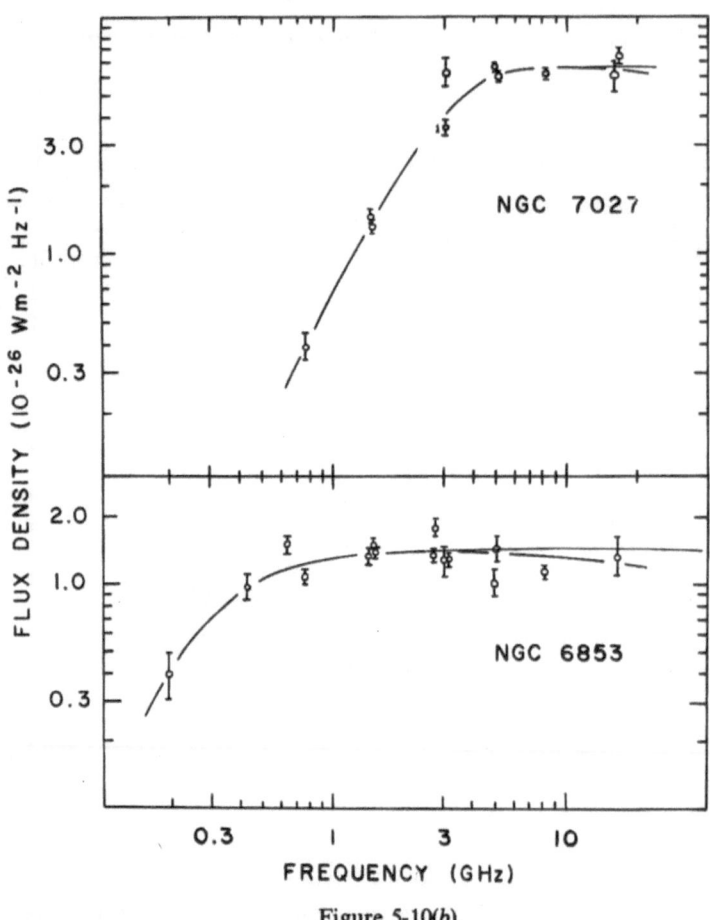

Figure 5-10(*b*)

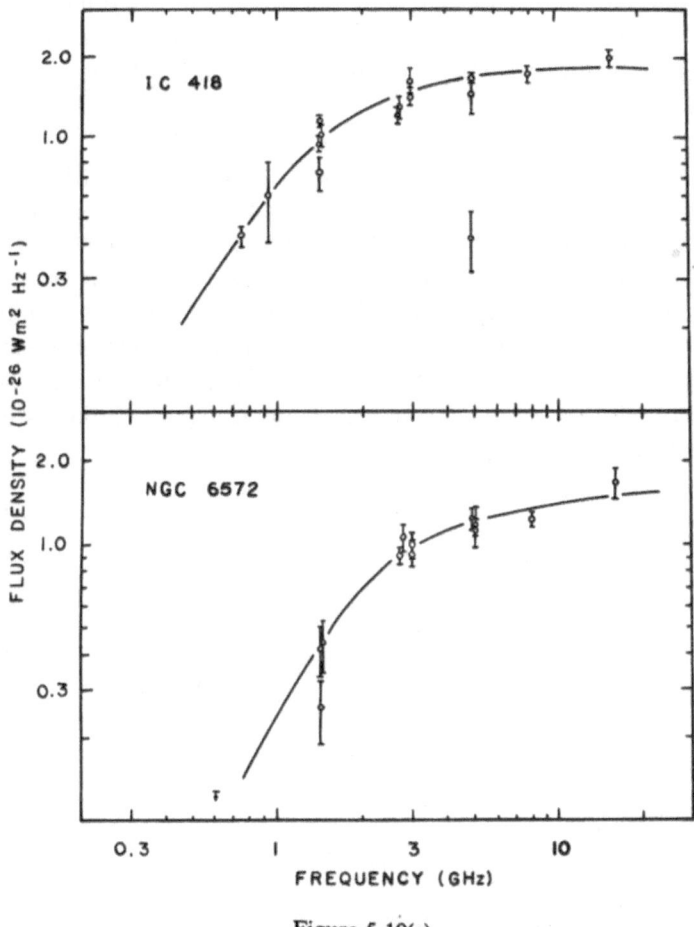

Figure 5-10(c)

The emission of synchrotron radiation, once established beyond all reasonable doubt, will further indicate the existence of magnetic fields in planetaries. It is in this way that we first arrive at the hypothesis of magnetic fields in planetaries (see Chapter IX).

Although the radio emission of most planetaries is of thermal origin, we must not rule out the possibility of nonthermal radiation. In certain periods of their active life, the central stars probably emit intense streams of high-energy electrons, and the synchrotron radiation temporarily prevails over the normal thermal radio emission. If this hypothesis is true, both the flux and the radio spectrum of planetary nebulae can be expected to vary in time. On the whole, a higher variability of planetaries can be expected in the radio than in the optical spectrum. Another possibility is that synchrotron radiation and thermal radio waves are generated in different parts of the nebula. To test this hypothesis we need high-resolution radio spectra from different regions of one planetary.

Chapter VI

Radiative Equilibrium of Planetary Nebulae

Planetary nebulae, as was pointed out in Chapter II, are transparent for radiation in forbidden lines and in the lines of subordinate series. Under these conditions the problem of the radiation field in the lines in the visible region of the spectrum can be solved very easily. The density of radiation, for example, in the lines of the Balmer series of hydrogen is proportional to the square of the electron concentration. Therefore the distribution of radiation density in these lines within the nebula is simply proportional to the distribution of the square of the electron concentration in it. Moreover, the intensity of the radiation emerging from a nebula in any line of this series can be determined simply by summing the radiation from each unit volume along the line of sight. In the same way one can find the intensity of radiation emerging from a nebula in a forbidden line. Nothing more need be said about the transfer of radiation in the lines of the visible part of the spectrum.

The situation is entirely different for radiation in the principal series, the so-called Lyman series, of hydrogen. For radiation in the continuous spectrum of this series (L_c-radiation) the optical thickness τ_c of the nebula is of order unity. Therefore processes which absorb and re-emit L_c-quanta may also play an essential role in the nebulae. In particular these processes produce quanta in the lines of the Lyman series, the absorption coefficient for which are some ten thousand times larger than that beyond the series limit. All Lyman quanta produced in this way are finally converted into L_α-quanta, while photons in the subordinate series, which escape the nebula, are also produced. Thus, only L_c and L_α photons are "imprisoned" in the nebula. Therefore the problem of radiation transfer or, in other words, of radiative equilibrium in nebulae will arise only for L_c and L_α-radiation. In contrast to L_c-radiation, for which the nebula is partially transparent, for L_α it is extremely opaque. Moreover both of these radiation fields interact with one another. The calculation of the radiation field in these conditions, generally speaking, is a very complex problem.

The problem of finding the radiation field in the L_α-line in planetary nebulae was solved for the first time by V.A. Ambartsumyan [1]. He developed a method of separating the L_c and L_α-fields, which greatly simplifies the solution of the problem. Although the results obtained by Ambartsumyan were subsequently refined and generalized by other workers, among whom a special place is occupied by V.V. Sobolev, the fundamental idea—that of separation—is preserved.

Considering the difficulty of the problem in its most general form—for a spherical nebula with a variable dilution coefficient—one usually resorts to studying a more simple geometrical model of the nebula. In the following, we assume that the nebula is bounded by two concentric spheres of radii r_1 and r_2 centered on the star which excites the nebula. The linear thickness is assumed to be small in comparison with its linear distance from the center, that is, $r_2 - r_1 \ll r_1$. In this case the nebula may be considered as a plane-parallel slab, in which the dilution coefficient is constant.

1. The L_c-Radiation Field

Let us denote by πS^* the number of photons in the Lyman continuum emitted by the nucleus per second from one cm^2 of its surface. Obviously the number of photons falling on 1 cm^2 of the inside surface of the nebula will be $\pi S = \pi S^*(r_*/r)^2$, where r_* is the radius of the nucleus.

Let κ be the mean absorption coefficient per neutral hydrogen atom* in the Lyman continuum and let τ be the corresponding optical depth, measured from the inner boundary of the nebula up to the layer being considered, that is

$$\tau = \int_{r_1}^{r} n_i \kappa \, dr, \tag{1}$$

where r is the linear distance from the center of the nebula to the layer in question.

The absorption of L_c-photons results in the ionization of hydrogen. Subsequently some of these photons will be re-emitted in recombinations into the ground state. Let p designate that fraction of all recombinations which go directly into the ground state. This depends only on the electron temperature of the nebula and is determined according to formula (48) of Chapter II. Obviously the probability of re-emitting the absorbed L_c-photons will be p and the probability of radiating L_α-photons will be $1 - p$.

Of the L_c-photons falling on the inner boundary of the nebula at the rate of πS per cm^2 per second, $\pi S e^{-\tau}$ photons reach optical depth τ. Of these, $n_1 \kappa \pi S e^{-\tau}$

---

* The mean absorption coefficient κ may be represented in this form

$$\kappa = \kappa_0 \phi(\tau, T_*)$$

where κ_0 is the absorption coefficient in the continuum at the series limit ($\lambda = 912$ Å) and $\phi(\tau_0 T_*)$ is some function depending on τ and on the temperature T_* of the star. Generally $\phi(\tau, T_*) \approx 1.1$, that is that κ only slightly exceeds κ_0 because of the rapid decrease of the absorption coefficient with increasing frequencies ($\kappa \sim \nu^{-3}$).

quanta will be absorbed per unit volume, where n_1 is the number of hydrogen atoms per unit volume. Diffuse L_c-radiation, arising from recombinations into the ground state, is also absorbed by hydrogen. The number of diffuse L_c-photons absorbed per unit volume per unit time will equal $n_1 \kappa \int I(\tau,\theta) d\omega$, where $I(\tau,\theta) h\nu_c$ is the intensity of the diffuse radiation directed with angle θ to the normal at optical depth τ and the integration is taken over all solid angles. Thus the total number of photons absorbed per unit volume per unit time at optical depth τ is

$$n_1 \kappa [\pi S e^{-\tau} + \int I(\tau,\theta) d\omega].$$ (2)

The number of photons emitted per unit volume per unit time in the Lyman continuum will be

$$4\pi n_1 \kappa C(\tau).$$ (3)

We are now able to write the condition of radiative equilibrium, equating (2) to (3), in the form

$$C(\tau) = \kappa \int I(\tau,\theta) \frac{d\omega}{4\pi} + p\frac{S}{4} e^{-\tau}.$$ (4)

The quantities $C(\tau)$ and $I(\tau,\theta)$ are also related to each other by the equation of radiative transfer, which for a plane-parallel layer has the form

$$\cos\theta \, \frac{dI(\tau,\theta)}{d\tau} = C(\tau) - I(\tau,\theta).$$ (5)

The combined solution of (4) and (5), together with the boundary conditions, gives us the quantities $C(\tau)$ and $I(\tau,\theta)$. In writing down the boundary conditions one must keep in mind that the inner surface of the nebula receives not only L_c-radiation from the central star, but also L_c-radiation from the opposite side of the nebula. Consequently the first boundary condition can be written in the form

$$I(0,\theta) = I(0,\pi - \theta),$$ (6)

that is, the diffuse radiation entering the nebula at a point on the inner boundary at an angle θ to the normal is not different from the radiation leaving the nebula at an angle $\pi - \theta$ on the opposite side (Figure 6-1).
The second boundary condition is written in the form

$$I(\tau_0,\theta) = 0 \text{ for } \theta > \frac{\pi}{2}.$$ (7)

It expresses the obvious fact that no radiation falls on the outer boundary from outside.

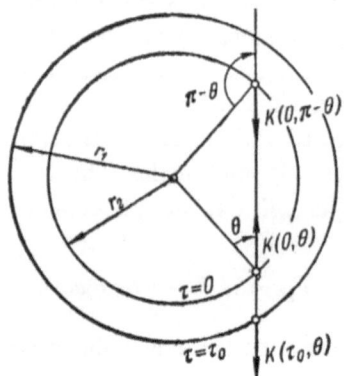

Figure 6-1

Geometrical scheme of a nebula in the computation of the radiation field.

The equations (4) and (5) may be solved with the help of one of the approximate methods known in the theory of radiative transfer (such as the Schuster-Schwarzschild approximation and the Eddington approximation). However, V. V. Sobolev [90] has given a more accurate solution of these equations based on the formulation of an integral equation. The integral equation for $C(\tau)$ has the form

$$C(\tau) = \frac{1}{2} p \int_0^{\tau_0} [E_1(|\tau - \tau'|) + E_1(\tau + \tau')] C(\tau')dt + p \frac{S}{4} e^{-\tau}, \tag{8}$$

where

$$E_1(\tau) = \int_\tau^\infty e^{-t} \frac{dt}{t}.$$

Since our goal is to ascertain the role of diffuse radiation in the transfer processes of L_c-radiation we may set $\tau_0 = \infty$ in (8). Then equation (8) may be written approximately in the form

$$C(\tau) = \frac{p}{2} \int_{-\infty}^\infty E_1(|\tau - \tau'|) C(\tau')d\tau'. \tag{9}$$

This equation is valid for large values of the optical depth τ.

Let us look for a solution of equation (9) in the form

$$C(\tau) = Ae^{-k\tau}, \tag{10}$$

where A is an arbitrary constant. Substituting this in (9), we find for k the equation

$$\frac{p}{2k} \ln \frac{1 + k}{1 - k} = 1. \tag{11}$$

TABLE 6-1

The Quantity p for Various Values of the Electron Temperature

$T_e(°K)$	5000	10,000	20,000	50,000
p	0.33	0.38	0.43	0.50

As regards the constant A, its value is to be determined by the condition that equation (8) be satisfied exactly on the average, treating the function (10) as only an approximate solution of the equation. Then we obtain

$$A = \frac{kpS}{4(1-p)}. \tag{12}$$

The values of p as obtained from formula (48) of Chapter II are shown in Table 6-1 for various values of the electron temperature.

Since the electron temperatures of nebulae are of order 10,000°K one may take $p \approx 0.4$ on the average. For this value of p we find from (11) $k = 0.958 \sim 1$. We may then write for the required function

$$C(\tau) = S\frac{p}{4(1-p)}e^{-\tau}. \tag{13}$$

Comparing this with formula (4), we see that the number of diffuse L_c-photons is approximately equal to the number of L_c-photons arriving directly from the star. Averaging equation (5) over the outward hemisphere we obtain

$$\frac{1}{2}\frac{dI(\tau)}{d\tau} = C(\tau) - I(\tau), \tag{14}$$

where $I(\tau)$ is the outward directed component of L_c-radiation. Substituting the value of $C(\tau)$ from (13) into (14) and integrating, we find, at the outer boundary of the nebula ($\tau = \tau_0$),

$$I(\tau_0) = \frac{pS}{2(1-p)}e^{-\tau_0} \approx 0.4Se^{-\tau_0}. \tag{15}$$

For the flux of diffuse L_c-quanta we have

$$H_c(\tau_0) = \int I(\tau_0)\cos\theta d\omega = \pi I(\tau_0) = 0.4\pi Se^{-\tau_0}.$$

Since the flux of direct L_c-radiation near the outer boundary equals $\pi Se^{-\tau_0}$, the total flux (diffuse + direct) is $1.4\pi Se^{-\tau_0}$.

The number of photons falling on the inner surface of the nebula is πS. The number of L_c-photons escaping the nebulae is $1.4\pi Se^{-\tau_0}$, so that $(1 - 1.4e^{-\tau_0})\pi S$ photons are converted into photons of other frequencies. Since one L_α-photon is formed from each L_c-photon, the number of L_α-photons at the outer boundary of the nebula must be $(1 - 1.4e^{-\tau_0})\pi S$ or, for $\tau_0 \gg 1$, this becomes just πS, that is, we obtained the result found in Chapter II from purely qualitative considerations.

2. The L_α-Radiation Field

Let us denote by k the absorption coefficient per hydrogen atom in the L_α line and define the corresponding optical depth t into the nebula by

$$t = \int_{r_1}^{r} n_1 k dr. \tag{16}$$

From (16) and (1) we have

$$\frac{t}{\tau} = \frac{k}{\kappa} = \frac{1}{q}. \tag{17}$$

The quantity k depends on the mean velocity of the hydrogen atoms. For the thermal velocities occuring in nebulae, the quantity q is of order $10^{-4} - 10^{-5}$. Since we assume that for the total optical thickness τ_0 of the nebula for L_c-radiation is of order unity or larger, we find that the optical thickness τ_0 of the nebula in the L_α-line is of order 10^4 or more, i.e., extremely large. Consequently one may talk about a completely opaque nebula for L_α-radiation.

On the other hand, we saw that one L_α-photon is formed from each stellar L_c-photon absorbed in the nebula. And since, because of the opacity of the nebula, the L_α-photons are unable to escape, an extremely large build-up of these photons is inevitable. In other words, the density of L_α-radiation in nebulae is very much larger than the density of L_c-radiation. Hence one may see a priori the very great importance of the L_α-radiation field in the dynamical problems of planetary nebulae.

Let us denote by $K(t,\theta)$ the intensity of L_α-radiation directed at an angle θ to the normal at optical depth t, and by $4\pi n_1 k G(t)$, the number of L_α-photons emitted per unit volume per unit time at the same depth. Then we may write the equation of transfer for L_α-radiation in the form

$$\cos \theta \, \frac{dK(t,\theta)}{dt} = G(t) - K(t,\theta). \tag{18}$$

In writing the equation of radiative equilibrium one must keep in mind that conversion of L_c-photons to L_α-photons occurs as well as scattering of the latter. As regards the L_α-radiation coming directly from the central star, it may be neglected in comparison with the L_c-radiation.

As we have seen in the preceding section, a unit volume of the nebula emits $4\pi n_1 \kappa C(\tau)$ L_c-photons per unit time. This is also the number of recombinations directly into the ground state per unit volume per unit time. Obviously, the number of recombinations to states other than the lowest, in which L_c-quanta cannot be emitted, will be equal to $[(1 - p)/p] \, 4\pi \kappa n C(\tau)$.

The number of diffuse L_α-photons absorbed per unit volume per unit time will equal $n_1 k \int K(t,\theta) d\omega$. Therefore the total number of L_α-photons, absorbed per unit volume per unit time at optical depth t, is given by

$$n_1 k \int K(t,\theta)d\epsilon \; + 4\pi \frac{1-p}{p} n_1 \kappa C(\tau). \tag{19}$$

The number of L_α-quanta emitted in the same unit volume equals $4\pi n_1 k G(t)$. Consequently we have for the condition of radiative equilibrium

$$G(t) = \int K(t,\theta) \frac{d\omega}{4\pi} + q \; \frac{1-p}{p} C(\tau), \tag{20}$$

where the value of the function $C(\tau)$ is taken from (13).

The solution of equations (18) and (20) gives us the values of the functions $G(t)$ and $K(t,\theta)$. Let us construct this solution in the Eddington approximation. As a result we obtain

$$\bar{K}(t) = a - 3bt - \frac{3}{4} \frac{S}{q} e^{-tq}, \tag{21}$$

$$F_\alpha(t) = b - \frac{S}{4} e^{-tq}, \tag{22}$$

where a and b are arbitrary constants and $\bar{K}(t)h\nu_\alpha$ and $4\pi F_\alpha(t)h\nu_\alpha$ are respectively the mean intensity and flux of L_α-radiation, that is

$$\bar{K}(t) = \int K(t,\theta) \frac{d\omega}{4\pi}, \; F_\alpha(t) = \int K(t,\theta) \cos\theta \frac{d\omega}{4\pi} . \tag{23}$$

In determining the constants of integration one must take into account the kinematic state of the nebula. In the particular case when it is motionless (stationary nebular model) or expands with a velocity not exceeding the mean thermal velocity of the atoms, we have the following boundary conditions, analogous to (6) and (7):

$$F_\alpha(0) = 0, \quad 2F_\alpha(t_0) = \bar{K}(t_0). \tag{24}$$

From (24) and (22) we find approximately (for $\tau_0 = tq \gg 1$)

$$a = \frac{3}{4} St_0, \quad b = \frac{S}{4} . \tag{25}$$

Substituting these values in (21) and (22), we find for the functions $F_\alpha(t)$ and $\bar{K}(t)$:

$$\bar{K}(t) = \frac{3}{4} S \left[(t_0 - t) - \frac{1}{q} e^{-tq} \right], \tag{26}$$

$$F_\alpha(t) = \frac{S}{4} (1 - e^{-tq}). \tag{27}$$

According to these formulae we are able to calculate the intensity $\bar{K}(t)h\nu_\alpha$ and the flux $4\pi F_\alpha(t)h\nu_\alpha$ of the L_α-radiation at any optical depth in the nebula.

For the density of L_α-photons we have:

$$\rho_\alpha(t) = \frac{1}{c}\int K(t)\,d\omega = \frac{4\pi}{c}\,\bar{K}(t). \tag{28}$$

We can use (26) and (28) at $t = 0$ to find the number of L_α-photons per cm^3 at the inner boundary of the nebula

$$\rho_\alpha(0) = \frac{\pi S}{c}\cdot 3t_0 \tag{29}$$

But $\pi S/c$ is just the number of L_c-quanta per cm^3 at $t = 0$. Consequently the density of L_α-radiation at the inner boundary of the nebula is $3t_0$ times larger, that is, tens of thousands of times larger than the density of L_c-radiation. In its turn the density of L_c-radiation for star temperatures of 40,000°K or 50,000°K is roughly $5 \cdot 10^4$ times larger than the density of L_α-radiation arriving from the star. Therefore the density of diffuse L_α-radiation at the inner boundary of the nebula is 10^9 times larger than the direct L_α-radiation from the nucleus. This density greatly exceeds in order of magnitude the radiation density in all other lines of the hydrogen spectrum. Nevertheless, if W is of order 10^{-13}, then the density of diffuse L_α-radiation is still 10^4 times smaller than the density of the surface of the stars.

An important case is that in which the nebula expands with a velocity significantly greater than the mean thermal velocity of the atoms. Essentially here again we have a stationary model of the nebula, but with different boundary conditions, signifying that a L_α-photon leaving a point on the inner boundary of the nebula will not be absorbed on the other side because of the Doppler effect. These conditions are:

$$2F_\alpha(0) = -\bar{K}(0), \quad 2F_\alpha(t_0) = \bar{K}(t_0). \tag{30}$$

Then for the constants a and b we will have from (21) and (22),

$$a = \frac{3}{4}\frac{S}{q}, \quad b = \frac{1}{4}\frac{S}{qt_0}. \tag{31}$$

As a result we obtain

$$\bar{K}(t) = \frac{3}{4}\frac{S}{q}\left(\frac{t_0 - t}{t_0} - e^{-tq}\right) \tag{32}$$

and

$$F_\alpha(t) = -\frac{S}{4}\left(e^{-tq} - \frac{1}{qt_0}\right) \tag{33}$$

For the flux of L_α-photons at the inner boundary ($t = 0$) of a nebula which is expanding with a large velocity, we have from (33):

$$F_\alpha(0) = - \pi S \left(1 - \frac{1}{qt_0}\right) = - \pi S \left(1 - \frac{1}{\tau_0}\right) \tag{34}$$

and at the outer boundary, assuming $qt_0 = \tau_0 \gg 1$,

$$F_\alpha(t_0) = \frac{\pi S}{qt_0} = \frac{\pi S}{\tau_0} . \tag{35}$$

For a motionless nebula we had respectively $F_\alpha(0) = 0$ and $F_\alpha(t_0) = \pi S$. Comparing these results with (34) and (35), we see, in the first place, that the L_α-flux is negative at the inner boundary of a rapidly expanding nebula, that is, that the flux is directed inward. Moreover in this case the majority of the L_α-quanta escape from the nebula through its outer boundary, the fraction escaping in this way decreases as the optical thickness τ_0 of the nebula increases. The remaining L_α-photons leave the nebula through the inner boundary. From these assertions it follows that somewhere within a rapidly expanding nebula, the flux of L_α-radiation must vanish. The optical depth of the layer where the L_α-flux changes sign is obviously determined from (33) by the condition $F_\alpha(\tau_1) = 0$ and equals

$$\tau_1 = \ln \tau_0. \tag{36}$$

Thus the character of the L_α-radiation field in a nebula depends strongly on its kinematic state.

3. The Radiation Field in an Expanding Nebula

The results obtained in the preceding section are valid for stationary nebulae, that is, those not moving at all or those expanding with large velocity. But in reality planetary nebulae are gaseous regions expanding with small velocities and the radiation fields in them will differ from those in nebulae at rest. Strictly speaking, the radiation field in a nebula expanding without velocity gradients will be exactly the same as that in a motionless nebula. However, even when the velocity of expansion is constant in all layers of the expanding nebula, there is always a velocity gradient along the line of sight, that, of course, provided the thickness of the nebula is not too small in comparison with its radius. Actually, the expression for the velocity gradient dv/ds in a given direction in an expanding region is

$$\frac{dv}{ds} = \frac{dv}{dr} \cos^2\theta + \frac{v}{r} \sin^2\theta, \tag{37}$$

where dv/dr is the velocity gradient along a radius and θ is the angle between the given direction and the radius vector. If the expansion velocity is constant and equal to v, then

$$\frac{\overline{dv}}{ds} = \frac{2}{3}\frac{v}{r}.$$ (38)

Moreover, even if at some moment at the expansion occurs without velocity gradients, radiation pressure inevitably sets up differences in the expansion velocities of various layers of the nebula. It is clear that for L_c-radiation, gradients in the expansion velocity are of no importance; the field is just the same as for a stationary nebula. But for L_α-radiation it appears that gradients of the expansion velocity have an enormous effect. Therefore, in the following, we limit ourselves to investigating the transfer problem for L_α-radiation in a nebula expanding with velocity gradients. This problem was posed originally by Zanstra [104], but its correct solution was first given by V.V.Sobolev [129, 90]. Without discussing the details of this solution, we quote here his final results.

The effect of gradients of the expansion velocity on the radiation field may be seen from the following argument. When the medium is stationary, that is, when the relative velocity of the atoms is zero (with the subtraction of thermal velocities), then a photon emitted by an atom at point A will be absorbed by an atom at point B, located some distance s from A. However, if the relative velocity of these points is not zero, the frequency of the photon emitted at A will be shifted due to the Doppler effect at B. If the shift is sufficiently large the atom at B will be unable to absorb the photon, which can then escape without hindrance from the medium. In other words, in this case the photon can escape from the interior regions of the nebula, as a consequence of the Doppler effect, as well as from the boundary regions.

The effect of a velocity gradient, therefore, is to reduce the radiation density in the nebula.

Let us denote by β the fraction of quanta leaving the medium because of the Doppler shift resulting from the velocity gradient. Obviously, the parameter β will depend on the kinematic and physical parameters of the medium: the velocity gradient dv/ds, the volume absorption coefficient k and the mean thermal velocity u of the atoms. The form of this dependence may easily be derived from the following considerations. The radiation emerging from point A, after arriving at point B a distance s away, has been attenuated by a factor e^{-ks} and shifted in frequency by an amount

$$v - v' = \frac{v}{c}\frac{dv}{ds}s.$$ (39)

Consequently, of the radiation emerging from A, only the following fraction will be absorbed in the medium:

$$\int_0^\infty e^{-ks}\left(1 - \frac{v - v'}{2\Delta v}\right)k\,ds = 1 - \frac{1}{2u}\frac{dv}{kds}, \tag{40}$$

where $\Delta v = 2(u/c)v$. The upper limit on the integral in (40) has only symbolic meaning; indicating that the product ks (i.e., the optical thickness) becomes substantially larger than unity after a small distance in the nebula.

The expression (40) gives the fraction of the radiation absorbed in the medium. Obviously, the fraction of radiation escaping from the medium will be

$$\beta = 1 - \left(1 - \frac{1}{2uk}\frac{dv}{ds}\right) = \frac{1}{2uk}\frac{dv}{ds}. \tag{41}$$

The problem now lies in introducing β into the equation of radiative transfer. The solution of this transfer equation together with the conditions of radiative equilibrium obviously gives us a representation of the radiation field in a nebula expanding with velocity gradients.

In the first place, substituting $k = n_1\kappa/q$ into (34) we find for β, in the case of L_α-radiation:

$$\beta = \frac{q}{2un_1\kappa}\frac{du}{ds} = \frac{q}{2u}\frac{dv}{d\tau}, \tag{42}$$

where τ is the optical thickness for L_c-radiation of the layer under consideration.

On the assumption that β is constant and positive inside the nebula, the solution of the transfer equation together with the condition of radiative equilibrium gives for the function $G(t)$:

$$G(t) = 3\frac{q}{\beta}\left\{\psi(\tau) - \left(1 - \frac{2}{3}\sqrt{\beta}\right)[\psi(0)e^{-t\sqrt{\beta}} + \psi(\tau_0)e^{-(t_0-t)\sqrt{\beta}}]\right\}, \tag{43}$$

where

$$\psi(t) = \frac{k}{4}Se^{-k\tau}, \quad (k \sim 1). \tag{44}$$

The L_α-flux may be expressed in the following form

$$F_\alpha(t) = 2\pi\left[\int_0^t E_2(t - t')G(t')dt' - \int_t^{t_0} E_2(t' - t)G(t')dt'\right], \tag{45}$$

where

$$E_2(x) = \int_1^\infty e^{-xz}\frac{dz}{z^2}.$$

Deep inside the nebula the formulae (43) and (45) give

$$F_\alpha(t) = -4\pi \frac{q^2}{\beta} \frac{d\psi}{d\tau}. \qquad (46)$$

At the boundaries the same flux is:

$$F_\alpha(0) = -2\pi\psi(0) \frac{q}{\sqrt{\beta}} \qquad (47)$$

and

$$F_\alpha(t_1) = 2\pi\psi(\tau_0) \frac{q}{\sqrt{\beta}}. \qquad (48)$$

Comparing formulae (47) and (48) with (27), on one hand, and with (38) and (26), on the other, we see that the occurrence of velocity gradients in the nebula reduces the flux and density of L_α-radiation by an extremely large factor. To obtain some quantitative estimates of these effects let us suppose, for example, that the velocities $dv/d\tau$ and u are of the same order of magnitude. Then formula (47) gives for the flux of L_α-radiation at the inner boundary a value some hundred times smaller than does formula (34). For the L_α-density deep in the nebula one obtains from formula (43) a value some ten thousand times smaller than that found from formula (32).

In spite of these differences, the L_α-radiation field in a nebula, expanding with a velocity gradient, has the same character as in a nebula expanding with a large constant velocity. In particular, in both cases the L_α-flux is directed inwards at the inner boundary and outwards at the outer boundary. The optical depth τ_1 of the layer where the L_α-flux vanishes is determined in the present case from the following relation, in analogy to (36)

$$e^{-k\tau_1} = e^{-\tau_1(\sqrt{\beta}/q)} + e^{-k\tau_0}e^{-(\tau_0-\tau_1)(\sqrt{\beta}/q)}, \qquad (49)$$

where it assumed that $\sqrt{\beta} \ll 1$.

The qualitative picture of the distribution of L_α-flux throughout the nebula for all cases considered is shown in Figure 6-2. These curves represent approximately the case in which the optical thickness τ_0 for L_c-radiation is of order three.

4. Diffusion of L_α-Radiation

There is still one more effect tending to reduce the density of L_α-radiation in nebulae. This is the so-called effect of redistribution of radiation in frequency, or non-coherent scattering.

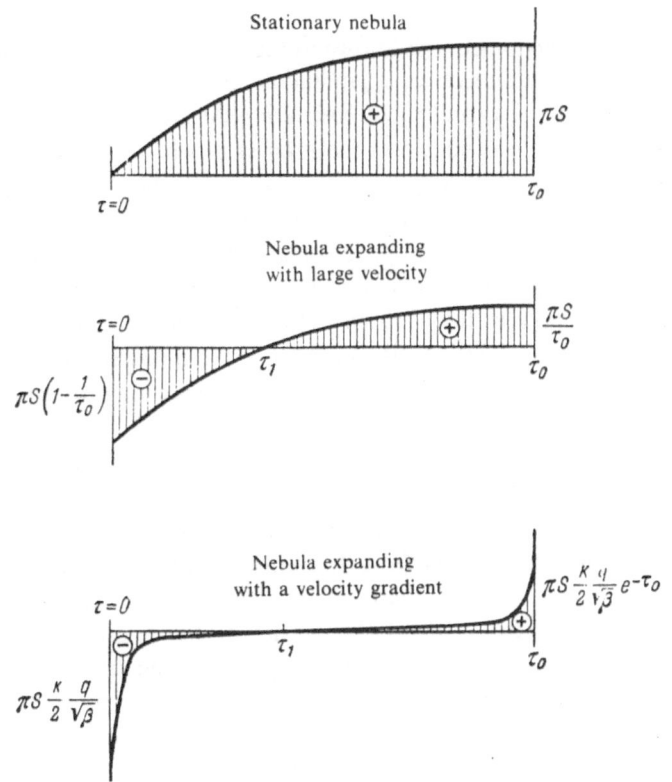

Figure 6-2

Distribution of flux of L_α-radiation throughout the nebula in different kinematic conditions, $\tau = 0$—on the inner boundary; $\tau = \tau_0$—on the outer boundary of the nebula.

The point is that in the problem of resonance scattering of L_α-photons (Section 2) it was assumed that diffusion of L_α-photons occurs without a change of frequency. In other words it was assumed that the absorption coefficient in the line is constant, that is, the line has a rectangular profile.

In reality the contour of the absorption coefficient in nebulae is due to the effects of radiation damping (natural width of the energy levels) and the Doppler effect arising form the thermal motion of the atoms. This contour takes the form of the Gaussian curve of probability distribution (at least when the radiation damping is negligible), rather than the rectangular shape used above. Thus, for example, when an atom absorbs a photon with the central frequency of the line, the photon will in general be emitted in another direction with its frequency slightly displaced from the center of the line (because of the thermal motion of the atoms) to a value where the absorption coefficient is smaller than at the center. In this way a fraction of the L_α-photons will pass from the center of the line into its wings, where the absorption coefficient is small and the probability that the photon escapes from the nebula is large. This phenomenon is known as the escape of quanta in the line wings.

The problem of the diffusion of L_α-radiation may be discussed with either of two assumptions; that of *actual* or of *complete* redistribution of radiation in frequency. In the first case one considers the process in which an elementary volume absorbs photons of one frequency and emits them at other frequencies. However the function characterizing this process is very complex [133] and consequently, one makes the assumption of complete redistribution in frequency. In this case it is assumed that the probability of emitting a photon of one frequency does not depend on the frequency of the absorbed photon. And, although this assumption is not entirely justified, it simplifies the problem very considerably since the energy emitted by an elementary volume is distributed in frequency proportionally to the absorption coefficient within the spectral line.

The problem of L_α-radiation diffusing with complete redistribution in frequency was investigated by Zanstra [130]. He assumed that the profile of an emission line is due to the thermal motion of atoms according to the Maxwell distribution. In this case the L_α-photons can escape from the central regions of the nebula by emission in the line wings where the optical thickness is small. According to Zanstra's calculations, this effect greatly reduced the density of L_α-radiation in nebulae from the values obtained for diffusion without change of frequency.

For a number of particular cases calculations or approximate solutions of diffusion of L_α-radiation in nebulae with complete redistribution have been given by Koelbloed [132]* and also by Miyamoto [132].

However, as was mentioned above, the assumption of complete redistribution is not in complete agreement with reality, while the real law of redistribution, which results from radiation damping and the thermal motion of the atoms, is very complex. Nevertheless, the investigation of the problem for the actual law of redistribution is of considerable interest.

This problem was studied by Unno [134] and V.V.Sobolev [135]. In particular it was shown that the values of the various physical quantities, such as density and flux did not differ greatly from the values obtained with the assumption of complete redistribution. Therefore the solution of the problem of diffusion of radiation with complete redistribution may be sufficient for many purposes.

The importance of redistribution in frequency in the diffusion of L_α-radiation can be decided more accurately after it is known which of the following two effects is more important in nebulae: the escape of photons from the nebula due to a velocity gradient or the escape of quanta in the line wings. Such an analysis has been performed by V.V.Sobolev [76]. To solve this problem, obviously one must de-

* Calculations performed on a computer.

termine the fraction of L_α-photons escaping from a given point because of the velocity gradient and the fraction escaping in the wings and compare them.

In the first case (effect of a velocity gradient) we have from (43) for the central parts of a nebula

$$G(t) = 3 \frac{q}{\beta} \psi(\tau), \quad (\tau = qt). \tag{50}$$

Here, obviously, $\beta/3q$ is the fraction of L_α-photons escaping from a given point in the nebula because of the effect of the velocity gradient.

In the second case (escape of quanta in the line wings) we assume that no velocity gradient exists and that the absorption coefficient in the line may be represented in the form

$$k_\nu = k_0 e^{-[(\nu - \nu_0)/\Delta\nu_0]^2}, \tag{51}$$

where $\Delta\nu_0$ is the Doppler half-width of the line. With the assumption of complete redistribution, the emission coefficient in the line is

$$\mathscr{E}_\nu = \mathscr{E}_0 e^{-[(\nu - \nu_0)/\Delta\nu_0]^2}$$

The function $G(t)$ for the present case is determined in the same way as in the case of diffusion of L_α-radiation in the presence of a velocity gradient, that is, by means of the approximate solution of the integral equation. By such a method one obtains for the central regions of the nebula [37]

$$G(t) = \frac{2\psi(\tau)q}{L(t) + L(t_0 - t)}, \tag{52}$$

where

$$L(t) = \frac{1}{\sqrt{\pi}} \int_{-\infty}^{\infty} e^{-x^2} E_2(te^{-x^2}) dx \ .$$

For large values of t we have approximately (for a three-dimensional medium),

$$L(t) \sim \frac{1}{2t\sqrt{\pi \ln t}} . \tag{54}$$

The quantity $1/2q \cdot [L(t) + L(t_0 - t)]$ in (52) obviously represents the fraction of the photons at the given point escaping in the line wings. Therefore, comparing (50) to (52), we may assert that when the inequality

$$\frac{\beta}{3} \ll \frac{1}{2}[L(t) + L(t_0 - t)] \tag{55}$$

is satisfied, the relative role of the velocity gradient will be small. It should be noted that if this inequality is satisfied in the central layers of a nebula, it will also be satisfied in the outer layers.

By assuming $t = t_0/2 = \tau_0/2q$ in (55) and utilizing (54), we obtain

$$\frac{1}{u}\frac{dv}{d\tau} \ll \frac{1}{\tau_0},$$

where we have used the value of β from (42). On the other hand we have $d\tau = n_1\kappa dr$ and $\tau_0 = n_1\kappa\Delta r$, where Δr is the thickness of the nebula. From (56) we find:

$$\frac{1}{u}\frac{dv}{dr} \ll \frac{1}{r}, \tag{57}$$

or approximately

$$\frac{v}{u} \ll \frac{r}{\Delta r}. \tag{58}$$

The expansion velocity v of the nebula is of the same order as the mean thermal velocity u and the thickness of a nebula amounts to a few tenths of its radius. In other words, the quantities v/u and $r/\Delta r$ in planetary nebulae are of the same order and therefore the inequality (58) is not satisfied for these objects. In this case both effects under discussion—the escape of quanta due to velocity gradients and the escape of quanta in the line wings—play approximately equal roles.

5. Relaxation Time

In view of the large density of L_α-radiation and the large optical thickness of the nebula in the L_α-line, we shall discuss some questions of interest in connection with the physical processes occurring in nebulae.

The large density of L_α-radiation may lead to the number of $1 \to 2$ transitions in hydrogen becoming large and hence to a greatly increased population n_2 of atoms in the second state. The density of diffuse L_α-radiation, as we have seen above, exceeds the density of the direct radiation from the star in this line by 10^9 times on the average for a motionless model of the nebula and by 10^7 times for a model expanding with a velocity gradient. Consequently, for the degree of excitation of the second state we will have (for the second model)

$$\frac{n_2}{n_1} \approx 10^7 \frac{W}{e^{h\nu/kT_*} - 1}. \tag{59}$$

For $T_* = 40,000°K$ and $W = 10^{-13}$ we have

$$\frac{n_2}{n_1} \sim 10^{-7},$$

that is, the degree of excitation of the second level is some million times larger than in normal conditions, when the population of the levels is determined by the processes of photo-ionization and subsequent recombination. (see Chapter III).

The question arises whether such a high degree of excitation leads to the nebula becoming opaque for radiation in the lines of the Balmer series.

Having denoted by t_i the optical thickness of the nebula in some line of the Balmer series, we will have

$$t_i = n_2 s_i \Delta r = 10^{-7} n_1 s_i \Delta r,$$

where s_i is the absorption coefficient in the lines of the Balmer series. For Doppler broadening corresponding to an electron temperature of 10,000°K, we have $s_i \sim 10^{-12} - 10^{-13}\,\mathrm{cm}^2$ for the first lines of the series. If we take $n_1 \approx 1\,\mathrm{cm}^{-3}$ and the thickness of the nebula $\Delta r = r_2 - r_1 \approx 10^{17}$ cm, we find $t_i \approx 0.01$ and less. Hence, one may conclude that real planetary nebulae, in which the effects of velocity gradients and redistribution of radiation are present, are in practice transparent to the lines of the Balmer series of hydrogen.

For a motionless model of a nebula, however, one obtains a somewhat different result. In this case $n_2/n_1 \sim 10^{-5}$ and $t_i \sim 1$.

In all probability, diffuse nebulae have optical thickness of order unity in the Balmer lines of hydrogen primarily because of their large linear thickness Δr. Since the optical thickness decreases rapidly in passing from the early lines of the series to the later lines, this fact will lead to an increase of the ratio H_α/H_β in diffuse nebulae in comparison with the "pure case" of the usual theoretical model. It is possible that in this way one can account for the deviations (even after correcting for the effect of interstellar selective absorption) in several nebulae of the quantity H_α/H_β from the theoretically expected value. Physically, fluorescence operates in establishing the ratios of the number of photons in H_β, H_γ and other lines; because of the opacity of the nebula, for example, in the line H_β a fraction of the photons in this line will be absorbed and some of these will be re-emitted in H_α and P_α. As a consequence the intensity of the H_β line will be reduced and the ratio H_α/H_β increased. As for the H_α-photons, their total number cannot decrease (if the absorption by the dust component in diffuse nebulae is neglected) no matter how large the optical thickness of the nebula in this line, for H_α-photons cannot be converted into photons of other frequencies under the conditions found in gaseous nebulae.

For the total number of L_α-photons in a nebula we may write on the basis of (28):

$$4\pi r_1^2\, \frac{4\pi}{c} \int_0^{t_0} \bar{K}(t)\, \frac{dt}{n_1 k},$$

or substituting the value of $\bar{K}(t)$ from (20), (23), and (43), which were derived allowing for the effect of velocity gradients, and performing the integration, we find approximately (for $\tau_0 > 1$, $dv/dt \sim u$):

$$\frac{4\pi r_1^2}{cn_1 k}\frac{3}{\beta}\pi S.$$

As for the number of L_α-quanta leaving the nebula, it is obviously equal to the number of L_c-quanta falling on the inner boundary of the nebula from the central star. But this number is just $4\pi r_1^2 S$. Therefore, allowing for the effect of a velocity gradient, we shall have for the mean time T_1 a photon remains in the nebula

$$T_1 = \frac{3}{n_1 kc}\frac{1}{\beta}. \tag{60}$$

It is of interest to observe that in this case, when one takes account of only the effect of a velocity gradient, the mean life-time of a L_α-photon diffusing in the nebula does not depend on its total optical thickness t_0. Assuming in (60) that $n_1 = 1 \text{ cm}^{-3}$, $k = 10^{-12} \text{ cm}^2$ and $\beta \sim q \sim 10^{-4} - 10^{-5}$, we find that the mean time for a L_α-quantum to remain in the nebula is on the order of several months.

Knowing T_1, one may determine the number of scatterings experienced by L_α-photons before emerging from the nebula. Obviously, to determine this number one must divide the quantity T_1 by the average interval of time between two successive scatterings. The mean distance between two successive scatterings equals $1/n_1 k$ and the mean interval is then $1/n_1 kc$. Therefore, having divided T_1 by $1/n_1 kc$, we find for the mean number of scatterings

$$N_1 \approx 1/\beta, \tag{61}$$

that is, each photon undergoes on the average several tens or hundreds of thousands of scatterings before it escapes from the nebula.

Analogous formulae can be derived as well for the case in which the density of L_α-radiation is determined by the effect of redistribution in frequency. From (52) and (54) one can write for the number $N_2(t)$ of scatterings, which incidentally depends on the optical thickness of the nebula,

$$N_2(t) = 2t\sqrt{\pi \ln t}. \tag{62}$$

For the time of diffusion of L_α-quanta we have:

$$T_2 = \frac{2t\sqrt{\pi \ln t}}{n_1 kc}. \tag{63}$$

With $t \sim 10^4 - 10^5$ and the same values of the remaining quantities we find for T a value on the order of one month to a year.

Finally, for a motionless nebula without the effect of redistribution of radiation in frequency, we have respectively

$$T_3(t) = \frac{3}{2} \frac{t^2}{n_1 kc} \tag{64}$$

and

$$N_3(t) = \frac{3}{2} t^2, \tag{65}$$

which gives $T_3 \approx 10^3$ years and $N_3 \approx 10^9$ for $t = 10^4 - 10^5$. The formulae (64) and (65), incidentally, may be applied to interstellar matter and to some extent to diffuse nebulae, since the physical conditions in them, one must assume, most likely approach the model of a motionless nebula.

Thus, even for motionless nebulae the time during which a L_α-photon remains in the nebula, that is, the relaxation time, is significantly smaller than the lifetime of the nebula. Therefore, one may assume that the observed nebulae are in a state of radiative equilibrium.

6. Radiation Pressure in Nebulae

The magnitude and character of the flux distribution in a nebula depend strongly on which kinematical model of the nebula is investigated. In a motionless nebula the flux of L_α-radiation, given by formula (27), is directed outwards and increases from the inner to the outer parts of the nebula. In this case the L_α-flux at the inner boundary is equal to zero and at the outer boundary equal to $+ \pi S$. If the nebula expands with a velocity gradient, then the flux at the inner boundary is directed inward and equals $-\pi S(q/\sqrt{\beta})$ and at the outer boundary is directed outward, with a value at $\tau \sim 1$ of approximately $\pi S(q/\sqrt{\beta})$ [see formulae (47) and (48)]. Since the magnitude of $q/\sqrt{\beta}$ is of order 10^{-2}, the flux at the outer boundary of the nebula is approximately one hundred times smaller than in the stationary case. Nevertheless in both cases the L_α-flux at the outer boundary is very great and consequently one must expect that the radiation pressure caused by L_α-radiation will also be great. The problem lies in determining by how much the radiation pressure will exceed the gravitational attraction of the central star.

Let us investigate first the case of a motionless nebula and determine the ratio of L_α-radiation pressure to the force of gravity. We take a unit volume near the outer boundary of the nebula and calculate the force due to radiation pressure on this volume. Obviously only those atoms which can absorb L_α-photons, that is, only ground state atoms, the number of which is n_1, will experience directly the force of L_α-radiation pressure. Subsequently this force is distributed by collisions and absorptions among all particles in the unit volume.

The radiation pressure due to L_α-radiation and acting on a unit volume will be equal to

$$R = \frac{n_1 k}{c} \pi S h v_c,$$ (66)

where $\pi S h v_c$ is the intensity of L_c-radiation falling on 1 cm^2 of the surface of the inner boundary of the nebula. For the determination of πS we have

$$4\pi r_2^2 \pi S = 4\pi r_*^2 \frac{2\pi}{c} \int_{v_0}^{\infty} \frac{v^2 dv}{e^{hv/kT_*} - 1}.$$ (67)

Consequently we obtain for R:

$$R = \left(\frac{r_*}{r_2}\right)^2 \frac{2\pi n_1 k h v_c}{c^3} \int_{v_0}^{\infty} \frac{v^2 dv}{e^{hv/kT_*} - 1},$$ (68)

where r_* is the radius of the central star.

In distinction from radiation pressure, which acts only on neutral atoms, the gravitational force of the central star acts on both neutral and ionized atoms, the concentration of which is n^+. Consequently we have for the gravitional force

$$G = \left(\frac{r_*}{r_2}\right)^2 g_* m_H (n_1 + n^+),$$ (69)

where g_* is the acceleration of the gravitional force on the surface of the star.

Let us form the ratio R/G. On the basis of (68) and (69) we have

$$\frac{R}{G} = \frac{2\pi k h v_c}{m_H c^3 [1 + (n^+/n_1)] g_*} \int_{v_0}^{\infty} \frac{v^2 dv}{e^{hv/kT_*} - 1}.$$ (70)

The value of k is determined by the Doppler halfwidth of the L_α-line for thermal velocities of order 10 km/sec. Assuming that $T_* = 40,000°K$ and $n^+/n_1 = 5000$, we find from (70):

$$R/G = 10^9/g_*.$$

Although we know very little about the nature of the nuclei of planetary nebulae, it is nevertheless difficult to believe that the surface gravity of these stars reaches 10^9 cm/sec^2. For this it would be necessary for the central star to be several thousand times more massive than the sun. But if this were the case, one would observe a red shift in the absorption lines in the spectrum of the nucleus corresponding to velocities of the order of 10,000 km/sec. Although observations give indications of a red shift in the spectra of nuclei, it is not more than 100 km/sec, which indicates that the masses of nuclei, in fact, are only approximately ten solar masses.

Thus we must conclude that the outer layers of the nebula are not in equilibrium and therefore that they recede under the influence of the repulsive force of radiation pressure, leading to the expansion of the nebula.

However, we have seen above that the L_α-flux at the outer boundary of a real nebula expanding with a velocity gradient is two orders of magnitude less than for a stationary nebula. Therefore the relation (71) in this case may be written in the form

$$R/G = 10^7/g_*. \tag{72}$$

A surface gravity of order 10^7 cm/sec^2 for the nucleus, although possible, is improbable. In some way, the L_α-radiation pressure in real nebulae expanding with a velocity gradient no longer plays as large a role as in the case of a motionless nebula. This does not exclude the possiblity that in isolated cases the force of radiation pressure may be comparable with the gravitational force of the central star.

The fact that planetary nebulae do expand with a velocity noticeably exceeding the thermal velocity of the atoms does not essentially contradict the idea that radiation pressure plays a vital role in them. However, the initial force, leading to the evolution of a nebula from the nucleus and its expansion, may be of a different nature.

In the above analysis of the role of L_α-radiation in a planetary nebula, it was assumed that the quantity β was constant throughout the nebula. In other words the problem was investigated with the assumption of a constant degree of ionization inside the nebula. Now the degree of ionization is not only not constant throughout the nebula; it also changes as the nebula expands.

A calculation of the degree of ionization varying throughout the nebula and an analysis of the radiation field in the inner parts of the nebula lead to interesting new consequences. In particular this work reveals a way to explain the formation of a second shell around the primary shell of the nebula at a definite stage of its evolution.

7. The Expansion of Nebulae

In the preceding section we have seen that planetary nebulae do not exist in a state of mechanical equilibrium and that they must expand under the influence of L_α-radiation pressure. This theoretical result appears to agree with the observational data; the measured expansion velocities of nebulae are found to exceed considerably the escape velocities of central stars at the boundary of the nebula (Chapter I).

In this connection the question arises: what would be the expansion velocity of a nebula if its motion were due only to the action of L_α-radiation pressure. One may expect two results. In the first case the expansion velocity will be very large if, in reality, the L_α-radiation pressure exceeds the gravitational attraction of the central star to the extent obtained for stationary nebulae [formula (71)]. In the second case the expansion velocity will be comparatively small, if the radiation pressure is actually only slightly larger than the gravitational attraction [formula (72)].

It is necessary to remark that L_α-photons are produced inside the nebula with their momenta in all directions, with the consequence that their mean momentum is zero. Therefore, the nebula as a whole can neither expand nor shrink from the action of L_α-radiation. But since the mean momentum of L_α-photons is directed outward near the outer boundary of the nebula and inward at the inner boundary, the nebula must expand intrinsically in both directions. In other words, the L_α-radiation pressure leads only to an increase in the linear thickness of a nebula. And when one speaks about the expansion of a nebula under the influence of L_α-radiation pressure, this refers only to the outer layers.

The nebula must also experience the pressure of L_c-radiation from the central star. However, in distinction from L_α-photons, the mean momentum of L_c-photons in the nebula is different from zero and is directed from the star to the nebula. Therefore the effect of L_c-radiation must be to increase the expansion velocity of the nebula as a whole. The difference between the L_c-fluxes at the inner and outer boundaries of the nebula, caused by absorption of L_c-radiation, will produce differences in the expansion velocities at these boundaries. This difference will tend to decrease the linear thickness of the nebula. However, from the available data it appears that the effect of L_c-radiation in decreasing the thickness of the nebula is considerably smaller than the effect of L_α-radiation in increasing the thickness.

Let us estimate first of all the expansion velocity of the outer layer of the nebula under the influence of L_α-radiation.

For the force on a unit volume of the nebula in its outer regions due to L_α-radiation pressure, one may write

$$P_\alpha = \frac{1}{c} \int n_1 k_v F_v dv, \qquad (73)$$

where F_v is the flux of L_α-radiation, k_v is the absorption coefficient per ground state atom and n_1 is the number of ground state hydrogen atoms per unit volume. If we assume that $n_1 \ll n^+$, we have for the equation of motion of the mass contained in a unit volume,

$$n^+ m_H \frac{dv}{dt} = P_\alpha, \qquad (74)$$

where n^+ is the number of ions per unit volume. On the basis of (74), we can write for the acceleration of the outer layers of the nebula:

$$\frac{dv}{dt} = \frac{P_\alpha}{n^+ m_H}.$$ (75)

Zanstra [136], using Koelbloed's solution of the integral equation for diffusion of L_α-radiation with complete redistribution in frequency, found that the expansion velocity of the outer layers of the nebula increases by 18.7 km/sec after 10,000 years with $t_0 = 10^4$ and by 6.7 km/sec in the same time with $t_0 = 10^2$, where t_0 is the optical thickness of the nebula in the center of the L_α-line.

Keeping in mind that in planetary nebulae both effects—redistribution in frequency and velocity gradients—play approximately equal roles in determining the character of the L_α-radiation field, we can say that, if a velocity gradient exists in the nebula, the expansion velocity of the outer layers will increase by nearly 10 km/sec after 10,000 years, that is, in a period comparable with the lifetimes of planetary nebulae. This estimate agrees with the observational data.

If one now proceeds from the value of F_α obtained for a stationary nebula, neglecting the redistribution in frequency and the influence of velocity gradients, then the increase of the expansion velocity will be on the order of 1000 km/sec after 10,000 years. In other words, in this case we would observe a planetary nebula of average dimensions with an expansion velocity comparable in magnitude with that of the shells of novae in their outburst. Moreover, the expansion velocity would have to increase with the size of the nebula. Observations do not confirm these inferences. Therefore, one may conclude that in real planetary nebulae the L_α-radiation pressure actually exceeds the gravitational attraction of the central star, but not to the extent obtained with the stationary model of a nebula.

Let us turn to the investigation of the dynamical effects of L_c-radiation. The L_c-radiation pressure, as was noted above, leads to an increase of the expansion velocity of the nebula as a whole. If one assumes that the nebula absorbs all of the stellar radiation beyond the Lyman limit, then the change of the expansion velocity is determined by the equation

$$\mathfrak{M} \frac{dv}{dt} = \frac{1}{c} E_c,$$ (76)

where E_c is the energy radiated in the Lyman continuum per second and $\mathfrak{M}$ is the mass of the nebula. From (76) we find:

$$v^2 - v_0^2 = \frac{2E_c}{c\mathfrak{M}} (r - r_0),$$ (77)

where v_0 is the value of the expansion velocity of the nebula with radius r_0. A calculation by I.N. Minin [137], performed according to formula (77), showed that when the radius of a nebula becomes of order 10,000 a.u. the expansion velocity of the nebula increases by approximately ten times in comparison with the original velocity.

The question of the L_c-radiation pressure in a nebula was studied also by Zanstra [136], but from a different point of view. We have for the total number of recombinations per unit volume per unit time $n_e n^+ \Sigma_1^\infty C_n(T_e)$, where $C_n(T_e)$ is the coefficient for electrons going directly into the n-th level (see Chapter II). The total energy radiated per unit volume in L_c-radiation is obviously $h\nu_0 n_e n^+ \Sigma_1^\infty C_n(T_e)$, where $h\nu_0$ is the mean energy of a L_c-photon, and the total momentum is $(h\nu_0/c) n_e n^+ \Sigma_1^\infty C_n(T_e)$. Using the fact that in each unit volume of the nebula the number of ionizations equals the number of recombinations, then, having written the equations of motion of the mass, one may derive the following relation for the change of velocity Δv of the nebula after a time Δt (acceleration):

$$\frac{\Delta v}{\Delta t} = n_e \frac{h\nu_0}{cm_{\mathrm{H}}} \sum_1^\infty C_n(T_e). \tag{78}$$

Assuming $n_e = 5 \times 10^3 \, \mathrm{cm}^{-3}$ and $T_e = 10{,}000°\mathrm{K}$, we find from (78) that the expansion velocity of the nebula increases by 3.2 km/sec after 10,000 years.

Thus the pressure of L_c-radiation exerts on the expansion of the whole nebula an effect of the same order as or slightly smaller than that which the L_α-pressure exerts on the outer layers, although L_α-radiation is completely absorbed in the nebula and L_c-radiation, only partially. This is explained by the fact that the absorption coefficient per hydrogen atom for L_c-radiation is significantly smaller than in the L_α-line.

8. The Effect of Thermal Expansion. "Vacuum" in the Centers of Nebulae

There is still one more factor leading to the expansion of nebulae, the diffusion of heated gas. Even a static, non-expanding nebula will with time increase its dimensions because of the diffusion of gaseous matter into space with a velocity on the order of the thermal velocity of hydrogen atoms. The latter has a value of order 10 km/sec. Therefore $v_0 = 10$ km/sec will be the minimum value of the expansion velocity of nebulae. However the real expansion velocity is of order 20 km/sec, that is, somewhat different from the thermal velocity. But the existence of ring-shaped planetary nebulae, that is, of formations half empty inside, is incompatible with their relatively slow expansion [138].

The point is that if the nebula was formed by discrete outbursts or emissions of gas from the nucleus in the form of spherical shells, then there is no reason why inward expansion of the inner boundary should not occur. As a result, the thickness of the shell must increase from both the inside and the outside. If, for example, the original expansion velocity of a shell is 10 km/sec, then the actual velocity at the outer layers of such a shell would be 20 km/sec and at the inner layer, zero. In other words, as the outer edge advances farther from the center, the inner edge remains motionless and will always be found in the center. As a result we would obtain, not a ring-shaped nebula, but one filling the whole volume enclosed by its outer boundary.

In reality the expansion velocity of planetary nebulae are larger than 10 km/sec; therefore a "vacuum" must be formed in the center of planetary nebulae. The whole problem lies in the fact that observations give a significantly larger value for the ratio r_1/r_2 (r_1 and r_2 are the radii of the inner and outer boundaries of the ring, respectively) than we might have expected from the "free" expansion of the nebula. Let us analyse the problem somewhat more thoroughly.

Let v_0 be the observed velocity of the "hollow" inside the nebula. Obviously v_0 represents the sum of the actual expansion velocity v and the thermal velocity u of the hydrogen atoms. Let us assume v and u constant over some interval of time. Then for the equations of motions of the outer and inner boundaries of the nebula, relative to the center, we will have

$$r_2 = (v + u)t + h = v_0 t + h,$$

$$r_1 = (v - u)t \qquad = (v_0 - 2u)t, \tag{79}$$

where h is the initial thickness of the shell. For sufficiently large values of t, from (79) we obtain approximately

$$\frac{r_1}{r_2} = \frac{v_0 - 2u}{v_0}. \tag{80}$$

Let us assume the electron temperature is, on the average, the same for all planetary nebulae. The expansion velocity v_0 may be different for various nebulae. Then we will have some theoretical relation between r_1/r_2 and v_0, given by the expression (80). In Figure 6-3 appears a curve showing this relation for the case $u = 10$ km/sec. In this figure the points represent the data for 13 planetary nebulae for which the expansion velocity v_0 and the ratio of radii r_1/r_2 are known. The first are taken from Table 1-6 (Wilson's measurements), the second, from the estimates of Curtis and from Wilson's measurements.

From Figure 6-3 it appears that the observations give significantly larger values for r_1/r_2 and therefore significantly smaller values for the relative thickness of the nebulae than follow from the "freely" expanding nebula—almost all of the points lay above the theoretical curve. And what is more, none of the nebulae for which the expansion velocity is less than 20 km/sec could have a "vacuum" in their centers. Nevertheless they make up almost half of the nebulae represented in Figure 6-3 (to the left of the broken line).

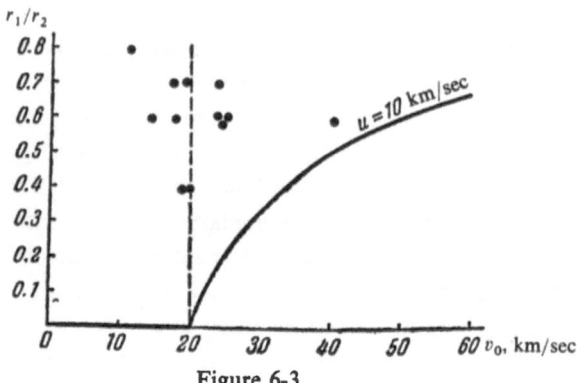

Figure 6-3

Theoretical curves of the dependence of r_1/r_2 on the expansion velocity v_0 for "free" expansion of the nebula. The points represent the observations of particular nebulae.

One can attempt to explain the formation of a "vacuum" of high electron temperature in the central region of a planetary nebula (see Section 5, Chapter IV). In fact, because of the high electron temperature, the gas pressure in the central region would be larger than in the outer part (where the electron temperature is two or three times smaller), if the gas density were everywhere the same. Then the gas would be ejected from the central region, with the result that the gas density there decreases.

Let us assume that the boundary dividing the "vacuum" from the shell itself is in a state of equilibrium. This is possible only when the gas pressure on the boundary from both sides—on the shell side and on the "vacuum" side—is equal. Neglecting the effect of radiation pressure, we have at the boundary

$$n_1 T_{e_1} = n_2 T_{e_2}, \tag{81}$$

where T_{e_1} and T_{e_2} are the electron temperatures in the "vacuum" and in the shell, respectively. The latter is of order of magnitude 10,000°K.

However, to judge from all the data (the brightness and therefore the matter density increases inside the inner boundary in the majority of ring-shaped nebulae), the inner boundary does not exist in a state of equilibrium, but travels outwards with some acceleration. In other words at this boundary we have the condition

$$n_1 T_{e_1} > n_2 T_{e_2}. \tag{82}$$

Hence we may write for the temperature of the "vacuum" region:

$$T_{e_1} > \frac{n_2}{n_1} T_{e_2}. \tag{83}$$

In the last section of Chapter VII we will show that $(n_2/n_1) \sim 3$ for some ring-shaped nebulae. Therefore for the electron temperature in the central regions of the nebulae we have:

$$T_{e1} > 30,000°K.$$

Thus we obtain still another demonstration of the existence of high electron temperatures in the central parts of planetary nebulae. Actually, however, taking account of the L_α-radiation pressure, directed into the nebula (on the inner boundary) one must decrease the degree of inequality in (82), as a result of which the electron temperature in the central parts of the nebula must not be much in excess of 30,000°K.

9. Deceleration of Nebulae

Planetary nebulae must also be subject to decelerating forces as they expand. Such a force, for example, is that due to the inward-directed (against the motion) L_α-flux found in the inner and central parts of an expanding nebula. Consequently, as the nebula expands, the outer layers of the nebula receive a positive acceleration, while the central and inner layers experience a negative acceleration, which tends to increase the depth of the nebula. The whole question, then, is just how effective this mechanism of deceleration will be.

In addition to radiation pressure, there are other reasons for the retardation of the nebular expansion, which we consider below.

The problem of deceleration occupies a particular place in the dynamics of planetary nebulae because of its direct relation to the problem of their origin. The point is that planetary nebulae are sometimes considered to be the result of ejections of gaseous matter from the central stars in the same way as is observed in the flareup of novae. But in the known cases of observed flare-ups, for example, in novae and Wolf-Rayet stars, the expansion velocity of the shell, judging from the broadening of the spectral line, is of order 1000 km/sec. On the other hand in planetary nebulae it is of order 10 km/sec. In this connection some alternatives arise, the resolution of which is of vital importance in explaining the mechanisms for the origin of planetary nebulae. Two hypotheses are possible:

a) the original velocity of ejection and the velocity of the subsequent expansion of planetary nebulae are quantities of the same magnitude;

b) the original velocity of ejection is very large, of order 1000 km/sec. Subsequently the velocity decreases because of the retarding effect of a number of factors acting on the nebula during its expansion.

Since up to the present the second alternative has received the most attention, we shall consider it first.

In the hypothetical explanation of the origin of planetary nebulae by means of the ejection of gaseous matter from novae and Wolf-Rayet stars, it is assumed that the original large expansion velocity subsequently decreases because of some retarding forces. Possible mechanisms of retardation are: gravitational deceleration, drag from the interstellar medium and radiation pressure. We shall dwell on these mechanisms in some detail.

(1) *Gravitational deceleration.* In order that ejection by impulsive forces takes place, it is necessary that the original velocity of the shell exceed the escape velocity, that is, the parabolic velocity at the surface of the star. Although the physical parameters of the nucleus of planetary nebulae are at present known with insufficient accuracy, there is reason to assume that the parabolic velocity for them is extremely high (of order 1000 km/sec). Then if the shell is detached suddenly from the surface of the star (ejection), its velocity must exceed this value.

At distances close to the star the shell is subject to deceleration as the result of gravitational attraction by the central star. Because of this the expansion velocity of the star is somewhat diminished. The dependence of velocity on distance from the star is given by the relation

$$v(r) = \sqrt{ v_0^2 - \left(1 - \frac{R}{r}\right) v_p^2 }, \tag{84}$$

where $v(r)$ is the velocity of the shell at a distance r from the center of the shell, v_0 is the original velocity, v_p is the parabolic velocity at the surface of the star and R is the radius of the star.

For large values of r $(r \gg R)$, the velocity of the shell approaches a constant value $v(\infty)$, given by

$$v(\infty) = \sqrt{ v_0^2 - v_p^2 }. \tag{85}$$

Obviously, the difference $v_0 - v(\infty)$ is the maximum change of velocity produced by the gravitational deceleration. If, for example, $v_0 = 1500$ km/sec and $v_p = 1000$ km/sec, then the minimum velocity of the shell will be equal to 1120 km/sec. Hence it appears that although the original velocity is somewhat reduced, it is nevertheless far in excess of the usual expansion velocities of nebulae. In order that $v(\infty)$ be of order 20 km/sec, it is necessary that v_0 be very close to v_p, which, as has been remarked earlier, is quite unnatural. Consequently gravitational deceleration must be considered insufficient to explain the small expansion velocities observed in planetary nebulae.

(2) *Deceleration by the interstellar medium.* The expansion of a gaseous shell, generally speaking, occurs in the interstellar medium which has a definite and far from negligible density. This medium will offer resistance to the moving shell and thereby decrease its velocity.

The mechanism of deceleration, first investigated by Oort [199], involves the following ideas. As the shell expands, the particles of the interstellar medium fall on it from all sides. These particles must be absorbed in the shell itself, since the thickness of the shell significantly exceeds the mean free path of the particles within it. At the same time part of the momentum of the shell is transferred to these new particles. The effect of these interstellar particles is then to decrease the expansion velocity of the shell and to increase its mass.

Let us write down the condition of conservation of momentum for the shell. We have

$$v_0 \mathfrak{M}_0 = v(r) \mathfrak{M}_0 + v(r) \frac{4\pi}{3} r^3 \rho_0, \qquad (86)$$

where $\mathfrak{M}_0$ and v_0 are the original mass and expansion velocity of the shell, ρ_0 is the density of the interstellar medium and $v(r)$ is the expansion velocity of the shell at radius r. The first term on the right side of the equation refers to the shell and the second, to the additional mass received from the medium.

From (86) we have for $v(r)$

$$v(r) = v_0 \frac{1}{1 + \sigma r^3}, \qquad (87)$$

where

$$\sigma = \frac{4\pi}{3} \frac{\rho_0}{\mathfrak{M}_0} \qquad (88)$$

Formula (87) gives the law of change (decrease) of the expansion velocity of the shell due to the resistance of the interstellar medium.

From (87) it is evident that a significant change in velocity can occur only when σr^3 is comparable with unity. For what value of the radius r_0 of the shell is this condition able to occur? From $\sigma r_0^3 = 1$ we find

$$r_0 = \left(\frac{4\pi}{3} \frac{\mathfrak{M}_0}{\rho_0} \right)^{-1/3} \qquad (89)$$

Taking $\rho_0 = 3 \cdot 1.67 \cdot 10^{-24} = 5 \cdot 10^{-24}$ gm/cm^3, we obtain:

$$r_0 = 70{,}000 \text{ a.u. for } \mathfrak{M}_0 = 0.01 \ \mathfrak{M}_\odot$$

$$r_0 = 140{,}000 \text{ a.u. for } \mathfrak{M}_0 = 0.1 \ \mathfrak{M}_\odot.$$

From these findings we are able to reach the following conclusions: the expansion of gaseous shells (nebulae) is indeed decelerated by the interstellar medium, but it can be effective only for nebulae of extremely large dimensions—for radii of order 100,000 a.u. This mechanism plays no role on the original state ($r \ll r_0$) of ejection of gaseous shells.

(3) *Deceleration by radiation pressure.* For a long time L_α-radiation pressure appeared to be the only significant method of deceleration of gaseous shells. In the central and inner parts of the expanding shell the pressure is directed inward, which causes retardation of the motion.

However, in a correct formulation of the problem, as we have seen above, the acceleration due to L_α-radiation pressure is comparable in magnitude with that of the force of gravity. Consequently the deceleration caused by radiation pressure produces an effect of the same order as the gravitational deceleration, that is, in reality quite insignificant. Essentially the same result was obtained both for planetary nebulae and for shells of small dimensions [90]. Therefore we may assert that radiation pressure is not effective even in the original state of expansion of the nebula and consequently that a velocity of the order 1000 km/sec cannot be reduced by two orders of magnitude in this way.

One can show that the relative role of radiation pressure in shells of small diameters is larger than for large diameters. This is seen from the following argument. For the flux of L_α-radiation near the inner boundary of the nebula [formula (47)] we have:

$$F_\alpha(0) = - 2\pi\psi(0) \frac{q}{\sqrt{\beta}}, \tag{90}$$

where the quantity β, appearing as a function of r, is expressed in the form

$$\beta = \frac{q}{2u} \frac{dv}{kds}. \tag{91}$$

Neglecting the change of velocity along the radius of the nebula and substituting $dv/ds = \overline{dv}/ds = (2/3)v/r$ from (38) into (91):

$$\beta = \frac{q}{3u} \frac{v}{kr} \approx \frac{q}{3u} \frac{v}{\tau}, \tag{92}$$

where τ is the total optical thickness of the nebula for L_c-radiation.

The optical thickness τ decreases as the nebula expands. In general the dependence of τ on r has a quite complex form (see Chapter VII). But in the particular case when $\tau < 1$, this dependence has a simpler form: $\tau \sim r^{-3}$.

For our purposes it is sufficient to know that τ decreases as the size of the nebula increases. Consequently, using the formal cubic law for τ, one may re-write (92) in the form

$$\beta = \frac{q}{2u} \frac{v}{\tau_0} \left(\frac{r}{r_0}\right)^3, \tag{93}$$

where r_0 is the radius of the nebula for which $\tau = \tau_0$.

Let us assume in (93) that if the expansion velocity v decreases as r increases, then in all cases it does so more slowly than with a cubic law. Then it follows from (93) that for small r the quantity β will also be small. But for small values of β, the flux of L_α-radiation will be larger, as follows from formula (93). Hence we may infer that the effect of deceleration due to radiation pressure will be largest when the dimensions of the nebula are small. This also implies that on the average a nebula of small dimensions must expand with a somewhat larger velocity than a larger nebula (of course, under the condition that other factors of deceleration or acceleration are absent).

Thus, although radiation pressure cannot slow down a shell ejected with enormous velocities, it can, however, bring about some redistribution of the expansion velocities of the nebulae along their diameters.

In summing up, one may conclude that *the original velocity of escape or ejection of a planetary nebula can in no case be of order* $1000 \, km/sec$. On the basis of the available observational data on the expansion velocities of nebulae, and also from the circumstance that the deceleration in general can lead to only insignificant changes of velocity, we conclude that, *the original velocity of ejection or escape of a nebula must be of an order of magnitude not larger than* $50-100 \, km/sec$.

A velocity of order $50-100 \, km/sec$ is significantly smaller than the escape from a surface of a star. Hence it follows that the origin of planetary nebulae cannot be related to phenomena having an explosive or catastrophic character. The actual process advances significantly slower and is characteristic of inflation. This conclusion automatically points the way to a correct discussion of the problem of the origin of planetary nebulae, on which we dwell in the last chapter.

Regarding the question of which of the types of force considered above predominates in nebula—forces leading to the acceleration of its expansion or forces leading to its retardation—the correct answer can be obtained only by knowing the law of expansion of the nebulae, that is, the dependence of the expansion velocities of nebulae on their diameters. A similar analysis, containing the most probable qualitative features of the nebulae, has been carried out in [51] (see Chapter IV).

On the basis of both this analysis and the conclusions obtained in this section, an attempt has been made, in particular, to estimate the lifetime of planetary nebulae. By the lifetime of a nebula we mean the interval of time from the first appearance of the nebula to its complete disappearance (for observers, of course). According to this estimate one obtains for the mean lifetime of the majority of planetary nebulae (diameters less than 100,000 a.u.) a value on the order of some tens of thousands, or rarely, a hundred thousand years. For gigantic nebulae it may be of the order of one million years.

Chapter VII

Double-Envelope Nebulae

1. Observations of Double-Envelope Nebulae

One often talks about a planetary nebula as consisting of one gaseous envelope which surrounds a central star—the nucleus. However, the nucleus is sometimes surrounded simultaneously by two concentric envelopes. In such cases we talk about *double-envelope planetary nebulae*. The outer (second) envelope is, as a rule, considerably less bright; in several cases it is very weak and can be discovered only with great difficulty. Examples of double-envelope planetary nebulae are given in Chapter I, and Plate I at the beginning of the book contains photographs of several of them.

At the time when attention was first drawn to double-envelope planetary nebulae (in the year 1953 [154]) the number of known objects of this type did not reach twenty. At the present time about thirty of these objects are known. These data show that double-envelope nebulae are not an isolated or chance phenomenon but rather, in all probability, that they are directly related to the physical processes which occur in the planetary nebulae. However, before reaching such conclusions, let us see to what extent the presence of two envelopes around the nuclei of several nebulae is real.

Planetary nebulae exhibit the phenomenon called "stratification of radiation" from which it follows (see Chapter I) that monochromatic images of a nebula have different sizes. One might think, therefore, that the observation of a second envelope is related to this effect. This idea, however, can be rejected for the following reasons:

(1) The aforementioned effect should lead to the observation of not two but several envelopes, since the wavelength interval usually covered by the photographs contains many lines that produce intense monochromatic images of the nebula.

(2) In those cases when slitless spectra have been obtained of double-evelope nebulae, the monochromatic images show two circles (for example, NGC 2392, 7662 and others [23, 24]). This shows that the emission of both envelopes has identical spectral composition.

The preceding facts are sufficient to show that the presence of two envelopes around certain nuclei is real, and does not depend on the observing conditions.

2. Basic Properties

In double-envelope planetary nebulae, the edges of both envelopes tend to be sharp; there is no gradual transition in brightness between the first and second shells. This is so in all known double-envelope nebulae. It shows, by the way, that the second envelope originates suddenly. It also allows the evaluation, without undue difficulty, of the apparent diameters of the inner (d_1) and outer (d_2) envelopes and, at the same time, the ratio d_2/d_1. This quantity, which is independent of the distance to the nebulae, is one of the basic parameters of two-envelope nebulae.

The numerical value of the quantity d_2/d_1 covers a rather wide interval. Values of d_2/d_1 for several double-envelope nebulae are shown on Table 7-1. The largest value of d_2/d_1 among the double-envelope nebulae known at the present time corresponds to the spiral nebula NGC 6543, and it is of order 15. The smallest value of d_2/d_1 corresponds to the bipolar nebula NGS 3587 and is equal to 1.16.

The brightness of the second envelope is always considerably less than that of the main envelope. In the majority of cases the brightness distribution over the second envelope is far from uniform (NGC 1535, 2022 and others), which shows the relatively small thickness of the second envelope. In several cases, although the mean surface brightness is constant within the boundaries of the second envelope, shows fluctuations in its distribution. Interesting objects from this point of view

TABLE 7-1

The Ratio (d_2/d_1) of the Outer and Inner Apparent Diameters of Several
Double-Envelope Planetary Nebulae

Nebula	d_1	d_2	d_2/d_1
NGC 1514	120″ × 90″	180″ × 150″	1.5
NGC 1535	20″ × 17″	40″	2
IC 2149	6″ × 12″	10″ × 15″	1.4
NGC 2392	19″ × 15″	47″ × 43″	2.6
NGC 2610	38″ × 31″	60″ × 50″	1.6
NGC 3242	26″ × 16″	40″ × 35″	1.8
NGC 3587	175″	203″	1.16
NGC 6445	38″ × 29″	150″	4
NGC 6543	22″ × 16″	∼300″	∼ 15
NGC 6826	27″ × 24″	135″	5
NGC 7662	17″ × 14″	32″ × 28″	2

are the double-envelope nebulae NGC 2392 and NGC 6543. Finally, there are cases when instead of a second envelope the photographs show only a bright, narrow arc (NGC 7293).

If we have the brightness distribution over both envelopes (isophotes) we can determine the mass ratio of the second $(\mathfrak{M}_2)$ to the first $(\mathfrak{M}_1)$ envelope—the second important parameter of two-envelope planetary nebulae—by means of the equation:

$$\frac{\mathfrak{M}_2}{\mathfrak{M}_1} = \frac{(L_2 V_2)^{1/2}}{(L_2 V_1)^{1/2}} = \left(\frac{L_2}{L_1}\right)^{1/2} \left(\frac{d_2}{d_1}\right)^{3/2},$$

where L_1 and L_2, V_1 and V_2 are the luminosity and volume of the first and second envelopes respectively. The ratio of the volumes of the two envelopes can easily be determined by measuring the relative sizes of the envelopes directly from the photographs of the nebula. The determination of L_2/L_1 is considerably more difficult since we have to deal with the effect of the projection of the outer envelope on the inner one. Therefore, the value of the ratio $\mathfrak{M}_2/\mathfrak{M}_1$ will be rather uncertain, particularly in those cases when the shape of the double-envelope nebula differs considerably from a homogeneous spheroid.

Nevertheless, an effort has been made [155] to obtain at least rough values of the ratio of the masses of the two envelopes for several double-envelope nebulae for which isophotes exist [8, 156], using also the results of direct measurements of the surface brightness of the second envelope [220]. The results are given in Table 7-2.

For the majority of the nebulae contained in the table the ratio is of the order of unity. In several cases the mass of the outer envelope even exceeds the mass of the inner one.

However, we should point out that in the case of the first seven nebulae in Table 7-2 the ratio $\mathfrak{M}_2/\mathfrak{M}_1$ was obtained without taking into account brightness fluctuations, *i.e.* fluctuations in the electron concentration. Besides, the brightness fluc-

TABLE 7-2

The Ratio of the Masses $(\mathfrak{M}_2/\mathfrak{M}_1)$ of the Outer and Inner Envelopes for
Several Double-Envelope Nebulae

Nebula	$\mathfrak{M}_2/\mathfrak{M}_1$	*Nebula*	$\mathfrak{M}_2/\mathfrak{M}_1$
NGC 650–1	0.8	NGC 7662	1.3
NGC 2392	1.5	NGC 6826	0.3
NGC 3440	1.0	NGC 6804	0.8
NGC 3242	1.8	NGC 7293	0.04
NGC 3587	0.2	NGC 6720	0.03
NGC 7009	1	NGC 6543	0.3

tuations are, as a rule, larger in the outer than in the inner envelope. At the same time, values of $\mathfrak{M}_2/\mathfrak{M}_1$ of the order of 0.1 and even smaller must be considered real. Therefore we are led to the important conclusion that the ratio in planetary nebulae varies between the limits 1 to 0.1, *i.e.* it can differ by a factor of ten or more from one nebula to the next.

It is of some interest tó make a statistical comparison of the linear thickness of the second envelope (the difference between the radii of both envelopes) with the linear thickness of the double-envelope nebulae. Using the data available in [16] one can apparently show that the linear thickness of the second envelope increases when the size of the double-envelope nebula increases [154]. This gives us a clear indication that the second envelope expands with a velocity somewhat higher than the expansion velocity of the inner envelope. If this is so we can reach the conclusion that there is a real link between the envelopes from the point of view of their origin, *i.e.* that there was a separation of the first envelope from the second at some stage during the life of the nebula. However, before we draw such a conclusion, which relies only on that observed regularity, it is necessary to collect more data on the linear dimensions of planetary nebulae.

We should add that from the point of view of the physical processes that take place within them there is no difference between the single- and double-envelope planetary nebulae. A convincing argument in this respect is the relative intensities of the emission lines in their spectra; it is identical in both types of planetary nebulae. This can be used as a proof that the formation of a second envelope in some planetary nebulae is not the consequence of some exceptional physical process or condition that occurs in one nebula and is absent in others.

Beyond what minimum linear dimensions can planetary nebulae have a second envelope? This is a difficult question to answer since our knowledge about the dimensions of planetary nebulae is so poor. According to the data in the catalogue [16], a second envelope is observed in nebulae whose inner envelopes have diameters of the order of 6000 to 9000 a.u. (IC 2149, NGC 2392, 6537, 6543, 7662). For these nebulae the second envelope has a thickness of 2000 a.u. or more. It follows, then, that a second envelope could be present in smaller nebulae, probably in those with diameters of just a few thousand astronomical units.

In connection with the properties of two-envelope nebulae we should also add that in the majority of cases the shape of the second envelope is similar to that of the first.

3. The Origin of the Second Envelope

The idea that the double-envelope planetary nebulae may originate as the result of two successive ejections of gaseous matter from the central star seems to be most attractive. It is corroborated by the well known fact that nova outbursts seem to be recurrent. The mean time between successive ejections for several nova-like objects is different, but for a given object it is more or less constant. On the average, the larger the amplitude of the brightness change, the longer the mean time between ejections. For the common novae, although we have no data, it seems that the mean time between two successive ejections is of the order of several hundreds or thousands of years. For the special group of recurrent novae the mean time between successive ejections is of the order of several tens of years. For these objects the amplitude of the brightness change is relatively small.

However, it is easy to argue that the origin of double-envelope nebulae cannot be due to successive ejections from the nucleus of the nebula. If this were true one would expect to find in several cases three, and once in a while four, envelopes simultaneously surrounding the nucleus.

If the mean time between successive ejections by the nucleus of a given planetary nebula were constant, then IC 2149, for example, should have three envelopes since the width of a second envelope ($\sim 4''$), which is proportional to the time between successive ejections, is smaller than the radius of the inner envelope ($\sim 13''$) by a factor greater than three. In the same way the nebulae NGC 6804 and NGC 7354 should have three envelopes, the nebula IC 2022, four envelopes, and NGC 3587, even five envelopes! In all these cases, however, only two envelopes have been observed arond the central star. *Up to the present time not a single planetary nebula having three envelopes has been observed.* This is a fact that should be kept in mind.

The only conclusion that can be reached within the framework of the hypothesis of repeated ejections, is the clearly inadmissible one that the nuclei of planetary nebulae are stars which eject matter only twice in their lives.

The hypothesis of successive ejections as a possible explanation of the origin of double-envelope nebulae can not be accepted on the basis of dynamical arguments, either. The second envelope, moving through the gaseous medium of the first, should quickly loose its original shape. Nothing of the sort is observed. On the contrary, the breaking-up of the shape and, as a consequence, the appearance of a spotty structure occurs in the outer envelope (second envelope in NGC 2392, 6543 and others).

Finally, the very idea of the origin of planetary nebulae as the result of ejections of a central star, as we shall see in Chapter X, is unacceptable in principle, and therefore further consideration of this process is a waste of time.

There is, however, another possibility: the second envelope appears as a consequence of the ionization of a region of interstellar gas. Since the optical depth beyond the Lyman limit remains smaller than unity, because of the expansion of the nebula, a fraction of the ultraviolet energy can penetrate the medium surrounding the nebula, excite the interstellar gas, and give rise to a second envelope around the first. This assumption, though, cannot explain many of the properties of double-envelope nebulae and, in addition, if this explanation were correct we would expect a strong galactic concentration of this type of nebula. Actually, however, nothing of the sort is observed; double-envelope nebulae are found around the galactic equator as well as at high galactic latitudes (up to $|b| = 60°$), corresponding to distances of the nebulae from the galactic plane of 500 and even 1000 parsecs.

Thus, double-envelope nebulae do not seem to be the result either of successive ejections from the nucleus or of a "continuation" into the interstellar medium. To us it seems that the second envelope is most probably the result of a certain a-mount of matter being torn off from the main envelope. Below we shall argue that duplicity of the envelope seems to be property shared by all planetary nebulae, and that each nebula, at a given stage of its life, should develop a second envelope.

4. The Evolution of Planetary Nebulae

The basic successive steps in the evolution of a planetary nebula present the following pattern. At first, when the nebula is small, it consists of just one envelope. During this stage the nebula appears as a disk, either round or oval. This is the *planet-like* nebula. Thus it remains up to a certain moment (some particular dimension for a given nebula) after which a fraction of the mass separates from the basic envelope; this mass moves away with small relative velocity, and after some time it becomes a second envelope arond the first. The nebula has reached the *double-envelope* phase.

The second envelope remains unobservable for a long time. As both envelopes expand their surface brightness decreases. For the second envelope the surface brightness decreases more rapidly, since it expands with a velocity somewhat larger than the expansion velocity of the basic envelope. As a consequence, the surface brightness of the second envelope reaches the limiting magnitude of our instruments earlier than the first, and therefore ceases to be visible.

At this time only the inner envelope can be seen, but by now it has acquired a ring-like shape since, as far as we can tell, no further emission of gaseous matter from the nucleus is taking place. Thus originate the *ring* nebulae. Ring nebulae follow the double-envelope type which, in turn, are preceded by the planet-like nebulae. In other words, the evolution of planetary nebulae goes as follows:

$$planet\text{-}like \quad \rightarrow \quad two\text{-}envelope \quad \rightarrow \quad ring$$

As this evolution takes place the nebulae expand, that is, their dimensions increase. It follows, then, that on the average the planet-like nebulae should have the smallest dimensions and the ring nebulae, the largest. The dimensions of the double-envelope nebulae should fall between the other two.

One may attempt to check the above conclusions by making a statistical comparison of the *mean* size of the nebulae of each type with the others. For this purpose one should use as far as possible homogeneous material in classifying the

TABLE 7-3

Diameter D in a.u. of Several Planet-Like, Double-Envelope and Ring Nebulae

| Planet-like | | Double-envelope | | | Ring | |
NGC	D	NGC	D_1	D_2	NGC	D
II 2165	20	1535	13	29	I 418	27
4593	12	2022	62	78	2438	129
5572	9	2149	6	11	2610	145
II 4476	13	2392	8	21	6058	54
6803	7	3242	11	22	6563	70
6879	9	6804	71	130	6565	27
6881	14	7354	21	32	6894	100
6884	22	7662	9	18		
6886	16	I 3568	25	50		
		6720	54	110		
II 2890	20	6826	15	30		

nebulae since in photographs obtained by one telescope a nebula may appear ring-shaped and, observed with another, it may appear as double-envelope nebula.

Table 7-3 is an incomplete list of dimensions of planet-like, double-envelope, and ring nebulae. The dimensions are taken from [16]. The nebulae were classified from plates obtained by Curtis [15] with the 36″ reflector at the Lick Observatory. We should point out that two of the nebulae included in Table 7-3, NGC 2610 and NGC 6058, have turned out to be of the double-envelope type when observed with more powerful telescopes (see below).

On the basis of the data shown in Table 7-3 we have prepared Table 7-4 where the mean diameters $\bar{D}$ of the nebulae of given types are shown in the second column. As can be seen from this table the dimensions of the nebulae increase considerably

TABLE 7-4

Average Values of the Diameter $\bar{D}$ in a.u. and of the Dilution Coefficient $\bar{W}$
for Various Types of Planetary Nebulae

Type of nebula	$\bar{D}$	$\bar{W}$
Planet-like	15	10^{-14}
Double-envelope	27, 48	0.4×10^{-14}, 10^{-15}
Ring	80	0.8×10^{-16}

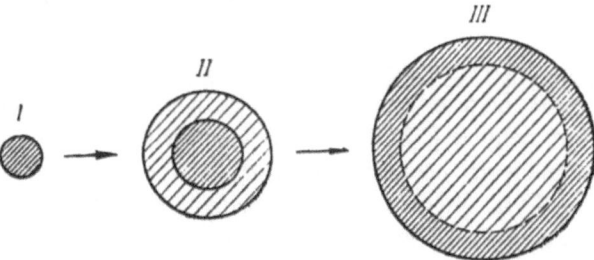

Figure 7-1.
Evolution of planetary nebulae. I, planet-like; II, double-envelope; III, ring.

as we go from one type to the next. For better illustration the data of Table 7-4 are shown graphically in Fig. 7-1.

As a further argument we have computed the dilution coefficient for each type of nebula by means of Eq. (37) of Chapter IV, using nebulae of known diameters D'' (in seconds of arc) and apparent magnitude of the nucleus m_*; the temperature and linear dimensions of the nuclei have been assumed the same for all nebulae. The computed values of W give the relative linear dimensions of the nebulae, completely independently from their distances to us.

The coefficient W was computed in this way for all the nebulae in Table 7-3, and an average was obtained for each nebular type. The results given in the last column of Table 7-4 show that the double-envelope nebulae have dimensions intermediate between the planet-like and ring nebulae.

Thus, in spite of a number of uncertainties, a comparison of the dimensions of the nebulae, determined by two different methods, confirm the assumption that the planet-like, double-envelope and ring nebulae are different stages of development of the planetary nebulae, and that they form an evolutionary sequence.

From what has been said it follows that the expression "ring nebulae," has to some extent a relative meaning and can be used only from the point of view of convenience in a formal classification of the shapes. We call ring-shaped those planetary nebulae where the second envelope is simply not visible, although it actually exists. In fact, instead of ring nebulae we should talk about double-envelope nebulae of *larger* size, to distinguish them from the double-envelope nebulae of normal size. If this is so, it is easy to predict the existence of a second envelope in the relatively large or in the ring-shaped planetary nebulae. To discover the postulated second envelope in such nebulae one must make a special search with very powerful telescopes and long exposures. Under such conditions the image of the nebula itself will be strongly over-exposed. In order to increase the contrast and reduce the influence of the sky background it is more efficient to obtain the photographs through a filter which transmits the $\lambda 3727$ Å line if, of course, the optics of the telescope have good transmission in this wave length region (for example, a Schmidt telescope).

Long before the above considerations regarding the nature of the double-envelope nebulae were made, Duncan made an attempt to discover second envelopes in a series of planetary nebulae [57]. He could detect (with the 100″ telescope) second envelopes in the planetary nebulae NGC 6826, NGC 6210, IC 3568 and also in the famous nebula in Lyra, NGC 6720. In the plates obtained by Curtis with the 36″ reflector all these were either planet-like or ring-shaped. Later on, second envelopes were found around the large nebula in Aquarius, NGC 7293 [158] and around NGC 6853 (the Dumb-bell Nebula).

The number of known double-envelope nebulae increased sharply after publication of the Palomar Sky Atlas (obtained with the 48″ Schmidt telescope). In this Atlas a second envelope was discovered in the following planetary nebulae [159]: NGC 1514, 2510, 4361, 6058, 6369, 6445, 6543; apparently, also in NGC 7008 and anon. $22^h 17^m$. The second envelope around NGC 6543, about which we talked earlier, is of special interest. At the present time it is the relatively most extended second envelope: the ratio d_2/d_1 is of order 15.

The fact that more and more nebulae turn out to have two envelopes poses a question: may not all planetary nebulae, without exception, have two envelopes? Could not it be that the planet-like nebulae are the result of an observational selection, determined by the limitations in our means of observation? The investigation of Duncan [160], in particular, gives a negative answer. He obtained long exposure plates of several *bright* nebulae with the 100″ telescope. However, no signs of a second envelope were found around these nebulae. The same conclusion can be reached after inspection of the Palomar Atlas. Many planetary nebulae, small but bright, or rather having a high surface brightness, do not show a second envelope, although the image of the nebula itself is strongly over-exposed. Thus, for example, the nebulae NGC 6894, A 39, A 54, NGC 2452, NGC 6778 and possibly also NGC 6842 and NGC 7048 are really planet-like.

Thus, all the facts presented here tend to show that the appearance of a second envelope around the basic one is a structural feature occurring in all planetary nebulae. *All planetary nebulae are born with one evelope but end their lives with two.*

5. The Theory of Envelope Separation

The second envelope originates at a given stage in the life of the nebula when a fraction of the gas splits off from the original envelope. The splitting occurs under the influence of selective pressure, due to L_α-radiation, which becomes very intense in the outer layers of the nebula. Below we shall show that this pressure is capable of producing a considerable difference between the expansion velocities of the outer layers of the nebula and the main mass.

As the nebula expands its total optical depth τ_1 beyond the Lyman limit decreases. When $\tau_1 < 1$ the dependence of τ_1 upon r can be easily obtained from the following argument. From the ionization equation we have:

$$\frac{n^+ n_e}{n_1} = Wf(T).$$

When $\tau_1 < 1$ the function $f(T)$ can be considered independent of r, and the electron (proton) concentration is constant and equal to the concentration of hydrogen atoms n. Then, since $W \sim r^{-2}$ we can write:

$$\frac{n^2}{n_1} \sim r^{-2}.$$

If during the expansion the mass of the nebula does not increase, we have $n \sim r^{-3}$ and, therefore, $n_1 \sim r^{-4}$. For the dependence of the optical depth τ_1 upon the radius of the nebula we get:

$$\tau_1 \sim n_1 r \sim r^{-3}.$$

When $\tau_1 \gg 1$ the dependence is more complicated and will be derived later.

Let us assume that τ_1 for a nebula of average dimensions is of order unity. Then, in nebulae of relatively small dimensions, τ_1 will be considerably greater than unity —of the order of several tens or even hundreds. The exponential dependence of the degree of ionization upon the optical depth and hence, on the linar distance from the nucleus, makes the number of hydrogen atoms undergo a sharp, almost discontinuous increase somewhere in the inner regions of the nebula. The nebula divides into two regions: an inner one, almost fully ionized, where $n_1/n \approx 0$, and an outer one, where practically all the atoms are in the ground state, *i.e.* where $n_1/n \approx 1$. In other words, the outer layers of the nebula possess maximum selective absorption, and because of the strong flux of L_α-radiation in these layers, the magnitude of the radiation pressure can become very large.

The L_α-flux in the upper layers of the nebula is directed outwards. When $\tau_1 \gg 1$ the magnitude of this flux is practically equal to zero and, although the number of neutral hydrogen atoms n_1 capable of absorbing L_α-quanta is large, the magnitude of the radiation pressure is extremely small. As the optical depth τ_1 decreases, the flux increases and, therefore, the radiation pressure in the outer layers of the nebula also increases. When $\tau_1 \ll 1$ its magnitude is again small but this time because of the negligible number of neutral hydrogen atoms. Obviously, there must be some optimum value τ_1^0 of the order of a few units, at which the radiation pressure attains a maximum. This force communicates a sufficient velocity to the outer layers of the nebula such that they separate and move away from the main nebular mass. Since the optimum value τ_1^0 is attained at certain definite nebular dimensions, *i.e.* at a given period of the life of the nebula, this moment can be related to the origin of the second envelope in the nebula.

From the quantitative point of view the study of the separation process, that is, of the problem of splitting, becomes that of finding the distribution of radiation pressure inside the nebula and, in particular, at its edge, and also of determining how this distribution changes with time, *i.e.* during the expansion of the nebula.

We shall call $n_1(s)$ the number of hydrogen atoms in the ground state per unit volume at a distance s from the inner edge of the nebula and $F_\alpha(s)$ the L_α-flux at that distance. Then, for the average light pressure $P(s)$ acting on the mass contained in a unit volume we have

$$P(s) = \frac{k}{c} n_1(s) F(s), \tag{1}$$

where k is the absorption coefficient of one hydrogen atom at the frequency of L_α-radiation.

Let us first determine $n_1(s)$. If we assume the dilution coefficient to be constant within the nebula, but decreasing as the nebula expands, we can write the ionization equation in the following way:

$$\frac{n^+ n_e}{n_1} = W f(T_*) e^{-\tau}, \tag{2}$$

where

$$f(T_*) = \left(\frac{T_e}{T_*}\right)^{1/2} \frac{2(2\pi\mu k T_*)^{3/2}}{h^3} e^{-h\nu_0/kT_*}, \tag{3}$$

and τ is the optical depth beyond the Lyman limit at the distance s. Let us take:

$$n^+ = n_e = xn, \quad n_1 = (1 - x)n, \tag{4}$$

where x is the degree of ionization, and n is the total number of hydrogen atoms per unit volume (it is constant within the nebula but decreases with its expansion). We shall make the simplifying assumption that the linear thickness of the nebula s remains constant while the nebula breaks up. This is justified if the splitting occurs in a relatively short time. Then n will vary according to:

$$n = n_0 \left(\frac{r_0}{r}\right)^2, \tag{5}$$

where r is the radius of the nebula, and n_0 the hydrogen concentration for a nebular radius r_0.

Using (4) we can rewrite (2) as follows:

$$\frac{x^2}{1-x} = Ae^{-\tau}, \tag{6}$$

where

$$A = \frac{W}{n}f(T_*) = \frac{1}{4n_0}\left(\frac{R}{r_0}\right)^2 f(T_*), \tag{7}$$

and R is the radius of the nucleus. Notice that in our approximation the quantity A is constant and does not depend on r.

If we differentiate (6) and introduce $d\tau = n_1\kappa ds = (1 - x)n\kappa ds$, where κ is the mean absorption coefficient for L_c-quanta, we get

$$\left[\frac{2}{x(1-x)} + \frac{1}{(1-x)^2}\right] dx = -n\kappa ds. \tag{8}$$

After integration we find that the dependence of the degree of ionization x upon the linear distance s is given by*:

$$s = C - \frac{1}{n\kappa}\left[\frac{2}{\text{Mod}} \log \frac{x}{1-x} + \frac{1}{1-x}\right]. \tag{9}$$

The constant C is determined as follows. At the inner edge of the nebula, *i.e.* at $s = 0$ we have $\tau = \tau_0$ and $x = x_0$. Therefore:

$$C = \frac{1}{n\kappa}\left[\frac{1}{\text{Mod}} \log \frac{x_0}{1-x_0} + \frac{1}{1-x_0}\right], \tag{10}$$

where x_0 is the solution of the equation

$$\frac{x_0^2}{1-x_0} = A. \tag{11}$$

The dependence of the degree of ionization upon the optical depth can be determined from Eq. (6), which we rewrite in the following way:

$$\tau = \frac{1}{\text{Mod}}\left[\log A - \log \frac{x^2}{1-x}\right]. \tag{12}$$

Since n which enters in (9) and (10), is a function of r [according to (5)], then s and τ will also be functions of r. Solving (9) and (12) simultaneously we get the relation between τ and s.

It is also important for us to know how the total optical depth τ_1 of the nebula varies during the expansion. For this purpose we first determine x_1, from the known

* [*Ed. note*] Here Mod = $\log_{10} e = 0.43429 \ldots$.

linear thickness s_1, by means of the equation

$$s_1 = C - \frac{1}{n\kappa} \left[\frac{2}{\text{Mod}} \log \frac{x_1}{1 - x_1} + \frac{1}{1 - x_1} \right], \tag{13}$$

and then τ_1 from the relation

$$\tau_1 = \frac{1}{\text{Mod}} \left[\log A - \log \frac{x_1^2}{1 - x_1} \right]. \tag{14}$$

This completes the solution of the first part of our problem, namely to determine the distribution of neutral hydrogen atoms, i.e. $1 - x$ inside the nebula, and also the dependence of this distribution on the radius r or the total optical depth τ_1 of the nebula. Figure 7-2 contains curves showing the distribution of the quantity $1 - x$, the fractional concentration of neutral hydrogen atoms in the nebula, against nebular thickness s for different values of τ_1, that is, for different nebular radii r, since an increase of nebular radius corresponds to a decrease in τ_1. As can be seen from the graphs, the quantity $1 - x$, i.e. $n_1(s)$ experiences a sharp, almost discontinuous increase somewhere inside the nebula. As τ_1 decreases the location of the jump moves towards the outer boundary of the nebula. For a value of τ_1 of the order of unity, there is no discontinuity in the distribution of $1 - x$.

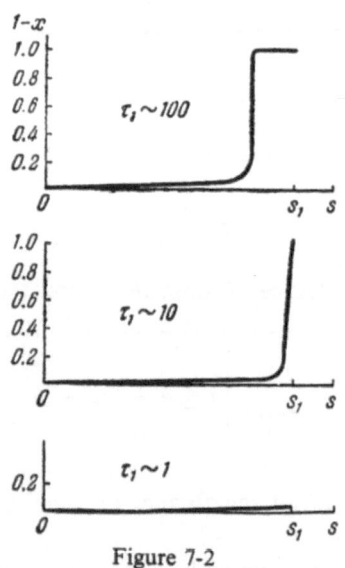

Figure 7-2

Distribution of the concentration $(1 - x)$ of neutral hydrogen with depth s in the nebula for various values of the optical thickness τ_1.

Let us turn now to the second aspect of the question: the determination of the L_α-radiation flux inside the nebula. To solve this problem we shall use the results of the theory of radiative equilibrium in real nebulae expanding with a velocity gradient (see Chapter VI, Section 3).

For the L_α-flux, $F_\alpha(t)$, as a function of the optical depth t at the frequency of L_α, passing through the inner boundary of the nebula, we have approximately:

$$F_\alpha(t) = -\frac{1}{3}\frac{dG(t)}{dt}, \tag{15}$$

where the function $G(t)$ has in general the following form [90]:

$$G(t) = Ae^{\sqrt{\beta}t} + Be^{-\sqrt{\beta}t} + 3\frac{q}{\beta}\psi(\tau). \tag{16}$$

Here A and B are constants, β and $\psi(t)$ are given by Eqs. (41) and (44) of Chapter VI and q—the ratio of the mean atomic absorption coefficient at the frequencies L_c and L_α—is of the order of 10^{-4}–10^{-5}.

In the use of Eq. (16) the following difficulty arises. The expressions which appear above are justified only if β is constant inside the nebula, i.e. β does not vary with τ. On the other hand, in our case β changes over a wide range, particularly at the transition from the ionized to the neutral region. We can resort to the following approximation: we write two expressions of the type of (16) with different constant coefficients and different values of β, one for the ionized and one for the neutral regions of the nebula. In this way β is constant within each region, and the constants (A, B, A_1, B_1) can be determined from the boundary conditions. However, it is more expedient to proceed differently; this will also imply an approximation but the problem will be solved more easily. And, in particular, it will be possible to use an expression for $G(t)$ that has already been derived [Eq. (38), Chapter VI]:

$$G(t) = 3\frac{q}{\beta}\left\{\psi(\tau) - \left(1 - \frac{2}{3}\sqrt{\beta}\right)\left[\psi(0)e^{-\sqrt{\beta}t} + \psi(\tau_1)e^{-\sqrt{\beta}(t_1-t)}\right]\right\}, \tag{17}$$

but taking β as varying continuously throughout the whole nebula. The numerical value of β at each point in the nebula can be determined in the following way. If we replace in (17) $d\tau = (1 - x)nk\,ds$ we find:

$$\beta = \beta_0\frac{1 - x_0}{1 - x}, \tag{18}$$

where β_0 is the value of β at the inner edge of the nebula:

$$\beta_0 = \frac{q}{2u}\left(\frac{dv}{d\tau}\right)_0.$$

Considering that $(dv/d\tau)_0$ and u are of the same order of magnitude we find that $\beta \approx 10^{-4}$.

If we substitute (17) in (15) and neglect $\sqrt{\beta}$ in comparison with unity, we get

$$F_a(t) = 2\pi \frac{q}{\beta} \left\{ \psi(\tau) - \frac{\sqrt{\beta}}{q} \left[\psi(0)e^{-\tau\sqrt{\beta}/q} + \psi(\tau_1)e^{-(\tau_1 - \tau)\sqrt{\beta}/q} \right] \right\}, \qquad (19)$$

where we have made the replacement

$$t = \tau/q$$

From Eq. (19) it follows that the L_a-flux in the nebula is not a monotonic function of τ and that it has a maximum somewhere inside the nebula. However, at present we are interested in the magnitude of the radiation pressure at the boundaries of the nebula and hence we shall leave the consideration of this point for later.

From (19) we can write at the boundaries of the nebula

$$F_a(\tau_1) = + 2\pi\psi(\tau_1)q/\sqrt{\beta_1} \qquad (20)$$

and

$$F_a(0) = - 2\pi\psi(0)q/\sqrt{\beta_0}, \qquad (21)$$

where β_1 is the value of β at the outer edge of the nebula. It can be determined from (18) upon setting $x = x_1$.

If we substitute the value of $\phi(\tau)$ from (44) of Chapter VI in (20) and (21) we get, respectively

$$F_a(\tau_1) = \pi S_c \frac{q}{2\sqrt{\beta_1}} e^{-\tau_1} \left(\frac{r_0}{r} \right)^2 \qquad (22)$$

and

$$F_a(0) = - \pi S_c \frac{q}{2\sqrt{\beta_0}} \left(\frac{r_0}{r} \right)^2 . \qquad (23)$$

The $+$ and $-$ signs show that the flux is directed outwards at the outer edge and inwards at the inner edge of the nebula.

We have already available all the necessary data for the determination of the magnitude of the radiation pressure at each point in the nebula. An example of such a calculation is given in [51] for the case: $T_* = 50,000°$, $r_0 = 5000$ a.u., $n_0 = 10^4$ cm^{-3}, $R = 0.1 R_\odot$. It turns out that the radiation pressure $P(s)$ at the outer boundary of the nebula will be negligible until its radius reaches a certain value, after which it increases and reaches a maximum rather quickly, and then gradually decreases with a further increase of the radius of the nebula (Figure 7-3). In this example the radiation pressure first becomes significant when $\tau_1^0 \approx 8$ and it attains its maximum at a nebular radius $r = 5600$ a.u. or $\tau_1 \sim 2\text{-}3$. The thickness of the layer which experiences the increase in radiation pressure is approximately 1/20 of the linear thickness of the nebula and hence, for the ratio of the masses of the two envelopes, we shall have $\mathfrak{M}_2/\mathfrak{M}_1 \approx 0.05$ which disagrees badly with the data shown on Table 7-2.

The magnitude of the expansion velocity Δv of the second envelope relative to the first can be determined from the following equation of motion:

$$nm \frac{d\Delta v}{dt} = \frac{n_1 k}{c} F_\alpha(r), \tag{24}$$

where m is the mass of the hydrogen atom.

If we put here $n_1 = (1 - x)n$, $k = \kappa/q$ and integrate, we get

$$\Delta v(t) = \frac{\kappa}{qmc} \int_{t_0}^{t} (1 - x)F_\alpha(r)dt = \frac{\kappa}{qmc} \int_{t_0}^{t} P(r)dt,$$

where $P(r) = (1 - x)F_\alpha(r)$. After some rearrangement we have

$$\Delta v(r) = \frac{\kappa}{qmc} \int_0^r P(r)\frac{1}{\frac{dr}{dt}}dr = \frac{\kappa}{v_0 qmc} \int_0^r P(r)dr, \tag{25}$$

where $dr/dt = v(r) = v_0$ is the expansion velocity of the nebula, here taken to be constant.

Equation (25) can be integrated graphically and the diagrams obtained are similar to that of Fig. 7-3. The relative expansion velocity for the second envelope is large at the beginning, slower later on and, for example, by the time the radius of the primary envelope reaches 8000 a.u., that is, approximately 2000 years after the beginning of the action of the radiation pressure, Δv attains a magnitude of a few kilometers per second. Let us now compute the ratio of the diameters of the two envelopes. For this purpose we first determine the linear thickness Δs of the

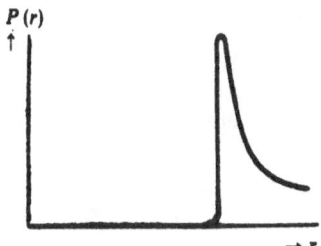

Figure 7-3

Variation of the radiation pressure $P(r)$ at the inner boundary of the nebula in relation to the diameter r of the nebula.

second envelope by means of the equation

$$\Delta s = \int \Delta v(r)dt = \frac{1}{v_0} \int \Delta v(r)dr, \tag{26}$$

and then in d_2/d_1 we substitute $d_2 = d_1 + 2\Delta s$. It appears that d_2/d_1 is strongly dependent upon the temperature of the nucleus and varies considerably with the expansion velocity of the nebula. In Table 7-5 we show the ratio d_2/d_1 for the example considered above, for several values of T_* and v_0 and different nebular sizes.

Minin [16] analyzed the role of the "inner" maximum of the light pressure, to which we referred earlier, in the process of separation of the second envelope. For this purpose he considered jointly the L_α-radiation field and the dynamics of the nebula, with the introduction of several approximations. He showed that the maximum is reached at $x = 2/3$, or $\tau \approx 6-7$ [the parameter A in Eq. (6) is of the order of 10^3]. Then he derived the following expression for the additional velocity acquired by the mass $\mathfrak{M}_2$ that splits off under the influence of the radiation pressure

$$\Delta v_2 = w \frac{m_2}{\mathfrak{M}_2} \frac{2\pi r^2}{\kappa} \int_1^{\tau_0} F(\tau,u)d\tau, \tag{27}$$

where

$$F(\tau,u) = \frac{2}{a^2(\tau)} \left[a(\tau)\sqrt{u} - b(\tau)ln\left(\frac{a(\tau)}{b(\tau)} + 1\right) \right], \tag{28}$$

$$a(\tau) = \frac{q}{2} \frac{\phi(\tau)}{\left(\int_0^\tau \phi(\tau)d\tau\right)^{1/2}}, \tag{29}$$

$$b(\tau) = \beta_0(1 - x)/(1 - x_0), \tag{30}$$

and

$$\phi(\tau) = (1 - x)e^{-\tau}.$$

Here u is a dimensionless parameter related to the time t through the expression

$$u = \frac{2\pi S\kappa}{wm_H c} \frac{v_c}{v_\alpha} t, \tag{31}$$

where w is the thermal velocity. The dependence of x on τ in (30) is given by Eq. (6).

The L_α-radiation pressure imparts an additional velocity also to the inner-ionized-region of the envelope, whose mass is $\mathfrak{M}_1$. It can be determined from the following

TABLE 7-5

Theoretical Ratio (d_2/d_1) of the Diameters of Double-Envelope Nebulae

Radius r of the nebula (a.u.)	$v_0 = 20\,\text{km/sec}$			$v_0 = 30\,\text{km/sec}$		
	$T_* = 50{,}000°$	$T_* = 75{,}000°$	$T_* = 100{,}000°$	$T_* = 50{,}000°$	$T_* = 75{,}000°$	$T_* = 100{,}000°$
6.5	1	1	1	1	1	1
10	1.2	1.9	4	1.1	1.5	2.5
20	1.4	3.6	9.7	1.2	2.9	6
30	1.7	5.4	16	1 4	3.7	10
40	1.9	7.2	23	1.6	4.9	14

expression:

$$\Delta v_1 = w \frac{m_\text{H}}{\mathfrak{M}_1} \frac{2\pi r^2}{\kappa} \int_0^1 F(\tau,u)d\tau. \tag{32}$$

Calculations show that the ratio $\Delta v_1/\Delta v_2$ is of the order of 0.5 when $\mathfrak{M}_2/\mathfrak{M}_1 = 1$ and 0.05 when $\mathfrak{M}_2/\mathfrak{M}_1 = 0.1$. Therefore, for the expansion velocity relative to the nucleus of the fraction of the nebula that splits off we can write

$$v_2 = v_0 + \left(1 - \frac{\Delta v_1}{\Delta v_2}\right)\Delta v_2, \tag{33}$$

where v_0 is the expansion velocity of the nebula itself, i.e. of the main envelope.

An analysis of Eq. (27) can be found in [202] where it is shown that in a number of cases the mass that splits off can be of the order of half the total mass of the nebula, i.e. $\mathfrak{M}_2/\mathfrak{M}_1 \approx 1$.

Table 7-6 shows the magnitude of $\overline{\Delta v}_2$, the expansion velocity of the outer envelope

TABLE 7-6

Relative Expansion Velocity $\overline{\Delta v}_2$ of the Outer Envelope for $\mathfrak{M}_2/\mathfrak{M}_1 = 1$

T_*	Δv_2 (cm/sec)		
	$R_* = 0.1_\odot$	$R_* = 0.25_\odot$	$R_* = 0.5_\odot$
40,000°	0.034	0.12	1.2
60,000°	0.45	1.5	5.2
80,000°	0.94	4	12.3
100,000°	1.5	8	25

relative to that of the inner one, i.e. $\bar{\Delta}v_2 = \Delta v_2 - \Delta v_1$ when $\mathfrak{M}_2/\mathfrak{M}_1 = 1$, as a function of the temperature T_* and the radius R_* of the nucleus. For the calculations we have taken: $\mathfrak{M} = \mathfrak{M}_1 + \mathfrak{M}_2 = 0.02\,\mathfrak{M}_\odot$, $w = 15\,\text{km/sec}$, $t = 10^4$ years, $\beta_0 = 10^{-4}$ and $q = 10^{-4}$. It was also assumed, in accordance with conclusions reached above, that the outer envelope begins to separate when the radius of the nebula is $r = 2000\,\text{a.u.} = 0.3 \cdot 10^{17}\,\text{cm}$.

We should point out that in the case of the "inner" maximum the absolute magnitude of the light pressure is roughly two orders smaller than the light pressure at the "outer" maximum. However, the "inner" maximum exists for a long time, beginning at the time when the optical depth τ_0 of the nebula is much larger than one, while the "outer" maximum exists for a relatively short time, when $\tau_0 \approx 8-10$. Therefore, for example, for small nuclear temperatures when the L_α-flux is small, the "inner" maximum cannot push away a large mass from the nebula, but the "outer" maximum is able to produce a second envelope of small mass ($\mathfrak{M}_2/\mathfrak{M}_1 \approx 0.01$). On the other hand, at very high nuclear temperatures the "inner" maximum will outstrip the outer maximum tearing off from the nebula an amount of mass comparable to the mass remaining behind ($\mathfrak{M}_2/\mathfrak{M}_1 \approx 1$). However, after a while, when the optical depth of the nebula becomes about 5 to 10, the "outer" (in the first case) and the "inner" (second case) maxima will coincide and a further development of the outer envelope will occur by means of the first process.

The account we have given here regarding the origin of the second envelope can also be extended to the envelopes of novae. Theoretically one would expect that a second envelope would originate in the gaseous envelopes produced by nova outbursts. On the other hand, "double-envelope" novae are not observed. The reason for this is clear. The magnitude of the radiation pressure at the outer boundary of the nova envelope is of the same order as at the outer boundary of a planetary nebula, since the temperatures of the central stars in both cases are about the same. But the expansion velocity of nova envelopes is almost two orders of magnitude greater than the expansion velocity of planetary nebulae. Therefore, in the case of novae the radiation pressure does not have time to impart a relative velocity to the second envelope (during the life of the envelope) comparable to the expansion velocity of the nova envelope.

The account presented in this section of the theory of splitting needs to be elaborated further, taking into account all aspects of the problem. In particular, it would be of special interest to determine a theoretical upper limit for the ratio $\mathfrak{M}_2/\mathfrak{M}_1$.

As early as 1932, when V.A.Ambartsumyan first considered the problem of radiative transfer in a stationary nebula, he reached the conclusion that radiation pressure plays an important role in planetary nebulae, and that it must be the main force present. Later on, in 1946, V.V.Sobolev solving the same problem for a nebula expanding with a velocity gradient, obtained for the momentum associated with radiation pressure a value two to three orders of magnitude smaller than the value obtained by Ambartsumyan. Accordingly it was concluded that the force of the radiation pressure was comparable to the attraction of the central star. Now in the case of a real nebula, if we take into account *the changes in the degree of ionization*, and also the gradient in the expansion velocity, we discover the very interesting, let us say, selective role of the radiation pressure. This selectivity is so strong that it causes a radical change in the shape of the nebula, namely the appearance of a second envelope. We can assert, therefore, that radiation pressure plays a very significant role in the dynamics of planetary nebulae.

6. On the Optical Depth

The determination of the correct value of the temperature of the nucleus of planetary nebulae, as we have seen in Chapter IV, is tied up with the knowledge of the total optical depth of the nebula beyond the Lyman limit.

The results obtained in the preceding section allow us to draw some conclusions about the order of magnitude of the optical depths for planetary nebulae of different types.

The process of formation of a second envelope begins when the optical depth beyond the Lyman limit τ_1 is of the order of five or ten. At this time, although the second envelope already exists, it cannot yet be distinguished since τ_1 still remains greater than unity for the whole nebula. The second envelope remains "concealed" until τ_1 decreases to a value of the order of unity because of the expansion of the nebula. By then the second envelope has had time to get some distance away from the main ring and hence, when $\tau_1 \approx 1$, we discover the second envelope already *at some distance* from the first; the moment at which the splitting takes place is not observable. From this fact we can draw the following conclusion: *The optical depth beyond the Lyman limit in the case of double-envelope nebulae is of the order of unity or less.*

As far as the ring nebulae are concerned, since their dimensions are even larger than those of double-envelope nebulae, τ_1 should be considerably smaller than one. Only in the planet-like nebulae τ_1 will be larger than one.

From the above it follows that the temperature of the nuclei determined by Zanstra's method will be on the average greater for planet-like nebulae; for double-envelope and, particularly, for ring nebulae they will be considerably lower.

Wurm and Singer [163] have also referred to the question of the incomplete absorption of radiation beyond the Lyman limit in many nebulae. They showed that in many cases the observed values of the ratio $I_{\text{He\,II}\,4686}/I_{H_\beta}$ was significantly greater than the theoretical value (determined by means of Ambartsumyan's equation, with the temperature calculated with Zanstra's method). Using the same basic material, but a different line of reasoning, Japanese astrophysicists have come to the same conclusion [164].

Some differences are observed in the magnitude of the ratio I_{4686}/I_{H_β} as we go from one type of nebula to another. For example, on the average this ratio is smaller for planet-like nebulae than for double-envelope or ring nebulae. This is further confirmation that the optical depth at the frequencies of the Lyman continuum is larger on the average for planet-like nebulae than for double-envelope nebulae.

One can get some ideas about the optical depth of the nebula at the frequencies of the Lyman continuum from the structure of slitless spectrograms of double-envelope nebulae. The monochromatic images of double-envelope nebulae in the most prominent emission lines (N_1, N_2, H_β and so forth) also show the two envelopes. But, beginning with certain lines belonging to ions with relatively high ionization potential, the second envelope vanishes from the monochromatic images. Thus, for example, in the planetary nebula NGC 2392 the second envelope is still visible in the lines λ 4684 Å He II, λ 3868 Å [Ne III] and barely visible in λ 3426 Å [Ne V]. In the nebula NGC 7009 the second envelope can be seen in the line λ 3868 Å [Ne III], but it is completely invisible in the lines λ 4686 Å He II and λ 3426 Å [Ne V]. The same situation occurs in NGC 7662, where the second envelope is well observed at λ 3868 Å [Ne III] but does not appear at λ 4686 Å He II and λ 3426 Å [Ne V].

From these examples it follows that the optical depth of the main envelope beyond the limit of the principal series of ionized helium ($\lambda < 228$ Å) and doubly ionized neon ($\lambda < 275$ Å) is of the order of unity in NGC 7009 and NGC 7662, and is smaller than unity in NGC 2392. In NGC 2392 the optical depth of the main envelope is of the order of unity at the frequencies of four times ionized neon ($\lambda < 65$ Å).

Knowing the optical depth of the nebula at the frequency of ionization of some ion or other, one can, in principle, determine the optical depth of the nebula (inner envelope) at the frequencies of the Lyman continuum. This requires a knowledge of the continuous absorption coefficients, which are well enough known for many ions (Tables 2-19 and 2-20), and of the relative abundance of these ions in a given nebula, which cannot always be determined with sufficiently high accuracy.

7. Ring Nebulae

Ring-shaped planetary nebulae are a more advanced stage of the double-envelope nebulae. They differ from the typical double-envelope planetary nebulae in that their second envelopes are weaker, to the extent that they do not show in the photographic plates with normal exposures, and in the fact that their main envelopes have extended to the point that they appear as rings in projection on the sky.

This latter circumstance leads us to pose the following question: are the ring nebulae spheres empty of gas, or is the density different from zero everywhere? In other words, we have to determine the spatial distribution of matter as a function of the nebular radius. This problem is not new, and was posed earlier (see Chapter

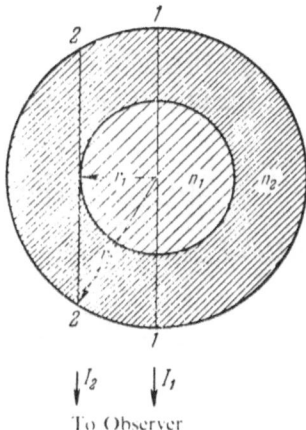

Figure 7-4

For the determination of the relative concentrations in the central regions of ring nebulae.
Arrows at bottom point to observer.

I). Here we shall consider a simple way to estimate the degree of rarefaction in the central regions of ring-like nebulae, *i.e.* their relative material density. To this effect it suffices to have available two observational facts: the ratio of the apparent radius of the outer (r_2) to the inner (r_1) boundaries of the "ring", and the ratio of the surface brightness at the center of the "ring" (I_1) to the surface brightness at the inner boundary (I_2).

Let us call n_1 the mean concentration of hydrogen ions in the central region of the sphere, and n_2 that of the nebula itself. Then, the surface brightness I_1 of the nebula in the hydrogen lines in the direction of the center—*i.e.*, along the line *1-1* (Fig. 7-4)—will include the emission of the central region due to the ions n_1, and the emission of the front and rear sections of the envelope, due to the ions n_2.

Along the line *2-2*, that touches the inner edge of the envelope at a distance r_1 from the center, the surface brightness I_2 will be due to the ions n_2 only. Since the amount of energy emitted per unit volume is proportional to the square of the ion concentration, it is easy to obtain the following expression for n_2/n_1, if we assume that the nebula is transparent to visual light:

$$\frac{n_2}{n_1} = \left\{ \delta \left[\left(\frac{r_2}{r_1} \right)^2 - 1 \right]^{1/2} - \left(\frac{r_2}{r_1} - 1 \right) \right\}^{1/2}, \tag{34}$$

where δ is the ratio of the surface brightness at the center of the ring to that at its inner edge, *i.e.*,

$$\delta = I_1/I_2 \tag{35}$$

and is determined from the observations. For a typical ring nebula δ should be smaller than one.

Let us consider the following extreme cases. Assume that the central region is completely empty, i.e. $n_1 = 0$. Then the value of δ will be a minimum and equal to

$$\delta_{min} = \frac{r_2/r_1 - 1}{[(r_2/r_1)^2 - 1]^{1/2}}. \tag{36}$$

When $n_1 = n_2$, δ will attain its maximum value,

$$\delta_{max} = \frac{r_2/r_1}{[(r_2/r_1)^2 - 1]^{1/2}}. \tag{37}$$

The observed value of δ, if the nebula is ring-shaped, should obviously satisfy the condition

$$\delta_{min} < \delta < \delta_{max}. \tag{38}$$

Let us apply this argument to several ring nebulae.

NGC 6720. From the surface brightness isophotes obtained for this nebula by Aller [8] in the line H_γ we find that the brightness at the center of the nebula (I_1) is equal to approximately 63, in arbitrary units, and at the inner edge of the ring (I_2) it is equal to 100. This gives $\delta = 0.63$. For the relative thickness of this nebula Curtis [15] gives the value 0.4. From this we get $r_2/r_1 = 1.7$. With these data and Eqs. (36) and (37) we find $\delta_{min} = 0.5$ and $\delta_{max} = 1.2$, that is, the condition (38) is satisfied and the inner regions of this nebula are not completely empty.

If we put in (34) $\delta = 0.63$ and $r_2/r_1 = 1.7$ we find

$$n_2/n_1 = 0.37.$$

According to more recent measurements $r_2/r_1 = 2$, which gives $n_2/n_1 = 0.30$.

Thus, the concentration in the central regions of the nebula NGC 6720 is about three times smaller than in the nebula itself. This difference, that may seem small at first sight, leads to an approximately tenfold decrease of the emission coefficient at the center of the nebula.

NGC 7293. For this nebula we have $\delta = 0.57$, $r_2/r_1 \approx 1.7$ [138]. With these data we find from (34)

$$n_2/n_1 \approx 0.3,$$

which shows a certain similarity between the nebulae NGC 7293 and NGC 6720 in the sense that they have the same internal structure.

IC 418. From the curves giving the distribution of surface brightness (in H_α) as a function of the radius obtained by Aller and Wilson [11] we find: $\delta = 0.73$ and $r_2/r_1 = 1.8$. It follows that $\delta_{min} = 0.54$ and $\delta_{max} = 1.2$ and hence condition

(38) is satisfied. From (34) we find:

$$n_2/n_1 \approx 0.54.$$

Using a different, more elaborate method Aller and Wilson [11] have obtained

$$n_2/n_1 \approx 0.46.$$

NGC 1501. From the isophotes shown in [8] we obtain the approximate value $\delta = 0.4$. It is more difficult to determine the value of the ratio r_2/r_1. Apparently it is of the order of 1.09. This gives from (36) and (37) $\delta_{min} = 0.29$ and $\delta_{max} = 2.6$, that is, condition (38) is satisfied. Furthermore we find

$$n_2/n_1 \approx 0.28$$

NGC 2392. From the very intricate isophotes of this nebula [8] we find approximately $\delta = 0.5$ and $r_2/r_1 \approx 2$. Therefore $\delta_{min} = 0.58$ and condition (38) apparently is not satisfied for this nebula; the nebula is practically empty inside. However, this conclusion has to be considered with caution since it is drawn from the analysis of highly confused isophotes.

Chapter VIII

The Stability of the Forms of Gaseous Envelopes

1. The Hydrodynamics of Planetary Nebulae

Up to now, in speaking of the forces acting in a planetary nebula or in any of its parts, we have considered the gravitational force of the central star and the radiation pressure of the nebula. Since the gas has a high temperature (of the order of 10,000°K) and is expanding with a speed on the order of 10 km/sec, it is clear that we should also consider the influence of the gas pressure on the nebula, with all of its hydrodynamic consequences. Finally, since planetary nebulae, as will be shown in the next chapter, have magnetic fields, the effect of the magnetic force also has to be considered.

In any nebula the four types of forces mentioned above can act simultaneously, but one of them may become dominant. We cannot say a priori which of the forces will predominate at any given stage of development of a planetary nebula. However, the problem can be solved in a comparatively simple way. Since by their nature the four forces are intrinsically different, the form and structure of the nebula will change radically according to which type of force has the dominant role. Therefore, we can sometimes guess the nature of the primary force responsible for the form and structure of the nebula simply by looking at the photographic plate that depicts it. For example, in the bipolar nebulae the dominant force responsible for its form and structure is the magnetic force (Chapter IX).

It follows that in special cases it is possible to describe the dynamics of a nebula on the basis of *one* type of force; this simplifies very much the treatment of the problem. Whether the hydrodynamic forces need to be taken into account or not when a magnetic field is present can be determined from the inequality

$$\frac{H^2}{8\pi} < \frac{1}{2}\rho v^2,\tag{1}$$

where H is the intensity of the magnetic field, v is the velocity of the gaseous masses and ρ is the density of the medium. For a planetary nebula, taking $v_0 \sim 20$ km/sec,

$\rho = 1.6 \times 10^{-20}$ gr/cm^3, we find that already for $H > 10^{-3}$ gauss the hydrodynamical forces can be neglected. The condition $H > 10^{-3}$ gauss can not always be fulfilled in a planetary nebula. In nebulae with large linear dimensions, and also in extended regions of normal nebulae, we can expect $H < 10^{-3}$ gauss. In this case the hydrodynamical forces can play a certain role in the development of the form and structure of the nebula.

Finally, the possibility of a dynamical interaction between the planetary nebula and the interstellar medium, in which the expansion and motions of the nebula take place, also points towards the consideration of the hydrodynamic forces in the dynamics of a planetary nebula.

Unfortunately, formidable obstacles are encountered in the quantitative elaboration of special problems of the hydrodynamics of planetary nebulae.

The main question, of great importance, concerns the validity of the application of the classical principles of hydrodynamics in the conditions that exist in cosmical objects. Even in the densest gaseous nebulae and galactic clouds the density of the medium is negligible compared with the density of the media for which the classical equations of hydrodynamics were deduced. Nevertheless, the usual laws of hydrodynamics can, under certain conditions, be also applied to cosmic objects. The determining factor is the ratio of the scale of the phenomenon or the size of the object (the "characteristic size") to the mean free path of the particles. We can assert that as long as the distance d between two points, such that the value of a parameter of the motion (for example, the velocity) or of the medium (for example, the density) changes by a significant amount, remains larger than the mean free path l of the particles, *i.e.*, while the condition

$$\frac{d}{l} \gg 1 \tag{2}$$

is satisfied, the usual laws of hydrodynamics are valid.

According to Spitzer's calculations [142, p. 31], even in the interstellar hydrogen clouds, which have a much smaller density than the planetary nebulae, the mean free path of the particles is just a few hundred astronomical units, while for the objects in which we are interested (planetary nebulae) the dimensions are huge and the distance over which there is a significant difference in the motion is of hundreds or thousands of astronomical units.

No significant investigation of the hydrodynamics of planetary nebulae has been carried out up to the present time.* Of some interest are the results obtained on the stability of gaseous envelopes emitted by the stars, among them the planetary nebulae [143, 51]. We shall summarize these results briefly in the following section.

---

* [Ed. note] The most recent work on the problem is that of Sofia and Hunter, *Ap. J.*, **152**, 405 (1968), who give references to other recent papers.

2. Statement of the Problem

It was noticed that planetary nebulae retain their shapes for a long time while, for example, the envelopes ejected by fast novae that have become very large, in the majority of cases, lack a true shape; they transform relatively quickly (during the time that they have been under observation) into amorphous nebulae. In other words, in the first case the gaseous envelope retains its shape while in the second it deforms and fragments.

According to these observations we call the envelope stable when it retains its shape, or changes it by a small amount, during the expansion, and we refer to it as unstable when it disperses relatively quickly. In the case of planetary nebulae we have stable envelopes and in the case of novae, unstable ones. Since these envelopes are characterized by different values of physical and kinematical parameters one may attempt to find conditions that must be satisfied by these parameters if the envelope is to be stable. If we can successfully tie together these parameters with some numerical magnitude characterizing the degree of instability ("instability coefficient") we shall be able to predict the future development of the gaseous envelopes ejected by the stars. In particular, we can determine the nature of the ejection mechanism giving rise to planetary nebulae. Finally, we may elucidate the role of novae, Wolf-Rayet stars, and others in the formation of planetary nebulae.

When we discuss reasons for the loss of stability (regular shape) of the gaseous envelopes we shall take into account the interaction of these envelopes with the interstellar medium. In this case we will speak of external causes of loss of stability. It is also possible that there exist internal causes leading to the same consequences. At the present time, however, it is hard to foresee them. Even the role of a general field of turbulence is not, apparently, very significant (see Section 8 of this Chapter). As regards the effect of an irregular or explosive ejection of gaseous matter from the central star, it is irrelevant to the question considered here; such envelopes would be unstable and irregular in shape from the very beginning of the ejection.

As it expands the gaseous envelope experiences a breaking action by the interstellar medium. Strictly speaking, the outermost layer of the envelope first experiences this resistance. The inner layers are subject to pressure with the result that a velocity gradient is established across the thickness of the envelope. This, in turn, leads to a redistribution of the gas density inside the envelope. If, at first, the density was constant at all depths, there will later on be an increase at the outer and a decrease at the inner boundary of the envelope. Similarly, the temperature of the gas cannot remain constant throughout the envelope. The departure of the density distribution from the homogeneous, isotropic case due to some volume force produces an acceleration within the nebula, *i.e.* related to the coordinate system of the moving envelope. Let us call it the "internal acceleration" and designate it

Obviously, the above considerations are valid until an ever increasing compression (at $r \rightarrow \infty$) of the envelope leads to the formation of a shock front.

Thus, because of the resistance of the medium and the resulting compression of the envelope, it acquires the properties of an incompressible fluid—it becomes "weighty." The rate of braking of the envelope changes with time, and the degree of compression changes accordingly. Therefore, the acceleration $g(t)$ will be a function of the time t. It can be shown [51], however, that the internal acceleration is approximately equal to the deceleration of the envelope itself in the interstellar medium, *i.e.*

$$g(t) = \frac{dv(t)}{dt}, \tag{3}$$

where $v(t)$ is the expansion velocity of the envelope [see Eq. (11) of Chapter VI].

$$v(t) = v_0 \frac{1}{1 + \sigma r^3}. \tag{4}$$

The direction of $g(t)$ is opposite to the deceleration of the envelope, *i.e.* $g(t)$ acts in the direction of the expansion. Under these conditions, that is, in the presence of an internal acceleration, the surface of contact between the outer edge of the nebula and the interstellar medium cannot remain stable for very long. Even the smallest irregularity at the outer edge of the envelope—an irregularity that can occur under for a variety of reasons—can grow during the expansion of the nebula and at a given moment bring about its total destruction.

The problem reduces, then, to answering the following question: through a gaseous medium of given density moves a gaseous envelope of considerably higher density; the size and mass of the envelope and its initial expansion velocity are known. One requires to establish under which conditions and at what stage during its expansion the envelope is stable as well as the time when it becomes unstable (as regards the conservation of its original regular shape). In view of the mathematical complexity of this problem, we can assume as a first approximation that the temperature of the envelope remains constant when the gas is compressed (*i.e.*, we limit ourselves to the examination of an isothermal medium); we neglect the effects of the compressibility of the gas, and also the fact that the expansion velocity of the envelope is supersonic. Then, using the well known methods used to investigate gravitational waves on the surface of an incompressible fluid, we can write down an equation for the outer surface of the envelope as a function of time or of its radius. We can then determine the changes in the shape of the outer surface with time and the degree of stability of the envelope. In particular, we wish to know when there will occur a relatively fast growth of the initial perturbation that will make the envelope lose its equilibrium shape and, on the other hand, when will this growth be very slow, or perhaps not occur at all, making the envelope stable.

The problem of the instability of the surface separating two liquids was first investigated theoretically by Rayleigh in 1899 (see, for example, [152]). More recently it was considered by Taylor [145], who obtained a theoretical relationship between the size of the perturbation (*i.e.*, the deformation of the surface) and the acceleration for incompressible fluids. Therefore, the instability of such surfaces are known as Rayleigh-Taylor instabilities.

Lewis [144] investigated experimentally the Rayleigh-Taylor instabilities for liquids. He showed that when two liquids of different densities are placed one above the other, an acceleration appears perpendicular to the surface of separation. This surface will be stable if the acceleration vector is in the direction from the lighter to the heavier fluid, and unstable when the vector is directed in the opposite sense.

In our case the role of the heavy fluid is played by the envelope, and the role of the lighter fluid by the interstellar gas; the "internal acceleration" vector is in the direction from the envelope to the medium. Therefore, in general terms we would expect the envelopes to be unstable. As we shall see later, the type of instability that arises is of the Rayleigh-Taylor type. However, for different envelopes with different kinematical and physical parameters the onset of the Rayleigh-Taylor instability occurs at different times. In other words, the duration of the life of different envelopes under equilibrium conditions is different.

To summarize, the question is, firstly, to determine the moment at which the envelope becomes unstable and, secondly, to clarify the character of the instability.

3. The Stability of the Forms of Gaseous Envelopes

Let us introduce a coordinate system xy moving with the nebula such that the x-axis is parallel to the surface of separation (the outer boundary of the envelope) and the y-axis is normal to the surface, pointing outwards from the envelope (Figure 8-1). We shall assume that there is no motion along the z-axis. The depth of the envelope will be considered very large (formally equal to infinity) in comparison with the small perturbation $\eta\ (x,y)$ of its outer boundary.

Our problem consists in establishing the form of the function $\eta(x,t)$ or $\eta(x,r)$ (where r is the radius of the nebula), *i.e.* the form of the outer edge of the nebula at different times or at different distances of this edge from the center of the nebula.

It can be shown that in our case we have potential flow; therefore we introduce the Eulerian coordinates v_x and v_y which represent the components of the motion of the disturbed particles relative to the coordinate system xy. They are connected to the velocity potential through the relations

$$v_x = -\frac{\partial \phi}{\partial x}, \quad v_y = -\frac{\partial \phi}{\partial y}.$$

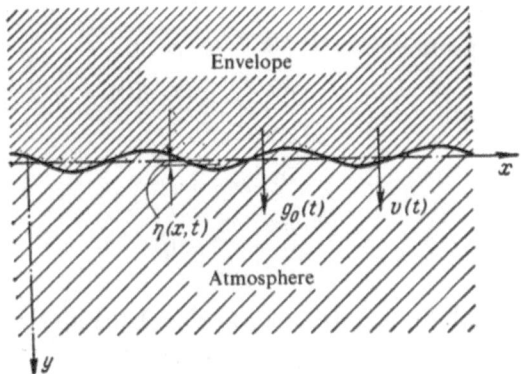

Figure 8-1.
For the stability problem of the forms of gaseous envelopes. $\mu(x,t)$—small perturbation of the lower boundary of the envelope, $v(t)$—expansion velocity of the envelope, $g_0(t)$—"inward acceleration."

The velocity potential satisfies the so-called equation of continuity, which for the potential flow of an incompressible fluid has the following form (Laplace's equation):

$$\frac{\partial^2 \phi}{\partial x^2} + \frac{\partial^2 \phi}{\partial y^2} = 0. \tag{5}$$

The equation of motion is given by the Cauchy-Lagrange integral

$$V + P + \frac{1}{2}\left[\left(\frac{\partial \phi}{\partial x}\right)^2 + \left(\frac{\partial \phi}{\partial y}\right)^2 \right] - \frac{\partial \phi}{\partial t} = F(t), \tag{6}$$

where V is the potential of the external forces and P the dynamical pressure. For V we have:

$$V = g(t)y, \tag{7}$$

where $g(t)$ is determined by (3) and (4).

After some manipulation we find from (6) the following expression for the motion of the outer edge of the envelope:

$$\eta = \frac{1}{g(t)} \left. \frac{\partial \Phi}{\partial y} \right|_{y=0}, \tag{8}$$

where $\Phi = \phi - \int P(t)dt$. To find the function $\Phi(x,y,r)$ we start from the condition that it satisfies Laplace's equation and the boundary conditions:

$$\frac{1}{g(t)} \left| \frac{\partial^2 \Phi}{\partial t^2} \right|_{y=0} + \left| \frac{\partial \Phi}{\partial t} \right|_{y=0} \frac{\partial}{\partial t} \frac{1}{g(t)} = - \left| \frac{\partial \Phi}{\partial y} \right|_{y=0}, \tag{9}$$

$$\left| \frac{\partial \Phi}{\partial y} \right|_{y=-\infty} = 0, \tag{10}$$

$$\left| \frac{\partial \Phi}{\partial x} \right|_{x=0} = 0. \tag{11}$$

From Eq. (4) it follows that as long as $\sigma r^3 < 1$ the motion of the envelope will be uniform, without acceleration, i.e. $g(t) = 0$. In this case we obviously will have

$$\eta(x,t) = \text{const.}, \tag{12}$$

that is, the surface of the envelope retains its shape as long as it remains inside a sphere of radius r_0 [see Eq. 13 of Chapter VI] determined by the condition $\sigma r_0^3 = 1$:

$$r_0 = \left(\frac{4\pi}{3} \frac{\rho_0}{\mathfrak{M}_0} \right)^{-1/3}, \tag{13}$$

where $\mathfrak{M}_0$ is the mass of the envelope and ρ_0 the density of the interstellar medium.

Outside the sphere of radius r_0 the condition $\sigma r^3 > 1$ is satisfied. Then, from (4) we can write

$$v(r) = \frac{v_0}{\sigma r^3} \tag{14}$$

for the expansion velocity of the envelope and

$$r = \left(\frac{4v_0}{\sigma} \right)^{1/4} t^{1/4} \tag{15}$$

for the equation of motion of the envelope. In this case $g(t)$ is different from zero,

$$g(t) = - \frac{3}{16} \left(\frac{4v_0}{\sigma} \right)^{1/4} t^{-7/4}. \tag{16}$$

It follows that $\eta(x,t)$ will no longer be constant: the surface of the envelope will change in shape with time or with increasing radius [according to Eq. (13)].

Substituting $g(t)$ from (16) in (9) and using Eqs. (5), (10) and (11) we find the function $\Phi(x,y,r)$ and from it the equation for the outer edge of the envelope $\eta(x,r)$ by means of (8). The final expression for $\eta(x,r)$—the small perturbation of the surface separating the two media—in a linear approximation is:

$$\eta(x,r) = \alpha_0 \left(\frac{r}{r_0} \right)^2 q(r) \cos kx, \tag{17}$$

where it has been assumed that the initial shape of the outer boundary of the envelope, when $r = r_0$ is given by the approximate equation

$$\eta(x,r_0) = \alpha_0 \cos kx, \tag{18}$$

where α_0 is the initial perturbation at time $t = t_0$ or $r = r_0$, assumed to be very small in comparison with the wavelength of the perturbation λ, and $k = 2\pi/\lambda$.

The form of $q(r)$ is given in [51, 143]. It can be expressed by means of Bessel functions of imaginary argument $I_3(v)$ and $K_3(v)$, where v is given by

$$v = 2\sqrt{6\pi} \left(\frac{r}{\lambda}\right)^{1/2} \tag{19}$$

In a first approximation we can take

$$q(r) \sim e^v. \tag{20}$$

Equation (17) is the equation of a bump whose height is given by the coefficient of $\cos kx$, *i.e.*, the magnitude of the perturbation at a given time and given size of the envelope. We are interested, of course, only in the relative value of the growth of the initial perturbation with time, *i.e.*, with the expansion of the envelope. Therefore, for the relative growth of the perturbation $\delta(r)$, that we shall call the "instability coefficient," we have from (17) and (18)

$$\delta(r) = \frac{\eta(x,r)}{\eta(x,r_0)} = q(r) \left(\frac{r}{r_0}\right)^2, \tag{21}$$

where r_0 is determined by (13). For $r = r_0$ we have $\delta(r_0) = 1$, and in this case the envelope is stable. For $r > r_0$ the instability coefficient is greater than one and, it follows, the envelope becomes unstable, the more so the stronger the inequality $\delta(r) > 1$. This is a Rayleigh-Taylor type of instability. In Figure 8-2 we show the development of the instability with increasing r. At the same time we show the order of magnitude of the instability coefficient $\delta(r)$.

Thus, every envelope, if it had a stable form after its ejection, will remain stable until the radius r_0 is reached [Eq. (13)] or until a time t_0, approximately equal to r_0/v_0. After this time the envelope begins to develop a Rayleigh-Taylor instability: the outer boundary adopts a periodic shape which steepens exponentially as the radius r increases.

The magnitude of r_0 is different for different envelopes and depends on their mass $\mathfrak{M}_0$ and on the density of the interstellar medium ρ_0. Then, within the framework of the scheme that has been used in posing and solving the question of the stability of the envelopes, the results obtained can be formulated thus: for each envelope there exists some sphere of radius r_0 —which we shall call "sphere of stability"—

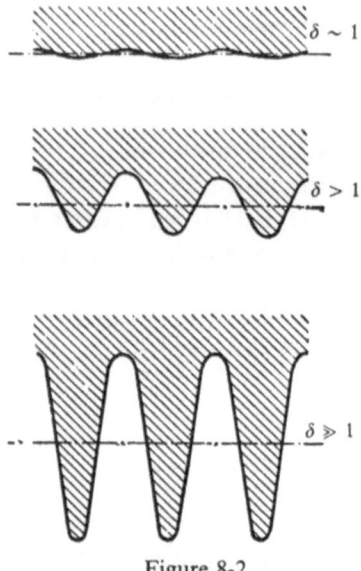

Figure 8-2

Development of the Rayleigh-Taylor instability in gaseous envelopes.

within which the envelope is stable; outside this sphere the envelope is unstable and after a certain time it will become an amorphous nebula.

The Rayleigh-Taylor instability has also been studied by Spitzer [140] in a purely qualitative manner (as applied to diffuse nebulae, see Section 7 of this Chapter). Frieman [147], using very approximate qualitative considerations, found an equation for $\eta(x,t)$ for the case of medium size perturbations (for an incompressible fluid), that had the following form:

$$\eta(0,t) = \exp\left[(gk)^{1/2}t\right], \tag{22}$$

where g is the acceleration, and k is the wavenumber of the periodic perturbation. If we compare (22) and (17) we see that the development of the instability for large values of η proceeds faster than for small values (i.e. for the initial phase of the loss of stability).

The further development of the hump or tongue instability is a simple "fall," i.e., its motion proceeds as the free fall of a body with acceleration g:

$$v \sim (g\eta)^{1/2}. \tag{22a}$$

This is the asymptotic behaviour of η for large values of the perturbation.

Obviously, for a more exact treatment of the problem of the stability of the shapes of gaseous envelopes and of other questions related to the hydrodynamics of planetary nebulae, it is necessary to abandon some of the restrictions imposed above and consider a real gaseous envelope or real nebula, taking into account the compressibility of the gas in the supersonic regime and the sphericity of the envelope (nebula). From a mathematical standpoint it is fruitful to apply the so-called perturbation method, which has been applied to different hydrodynamical problems; the basic idea of the method is to consider the motion of each volume element of gas as the sum of two motions—the unperturbed and the perturbed; the parameters of the unperturbed component are given. Then the basic equations of gas dynamics (equation of motion, equation of continuity and energy equation) constitute the differential equations for the perturbed motion and their solution for given initial and boundary conditions gives the parameters of the perturbation as a function of the coordinates and time.

Below we present examples of gaseous envelopes to which the Rayleigh-Taylor instability criterion has been applied.

4. The Stability of Planetary Nebulae

Using for the mass of planetary nebulae $\mathfrak{M}_0 \approx 0.01 - 0.1 \mathfrak{M}_\odot$ and for the density of the interstellar matter $\rho_0 \approx 5 \times 10^{-24}$ gm cm^{-3}, we find from (13) the following value for the radius of the equilibrium sphere:

$$r_0 \approx 10^5 \text{ a.u.}$$

This result must be interpreted as follows: while the radius of the planetary nebula remains smaller than 10^5 a.u. its form will be stable. Planetary nebulae can develop a Rayleigh-Taylor instability only after their radii exceed the above value. Since the radii of the overwhelming majority of planetary nebulae is smaller than 10^5 a.u. it follows that almost all planetary nebulae are within the zone of stability with the result that they should retain their shapes. The observations, as was shown before, confirm this conclusion.

5. The Stability of Nova Envelopes

The mass $\mathfrak{M}_0$ of the envelope of a fast nova is of the order of $10^{-5} \mathfrak{M}_\odot$. This gives for the radius of the sphere of stability

$$r_0 \approx 7000 \text{ a.u.}$$

As soon as the radii of nova envelopes exceed this value they should exhibit a Rayleigh-Taylor instability. But the envelopes attain this size very quickly—in a few tens of years if we take the expansion velocity of the envelopes to be of the order of 1000 km/sec. This means that nova envelopes find themselves outside the zone of stability almost immediately and should be destroyed quickly. The process of destruction is as follows: a perturbation initially small will begin to grow and, as a result, the flutes will turn into "tongues" which, in turn, break up leading to the total disintegration of the envelope.

There are relatively little data on the shape and structure of gaseous envelopes ejected by fast novae. Even less is known about the way the shape and structure change during the expansion. Nevertheless it can be considered as established that in most cases the nova envelopes, beginning at a given stage in their expansion, enter into an unstable, *i.e.* amorphous, condition.

As an example we can take the envelope produced by Nova Persei (1901). According to photographs published at different times, the envelope of this nova still had a fairly regular, stable form about 10″ in diameter in 1917. However, in later photographs it shows a partial loss of stability. At the present time the envelope of this nova has completely lost its stability and has broken up in different wisps and blobs. We can ascertain that this occurred because of the dynamical interaction between the envelope and the interstellar medium. We can even estimate an upper limit for the density of the interstellar matter in the vicinity of Nova Persei from the available data on the mass of the envelope, which is $2 \times 10^{-5} \, \mathfrak{M}_\odot$ [48], its distance—approximately 600 parsecs [149]—and the time at which the envelope became unstable (1917). This upper limit turns out to be $1.5 \times 10^{-22} \, gm/cm^3$, that is, fifteen times higher than the mean density of interstellar matter. This is the density of a diffuse nebula. Since then we have learned that this nova occurred inside a dust cloud.

Apparently, the Rayleigh-Taylor instability also occurs in the case of the Crab Nebula. In the numerous pictures ob.ained for this nebula, it appears to consist of many filaments and elongated blobs distributed radially relative to the center of the nebula. Assuming that after the explosion of the supernova (1054) the ejected envelope had initially a more or less regular shape, we can attribute the later loss of stability to the interaction of the envelope with the interstellar medium. From the fact that the envelope is badly disrupted we conclude that at the present time it is outside the sphere of stability. We can even make an attempt to determine the mass of the gaseous envelope of the Crab Nebula from the known density of the interstellar matter in the vicinity of this nebula. Unfortunately, data on this density are very poor. The interstellar absorption in the direction of the Crab Nebula as measured from the color index of B stars [150], is apparently not very large, about $1^m.2$, the normal value for regions free of dense absorbing clouds. However, other facts point to the presence of interstellar matter of significant density in the vicinity of the nebula; in plates obtained by Baade one can see distinctly (especially between $\lambda\lambda \, 7200$ and $8400 \, Å$) regions of dark matter both on the eastern and, especially, on the western sides of the nebula, clearly *interacting* with it.

As far as the density of interstellar hydrogen in the vicinity of the Crab Nebula is concerned, the data that we have are again not very reliable. Nevertheless, from the usual value of the density of interstellar matter and a more or less reliable estimate of the radius of the sphere of stability [51], we can reach the conclusion that if the disruption of the gaseous envelope of the Crab Nebula is tied to its dynamical interaction with the interstellar matter, then the mass of this envelope should be less than $0.1 \, \mathfrak{M}_{\odot}$.

We also have cases of gaseous nova envelopes with regular spherical shapes under stable conditions. Thus, for example, the envelope ejected by Nova Aquilae (1918). The envelope around this nova was discovered almost immediately after the explosion, in October of 1918; its diameter was then $0''.65$, and the expansion velocity 1700 km/sec. Within two months the diameter of the nebula had reached $1''.8$. In photographs obtained in 1926 it looked like a full circle and its diameter was $16''.4$ [151]. Since Nova Aquilae is at a distance of 360 parsecs, this corresponds to a linear diameter of 6000 a.u. In 1930 its diameter was 8000 a.u., and in 1940, 11,000 a.u.; the envelope was still stable.

The stability of the envelope of Nova Aquilae can be explained by saying that it has not yet come out of its sphere of stability. If we estimate this radius from the known mass of the envelope of Nova Aquilae,
we get from (13) $r_0 \approx 14,000$ a.u. (taking $\rho_0 = 5 \times 10^{-24}$ gm/cm^3). This is more than twice the radius of the envelope in 1940. The envelope will be able to retain its shape until its radius exceeds 14,000 a.u. (with the assumed density of the interstellar matter). With the expansion velocity quoted above this will occur some 50 to 60 years after the ejection.

The above considerations about the stability of gaseous envelopes which, of course, are only qualitative, are not sufficient to come to any final conclusion about the role of novae or supernovae in the process of formation of planetary nebulae. Nevertheless, these considerations point to the conclusion that the planetary nebulae cannot originate as the result of the ejection of gaseous matter from novae and supernovae.

In those cases the gaseous envelopes lose their stable forms relatively quickly and turn into small diffuse nebulae or disperse into space. This marks a separation between the processes related to nova and supernova explosions and processes related to the formation of planetary nebulae.

6. Stability of the Envelopes of Wolf-Rayet Stars

It has sometimes been assumed that planetary nebulae result from the continuous flow of gaseous matter from nonstationary stars, and in particular from Wolf-Rayet stars. If the star ejects matter at a rate of $10^{-5} \, \mathfrak{M}_{\odot}$ per year, continuously during several thousand years, then at some distance from the star an accumulation of mass equal to the mass of a planetary nebula may occur. The ejected matter would be slowed down by the resistance of the interstellar matter. This is the essence of the hypothesis.

Leaving aside other possible objections against this hypothesis, let us consider its validity from the point of view of the stability of the envelope: can an envelope produced by the continuous ejection of gaseous matter from a central star be stable and turn into a planetary nebula?

Let us call $\mathfrak{M}_0$ the intial mass of the ejected envelope. The mass will increase because of the matter ejected by the star with velocity v_0 and rate av_0 and of the matter accreted by the envelope from the interstellar medium and carried away with the envelope, *i.e.*

$$\mathfrak{M} = \mathfrak{M}_0 + \frac{4\pi}{3}\rho_0 r^3 + a \int_0^t (v_0 - v)dt, \tag{23}$$

where ρ_0 is the density of the interstellar medium, and r the radius of the envelope at time t.

According to the calculations, the loss of stability of the envelope begins when $\sigma r^3 = 1$, when the mass of the expanding envelope proper becomes comparable with the amount of interstellar matter inside a volume equal to the volume of the nebula. Therefore, we can write for the critical (maximum) mass and radius of the nebula that will still remain stable

$$\mathfrak{M} = \frac{4\pi}{3}\rho_0 r^3. \tag{24}$$

To determine $\mathfrak{M}$ or r from this relation we need to know the equation of motion of an envelope with variable mass and variable expansion velocity. This problem was investigated by Minin [137] who obtained the following expressions for r, $\mathfrak{M}$ and v:

$$r = \frac{\mathfrak{M}_0}{a}x, \tag{25}$$

$$\mathfrak{M} = \mathfrak{M}_0(1 + \alpha x^2)/y, \tag{26}$$

$$v = v_0(1 - y), \tag{27}$$

where x and y are related through the equation

$$\frac{1 - y}{y} = \frac{[2x + (\alpha/2)x^4]^{1/2}}{1 + \alpha x^3}. \tag{28}$$

In Eq. (28) we have used

$$\alpha = \frac{4\pi\rho_0\mathfrak{M}_0^2}{3a^3}. \tag{29}$$

Using (25), (26), and (28), Eq. (24) can be written:

$$\frac{4\pi}{3} \frac{\rho_0 \mathfrak{M}_0^2}{a^3} = \frac{1}{x^3} \left[(1 + \alpha x^3) + \sqrt{2x + \frac{\alpha}{2} x^4} \right] \tag{30}$$

From known values of $\mathfrak{M}_0$, ρ_0 and a we can determine x from (29) and then $\mathfrak{M}$ and r by means of (26), (28) and (25).

Taking, for example, $\mathfrak{M}_0 = 10^{-5} \mathfrak{M}_\odot$, the amount of matter ejected per year equal to $10^{-5} \mathfrak{M}_\odot$, the velocity of ejection $v_0 = 1000$ km/sec and $\rho_0 = 5 \times 10^{-24}$ gm/cm^3, we find as parameters of the stable nebula

$$r = 18,000 \text{ a.u.},$$

$$\mathfrak{M} = 2 \times 10^{-4} \mathfrak{M}_\odot,$$

$$v = 870 \text{ km/sec},$$

where r stands for the maximum size of the nebula. From these data it follows that a typical planetary nebula cannot be formed in this way. Furthermore, the lifetime of such a nebula would be of the order of hundreds of years; after this interval of time the envelope would begin to disperse. Therefore, we cannot talk about a further accumulation of mass within the boundaries of the envelope by continuous ejection from the central star.

Therefore, considerations of the stability of gaseous envelopes lead us to the conclusion that planetary nebulae cannot originate by continuous ejection of gaseous matter from nonstationary stars and, in particular, from Wolf-Rayet stars. They are more likely the result of one single violent ejection, since only in this case can the ejected envelope, containing sufficient mass, remain stable until it reaches the typical size of a planetary nebula.

7. Application of the Theory to Diffuse Nebulae

The necessary condition for the occurrence of a Rayleigh-Taylor type of instability is the appearance of an acceleration at the surface of separation of the two fluids. This acceleration can arise for different reasons in different situations. In the case of planetary nebulae and the gaseous envelopes ejected by novae we have seen that it is due to a compression at the surface of separation between the envelope, which is expanding under the influence of inertial forces, and the interstellar medium at rest. An analogous situation can occur through the interaction of diffuse nebulae with the surrounding small clouds of cold gas and dust, with the only difference that in this case, as was shown by Spitzer [146], the acceleration originates in the thermal expansion of the nebula. When the gas surrounding a hot star heats up and expands, it becomes less dense than the cold cloud around the nebula. During the expansion (with a velocity of the order of 10 to 20 km/sec) the hot light gas compresses the dense,

but cold, gas (cloud). This pressure is equivalent to an acceleration at the surface
of separation between the hot and cold gases, that in the end results in Rayleigh-
Taylor instability of the surface.

Spitzer discovered a region of Rayleigh-Taylor instability in the diffuse nebula
M 16 (NGC 6611). It had the appearance of a tongue or "elephant trunk," and could
be easily distinguished against the bright background of the nebula. The "elephant
trunks" are less luminous and in some cases they are completely dark: they consist,
obviously, of cold gas associated with a large amount of dust.

One should not forget the possibility that a perturbing acceleration, and even
a Rayleigh-Taylor instability, can be produced by gravitation, when relatively cold
condensations of gas and dust form in the nebula under the influence of its gravitational force. However, for typical diffuse nebulae with dimensions of the order of
a few parsecs and concentrations of the order of 50 atoms/cm^3, calculations show
that this acceleration is negligible. Therefore, a Rayleigh-Taylor instability arising
from gravitation will occur only either in a particularly dense diffuse nebula of small
dimensions or in a relatively cold nebula of gas and dust (with a temperature of
the order of a few hundred degrees) having a small thermal expansion velocity.

In Figure 8-3 we show schematically the development of a Rayleigh-Taylor instability due to any kind of acceleration. The heavy fluid (the cold cloud) is located
above, and the light fluid (the heated gas) is below. While the amplitude of the perturbation (irregularity) at the surface of separation of the two fluids remains small
compared to the wavelength of the perturbation, the amplitude grows according
to the Rayleigh-Taylor theory, *i.e.*, with an exponential law. For very large ampli-
tudes, the theory no longer holds and the instability becomes a simple "fall" of the
heavy fluid into the light one.

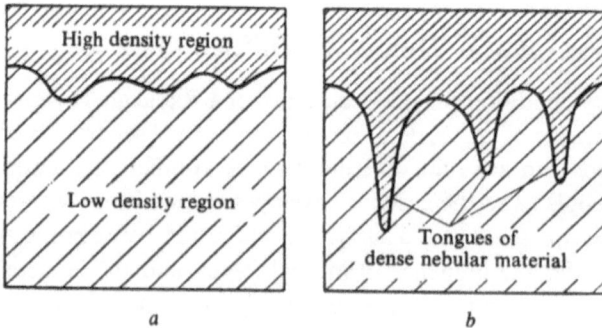

Figure 8-3
Diagram of the formation of Rayleigh-Taylor instabilities in diffuse nebulae: (*a*) onset
of the condition at the boundary between the cold envelope and the hot nebula;
(*b*) formation of "elephant trunks."

An examination of the photographs of diffuse nebulae shows that Rayleigh-Taylor instabilities are quite wide-spread among them. Many dark tongues and "elephant trunks" can be discovered in the diffuse nebulae NGC 2264, 2237–9, 7000, IC 405, 410, S 139, IC 1848. Especially remarkable in this respect are the nebulae NGC 6523 and IC 1396. In the majority of cases the "elephant trunks" develop in the periphery of a nebular region and, in particular, at its outer edge.

8. The Influence of Macroturbulence on the Stability of the Envelopes

Up to now we have talked about just one, external factor that can cause the instability of a gaseous envelope, namely, the resistance of the medium. But there is another, internal factor that can lead to the same consequences. We have in mind macroturbulence in the envelope, *i.e.* turbulence set up by the radial expansion of the nebula which produced a velocity differential between any two arbitrarily chosen points. As far as microturbulence is concerned, it was already mentioned in Chapter I (as the result of Wilson's work) that it plays no significant role in planetary nebulae.

In estimating the influence of macroturbulence on the stability of envelopes we encounter the difficulty that we do not know in which way and to what extent turbulence can influence the shape of the envelope. In other words, from the simple fact of the deformation or loss of shape of the envelope we cannot tell whether it is due to turbulence or to some other reason.

It is well known that the regime of motion of a fluid is characterized by the non-dimensional Reynolds number. For small values of this number laminar flow is stable, while for large values it is not stable and the motion becomes turbulent. Between these values there is a transition region, where the motion changes from laminar to turbulent. The corresponding Reynolds number is called *critical*. It can be determined experimentally and, for example, for the conditions present in terrestrial aerohydrodynamics it is of the order of one thousand. However, its order of magnitude in a cosmic object like a planetary nebula is not known. It cannot be simply assumed that the critical Reynolds number for cosmic objects is of the same order as in terrestrial conditions. Undoubtedly, some role must be played, for example, by the magnetic field interacting with the gas.

We can attempt to estimate the order of magnitude of the critical Reynolds number from observational data. Indeed, among planetary nebulae there are some that show quite regular and stable shapes (for example, IC 418, NGC 7610, and others). This means that the Reynolds number in these nebulae is smaller or of the order of the critical number. On the other hand, it is not difficult to compute their Reynolds numbers, which turn out to be of the order of 10^5.

Thus, the Reynolds number in real nebulae is one or two orders of magnitude larger than the critical Reynolds number of usual hydrodynamics. Nevertheless, turbulence is not present in the nebulae. One must conclude that there exists some force or mechanism which prevents the transition to the turbulent phase. We shall investigate whether the magnetic field associated with the nebula can act in this way. It can indeed play a stabilizing role in the sense that the internal motions of the gas can no longer have arbitrary directions but must conform to a given law, namely, they must, be directed along the magnetic line of force.

At the present time it is generally assumed that planetary nebulae are associated with an ordered magnetic field. Evidence of many types points in this direction (see Chapter IX). Here we shall accept that any general consideration of turbulence in an expanding gaseous envelope should take into account the presence of a regular magnetic field in planetary nebulae.

It is also of interest to determine how the Reynolds number varies as the nebula expands. First we need to establish the dependence of the Reynolds number upon the parameters of the envelope.

The difference in velocity between two arbitrary points in an expanding nebula changes with the points chosen. Therefore, the Reynolds number will also change. The maximum value that this number can attain is:

$$\text{Re} = \frac{2v_0 D}{v},\tag{31}$$

where v_0 is the maximum velocity difference, D the diameter of the envelope, and v the kinematic viscosity, which can be expressed in the following way:

$$v \approx ul,\tag{32}$$

where u is the thermal velocity, and l is the mean free path of the particles in the gas. We have $l \approx (\sigma n)^{-1}$ where n is the number of particles per unit volume and σ is the effective collision cross-section. Finally, for the Reynolds number in the outer layers of an expanding envelope we find

$$\text{Re} = \frac{2v_0}{u} Dn\sigma.\tag{33}$$

If we assume that the expansion velocity is constant and that the gas density decreases as the square of the radius of the envelope (for an envelope of constant thickness), we find from (33) that the Reynolds number decreases as the nebula increases in size. This conclusion leads to the following: if a gaseous envelope at some initial time is not in a turbulent state and is stable in shape, then later on, with increasing size, it will certainly be stable.

Therefore, there are two hydrodynamic factors leading to the loss of stability of the envelope and to its eventual destruction, that change in opposite directions. The first one, a resisting medium, plays a significant role only in envelopes of large dimensions: the second, on the contrary, is influential only for small envelopes.

9. On the Motion of Planetary Nebulae in the Interstellar Medium

Let us consider briefly the possible influence of the interstellar medium on the geometry of planetary nebulae as they move through this medium.

The overwhelming majority of planetary nebulae are distributed near the galactic equator, where the mean density of the interstellar matter is large. The gaseous envelope of the nebula, having collossal dimensions and a relatively small mass, must experience a drag opposing its motion through the medium and, after a certain time, it may "fall behind". Then the nucleus, which experiences practically no drag, should not appear in the center of the nebula but shifted somewhat to the front edge. If such a phenomenon does indeed occur we should quite frequently observe planetary nebulae with eccentric nuclei.

Actually, however, the nuclei of the majority of the nebulae are located quite symmetrically relative to the nebula. If any eccentricity is observed it never exceeds 0.1 or 0.2 of the radius of the nebula. Therefore we must conclude that for some reason or other the interstellar medium exerts no influence on the motion of the nebula.

In its motion through the interstellar medium only half the nebula experiences a drag. The envelope is decelerated because part of its momentum is transferred to the interstellar matter collected along the way and accreted by the envelope. It is quite probable that such a process does not occur because the linear thickness of the envelope of the planetary nebula is many times larger than the mean free path of the gas particles. Therefore we can write, neglecting the variation in size of the nebula,

$$\mathfrak{M}_0 w_0 = (\mathfrak{M}_0 + \pi R^2 \rho_0 s)w, \tag{34}$$

where w_0 is the initial relative velocity of the envelope (the velocity of the translational motion) and also of the nucleus; $2R$ is the diameter of the nebula, w is the velocity of the "front" side of the nebula (*i.e.* the side facing the motion upon which the drag force is applied) after the nebula has moved a distance.

From (34) we get

$$w = w_0 \frac{1}{1 + (\pi \rho_0/\mathfrak{M}_0)R^2 s}. \tag{35}$$

Substituting $w = ds/dt$ and integrating we find:

$$(s - R) + \frac{\pi\rho_0}{\mathfrak{M}_0} R^2(s^2 - R^2) = w_0 t, \tag{36}$$

where the constant of integration has been evaluated from the condition that at $t = 0$ the nucleus is at the origin of our coordinate system and the front side of the nebula is at the distance R.

Equation (36) is actually an equation of motion for the front side of the envelope. Since the nucleus is not decelerated, its equation of motion is simply:

$$s' = w't. \tag{37}$$

As an example let us calculate the time necessary to produce a "displacement" between the nucleus and the center of the nebula (eccentricity) of one tenth of the radius of the nebula. We have

$$s' = s_0,$$

$$s' = s_0 - 0.9R.$$

Comparing the above with (37), and also with (36), we find that the distance that the nebula must cover in order to reach the specified eccentricity is

$$s_0 = \left[R^2 + \frac{0.1\mathfrak{M}_0}{\pi\rho_0 R} \right]^{1/2}, \tag{38}$$

and the required time is

$$t_0 = \frac{1}{w} \left[R^2 + \frac{0.1\mathfrak{M}_0}{\pi\rho_0 R} \right]^{1/2} \tag{39}$$

Taking $\mathfrak{M}_0 = 0.01\,\mathfrak{M}_\odot$, $\rho_0 = 5 \times 10^{-24}$ gm/cm^3, $w_0 = 30$ km/sec and $R = 20{,}000$ a.u. we get $t \approx 10{,}000$ years.

Thus, the resistance of the interstellar medium can produce an asymmetry in the position of the nucleus relative to the nebula of one tenth of the radius of the latter in a time of the order of 10,000 years. Here we did not take into account the expansion of the nebula, which if included would lead to even longer times (since for nebulae of small dimensions the drag would be smaller). Therefore we can conclude that in order to produce a significant displacement between the nucleus and the nebula under the influence of the drag of the interstellar medium the time required is of the order of the life of the nebula itself. The consequences of such influence could only be observed in very large nebulae, and indeed have been observed.

The fact that in a few relatively large planetary nebulae we can observe an increase in brightness at one side shows that the drag exerted by the interstellar medium is already occurring. Examples of such nebulae are NGC 6888, A 16, NGC 7139.

Returning to the question of the possible capture of particles from the interstellar medium (hydrogen atoms) by the nebula, we should consider the repulsive influence of radiation pressure upon the medium. If, indeed, such an influence exists then, obviously, the interstellar hydrogen atoms not only will not be able to enter the nebula but, on the contrary, will be pushed away by the radiation pressure so that a certain space in front of the nebula will be free of them. However, it is easy to prove the small likelihood of such a process. The interstellar hydrogen atoms can be repelled only by the L_α-radiation from the nebula. However, because of the Doppler effect due to the difference in velocity between the nebula and the interstellar medium, which is of the order of 30 km/sec and larger, this radiation will go through by the latter practically without absorption.

The impossibility that the interstellar hydrogen is repelled by the radiation pressure of the nebula also follows from the fact that we do not observe the relatively bright arch that should form on *one side* of the nebula (the "trailing" side), if this mechanism were really effective. True, in a few cases, something resembling a condensation has been observed. However, apparently they are observed on *all sides* of the nebula and may be the remains of a second envelope.

Chapter IX

Magnetic Fields in Planetary Nebulae

1. The Role of Galactic Magnetic Fields

Bipolarity is one of the most common structural features of planetary nebulae. Bipolar nebulae have two bright "beads", or regions of enhanced brightness, arranged symmetrically in relation to the nucleus. Examples of bipolar nebulae are given in Chapter I.

No planetary nebulae with, say, one or three "beads" have been discovered to date. The highly pronounced bipolar structure of planetary nebulae can therefore be taken as evidence of the electromagnetic origin of the phenomenon.

We will first consider the possible contribution from the general magnetic field of the Galaxy. Since the regular galactic magnetic field may cause redistribution of the almost completely ionized nebular matter, the density may be nonuniformly distributed in different directions within the nebula. If this hypothesis is true, a certain correlation can be expected between the direction of the galactic magnetic lines of force and, say, the direction of the major axis of a bipolar planetary nebula.*

The direction of the galactic magnetic lines of force normally varies from one part of the sky to another. The magnetic field, however, is fairly homogeneous within certain parts of the Galaxy, and in regions of substantial apparent size the lines of force are nearly parallel. The average direction of the lines of force generally makes a small angle with the galactic plane; it is only occasionally that the inclination of the lines of force is not correlated with the direction of the galactic equator (these exceptional cases arise when the magnetically homogeneous region is situated along one of the spiral arms of the Galaxy). The above results were inferred mainly from polarimetric observations of numerous sufficiently distant stars, giants included. In particular, the direction of the galactic magnetic lines of force is determined from the plane of polarization of the electric field vector. The strength of the galactic magnetic fields, as estimated by a variety of methods, was found to be of the order of 10^{-5}–10^{-6} gauss.

---

* The minor axis of a bipolar planetary nebula passes through the two "beads", and the major axis is perpendicular to the minor axis.

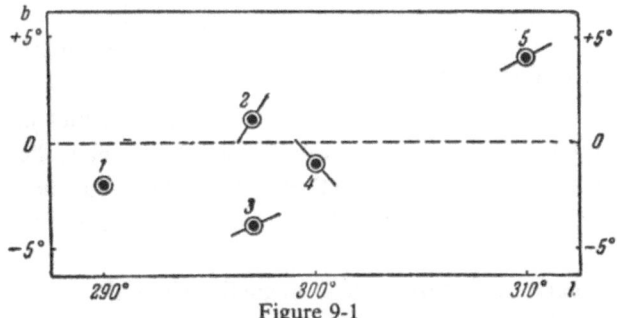

Figure 9-1

An example illustrating the lack of correlation between the direction of the major axis of bipolar planetary nebulae and the direction of the Galactic equator (dashed line): 1) anon. 15^h47^m, $-51°21'$; 2) anon. 15^h30^m, $-58°59'$; 3) anon. 16^h10^m, $-54°50'$; 4) anon. 16^h13^m, $-51°52'$; 5) NGC 6153.

Analysis of observational data does not reveal any correlation between the orientation of the "beads" and the direction of the galactic equator: planetary nebulae with their major axis parallel to the galactic equator are as frequent as those with the major axis at right angles to the equator [165]. An excellent example is provided by a group of five nebulae, four of which are bipolar, clustered in a relatively small region of the southern sky, near the galactic equator. A map of this region is shown in Figure 9-1. There are unfortunately no data on the polarization of the light of stars in this part of the sky, and the direction of the galactic magnetic field remains unknown. This group of nebulae is nevertheless of considerable interest (good photographs of these nebulae were published in [166]). First, not all the nebulae are bipolar: one of them is an ordinary ring nebula (anon. 15^h47^m), without any traces of bipolar structure. The presence of a fairly large ring nebula ($D = 72''$) near a cluster of bipolar planetary nebulae is puzzling if the galactic magnetic fields are regarded as being responsible for the bipolar structure. Next, the major axes of the four bipolar nebulae are not parallel to one another. Finally, there is no correlation between the orientation of the major axes and the direction of the galactic equator.

The scatter in the orientation of the major axes of the bipolar planetary nebulae and the lack of correlation with the direction of the galactic equator may be attributed to these nebulae being scattered at various distances along the line of sight, where the magnetic fields have different strength and direction, but this explanation is not very convincing. Stars for which polarimetric observations are available are also scattered at various line-of-sight distances, but their electric vectors nevertheless oscillate in a plane almost parallel to the plane of the galactic equator.

It follows from the preceding that at least the process of "bead" formation in planetary nebulae is independent of the general magnetic field of the Galaxy. This conclusion is further supported by the fact that bipolar planetary nebulae sometimes occur at very high galactic latitudes (*e.g.*, the typical bipolar nebula NGC 3587, $b = +58°$), where the magnetic field is too weak to have a marked influence on internal motions in the nebula.

The magnetic field of the central star—the nucleus—cannot affect the ionized matter of the nebula either. As this is a dipole field, it falls off as the cube of the distance, so that the magnetic field strength in the region of the "beads" is as low as 10^{-10} gauss, if a field of the order of 10^6 gauss at the surface of the central star is assumed.

We conclude therefore that the regular magnetic field of the Galaxy and the dipole magnetic field of the central star do not affect the fundamental structure and the shape of the planetary nebulae. And yet, the hypothesis of the electromagnetic origin of the bipolar structure of planetary nebulae can hardly be rejected. Hence we are left with the conclusion that *the nebula itself has an intrinsic magnetic field*, which is neither an extension of the field of the central star nor related to interstellar magnetic fields.

The magnetic field of a planetary nebula cannot be homogeneous (have equal magnitude and constant direction at all points of the nebula), since it would have then produced the same effect everywhere and no redistribution of matter within the nebula would have resulted. For the same reason this field cannot be random. The bipolar structure of the planetary nebulae indicates, first, that the strength and the direction of the magnetic field should vary in a definite regular manner from one point in the nebula to another and, second, that this field should be symmetric about one of the axes of the nebula. A *dipole* or a dipole-like magnetic field formally meets these requirements.

Thus, without making any assumptions as to the origin of the magnetic field in a planetary nebulae and the mechanism sustaining the field, we regard it as a dipole field. In the next sections of this chapter we will show that the hypothesis of a dipole magnetic field in planetary nebulae provides a satisfactory interpretation of the various shapes and structures observed.

2. Mechanical Equilibrium in Planetary Nebulae with a Magnetic Field

Planetary nebulae are expanding spheres of gas, large in comparison with stars, and as such they possess positive energy. How will a sphere of this kind behave in the presence of a magnetic field?

If the nebula has an intrinsic magnetic field, it also carries a certain quantity of magnetic energy. Let H be the average magnetic field strength (averaged over the entire nebula). The specific magnetic energy, *i.e.*, energy per unit volume of the nebula, is then $(\bar{H})^2/8\pi$. The total magnetic energy $\mathfrak{M}$ of a homogeneous spherical nebula is thus given by

$$\mathfrak{M} = \frac{1}{8\pi}(\bar{H})^2 V = \frac{1}{6}(\bar{H})^2 R^3, \tag{1}$$

where V and R are the volume and the radius of the nebula. The magnetic energy is clearly a positive quantity, independent of the polarity of the field. Its contribution to the mechanical equilibrium of a gaseous sphere is therefore identical to that of the kinetic energy: the magnetic energy opposes the negative potential energy and tends to destroy the equilibrium, *i.e.*, to cause expansion of the sphere. The problem is to find the field that causes expansion in spheres of known mass and radius.

This problem is best solved by means of the virial theorem. The most general expression for this theorem was derived by Chandrasekhar and Fermi [200]:

$$\frac{1}{2}\frac{d^2I}{dt^2} = 2T + 3(\gamma - 1)U + \mathfrak{M} + \Omega. \tag{2}$$

Here I is the moment of inertia of the gaseous sphere, T, U, $\mathfrak{M}$, and Ω are respectively its kinetic, thermal, magnetic, and potential energy, and γ is the adiabatic index.

The moment of inertia I varies as the nebula expands. Let the rate v_0 of expansion be constant; then for a homogeneous nebula of mass M_0,

$$I(t) = \int_0^{M_0} r^2 dm = \frac{3}{5}M_0 v_0^2 t^2, \tag{3}$$

where t is the time for the nebula to expand to radius $R (= v_0 t)$. Inserting for I in (2) its expression from (3) and seeing that $M_0 v_0^2/2 = T$, we find

$$\frac{4}{5}T + 3(\gamma - 1)U + \mathfrak{M} + \Omega = 0. \tag{4}$$

On the other hand, the total energy of the nebula (ignoring the radiative energy) is

$$E = T + U + \mathfrak{M} + \Omega. \tag{5}$$

Eliminating U between (4) and (5), we find

$$E = -\frac{3\gamma - 4}{3(\gamma - 1)}(|\Omega| - \mathfrak{M}) + \frac{15\gamma - 19}{15(\gamma - 1)}T. \tag{6}$$

The nebula will not expand, *i.e.*, will remain in equilibrium, if its total energy is negative, $E < 0$. Thus, putting $\gamma = 5/3$ (hydrogen), we have from (6)

$$\frac{6}{5}T + \mathfrak{M} < |\Omega|. \tag{7}$$

If $T = 0$, the nebula is in equilibrium when

$$\mathfrak{M} < |\Omega|. \tag{8}$$

Planetary nebulae expand, however; that is to say, they are not in equilibrium. Assuming for the moment that the lack of equilibrium is due to the action of the magnetic field alone, we replace (8) with

$$\mathfrak{M} > |\Omega|. \tag{9}$$

For constant-density spherical configurations we have

$$|\Omega| = \frac{3}{5} \frac{GM_0^2}{R}, \tag{10}$$

where G is the gravitational constant. Substituting (1) and (10) in (9), we find

$$\bar{H} > \left(\frac{18}{5} G\right)^{1/2} \frac{M_0}{R}. \tag{11}$$

For a typical planetary nebula, $M_0 \sim 0.1 \, M_\odot$, and $R \sim 10{,}000$ a.u. Inserting these figures in (11), we find

$$\bar{H} > 10^{-6} \text{ gauss}. \tag{12}$$

Magnetic energy is obviously not the only factor opposing the mechanical equilibrium of planetary nebulae: other forces, having a nonzero total kinetic energy T, may also contribute to this effect. We therefore return again to relation (7) (with the inequality sign reversed) and obtain an even smaller estimate for the right-hand side of (12). However, a field of 10^{-6} gauss is already comparable (or even less) in its order of magnitude to the general galactic magnetic field, which has been shown above to make no contribution to the local condensation of gaseous matter in the nebulae. We therefore need not bother to improve on the lower-limit estimate in (12), since it is clear that the magnetic field strength in the nebulae should be considerably larger than 10^{-6} gauss to have a noticeable effect.

This result is significant mainly in that it points to the important role of magnetic fields in the mechanical equilibrium of nebulae. Even if other forces capable of causing expansion are present (inertial forces, gas and radiation pressure), *a magnetic field of $\bar{H} > 10^{-6}$ gauss is sufficient to destroy the mechanical equilibrium of a nebula and make it expand.*

A realistic estimate of the field strength can be obtained if we assume that the motion of gases in certain parts of the nebula is controlled by magnetic fields. This implies that the magnetic energy in the relevant regions is comparable in its order of magnitude with the corresponding thermal energy of gas particles. In other words, if the formation of "beads" in planetary nebulae is indeed attributable to the internal magnetic field, we should have $U \sim M$ or

$$\frac{1}{6}(\bar{H})^2 R^3 \sim \frac{M_0 u^2}{2}, \tag{13}$$

where u is the mean thermal velocity of the gas particles. Taking $u \sim 10\,\mathrm{km/sec}$ we obtain for the average field strength (for the same mass and radius as above)

$$\bar{H} \sim 10^{-3}\,\text{gauss}. \tag{14}$$

Given magnetic fields of this magnitude, planetary nebulae will no longer be in mechanical equilibrium and must therefore expand, even if no other forces act in this direction.

We have considered one method for the determination of the average magnetic field in nebulae. Another, more precise, method, which can be applied to particular planetary nebulae, is described in what follows.

Table 9-1 gives a summary of the various contributions to the energy of a medium-sized planetary nebula of average mass. The last line gives the total energy (gravitational, thermal, ionization, *etc.*) for an average star (the Sun). The gravitational energy of a nebula, as we see from the table, is three orders of magnitude less than the kinetic, the thermal, or the magnetic energy, and the sum of the last three components is three orders of magnitude less than the total energy of the central star, if this is a star like the Sun. In other words, planetary nebulae contain only a negligible fraction of the total energy of the nucleus.

TABLE 9-1
Breakdown of Total Energy for a Homogeneous
Planetary Nebula with $M = 0.1\,M_\odot$
and $R = 1.5 \times 10^{17}\,\mathrm{cm}$

Form of energy		Energy (erg)
Kinetic $T(v_0 = 20\,\mathrm{km/sec})$		$4 \cdot 10^{44}$
Thermal $U\ (u = 10\,\mathrm{km/sec})$		10^{44}
Gravitational Ω		$2 \cdot 10^{40}$
	$H = 10^{-3}$ gauss	$6 \cdot 10^{44}$
Magnetic $\mathfrak{M}$	$H = 10^{-4}$ gauss	$6 \cdot 10^{42}$
	$H = 10^{-6}$ gauss	10^{39}
Total energy of an average star (Sun)		$5 \cdot 10^{48}$

3. Energy Equilibrium of Planetary Nebulae

A different picture is obtained, however, if we consider the energy equilibrium in a planetary nebula with magnetic fields. The possibility of radiative energy equilibrium was demonstrated in Chapter VI. The other types of energy are apparently also in equilibrium, since otherwise a difference in energy would be established between any two volume elements in the interior, resulting in a pressure gradient and corresponding large-scale motion. This is not supported by observations (see Chapter I). We therefore conclude that energy equilibrium is attained and that the relaxation time (time to attain equilibrium) is significantly less than the lifetime of the planetary nebulae.

These considerations, however, do not constitute a rigorous proof. In particular, the order of magnitude of the relaxation time deserves special consideration. It is therefore advisable to assume from the start that *the total specific energy is equal at all points in the interior of a nebula*, i.e.,

$$\sum E_i = C, \tag{15}$$

where the sum is taken over all energy types. We will show in what follows that this assumption, together with the hypothesis of symmetric, dipole-like magnetic fields in planetaries, provides a satisfactory interpretation of the various features of their structure and shape.

The kinetic energy of the thermal motion of gas particles and the magnetic energy are the principal components of the total energy of the nebula; gravitational energy is vanishingly small, and there is virtually no turbulence. For simplicity we will consider a stationary, nonexpanding nebula, although this is not a fundamental restriction; it only implies that the energy equilibrium in the nebula was attained before the nebula had significantly changed its size.

We therefore write the condition of energy equilibrium (15) in the form

$$\frac{H^2}{8\pi} + 2nkT = C, \tag{16}$$

where H is the magnetic field strength at a point, n the concentration of protons (electrons) at the same point, T the electron temperature, and k Boltzmann's constant. The first term in the left-hand side of (16) is the magnetic energy per unit volume, and the second is the thermal energy per unit volume. Both are local quantities varying with position. The factor 2 in the second term signifies that the gas (hydrogen) in the nebula is almost completely ionized, the number of protons being equal to the number of electrons, so that the total number of free particles is twice the number of free electrons.

In a regular, nonrandom magnetic field geometry, the relation (16) can be physically interpreted as follows. If the magnetic field is frozen, as we say, into the plasma (the highly ionized gaseous medium), the gas will move only along the magnetic lines of force; the flow of gas at right angles to the lines of forces is forbidden, unless special factors stimulating motion in this direction are present. It can be shown [168] that in the absence of volume forces (this is apparently the case everywhere in the interior of the nebula), all motion at right angles to the magnetic field decays exponentially in a time $t = c^2\rho/\sigma^I H^2$, which is less than 1 sec in planetary nebulae (σ^I and ρ are the forward conductivity and the density of the medium).

Now if the field strength varies along a line of force, the magnetic energy density (magnetic energy per unit volume) will also vary along that line, resulting in different magnetic pressures at different points. The magnetic pressure gradient will clearly cause the gas to drift along the lines of force from high to low magnetic pressure. This will destroy any initially uniform distribution of gas: the gas density will increase in regions of low magnetic pressure, and *vice versa*. The drift of gas along the line of force will cease when the magnetic pressure gradient $(1/8\pi)\nabla(H)^2$ is balanced by the gas pressure gradient $2kT\nabla(n)$, *i.e.*,

$$\nabla \left(\frac{H^2}{8\pi} + 2nkT \right) = 0. \tag{17}$$

Integrating, we obtain (16).

Unfortunately, the relation (16) does not apply to dipole fields since, first, it is valid only for parallel lines of force and, second, the magnetic dipole strength being balanced by the tension of the lines of force, the dipole field cannot alter the initial distribution of gas in the nebula. Despite these fundamental objections, we will adopt relation (16) as a working hypothesis for planetary nebulae, with the further additional assumptions:

(a) The magnetic lines of force in the main part of the nebular volume—the envelope—are vitually parallel, being directed along the envelope. The absolute magnitude of the field varies in a certain way from the magnetic pole to the magnetic equator.

(b) The structure of the magnetic field in the central part of the nebula and near the magnetic poles is ignored.

In (16), H and n are the only position-dependent quantities; the electron temperature T is generally determined by atomic absorption and emission processes, and it is henceforth assumed constant over the entire nebula. In functional notation, the relation (16) therefore has the form

$$\frac{1}{8\pi} |H(x, y, z)|^2 + 2kTn(x, y, z) = C. \tag{18}$$

The constant C is clearly determined from the constancy of the total mass along the line of force, *i.e.*,

$$\oint_L n(x, y, z)\, dl = n_0 \oint dl = n_0 L, \qquad (19)$$

where the integral is taken along the line of force; L is the length of the line, n_0 the number of atoms per unit volume of "unperturbed" nebula (*i.e.*, a nebula of uniform density without a magnetic field). Using (18), we find

$$C = 2n_0kT + \frac{1}{8\pi L} \oint |H(x, y, z)|^2\, dl. \qquad (20)$$

Thus, although C is constant for any given line of force, it may change from one line of force to another. As the field strength is a continuous function of position, C is also continuous at any point inside the nebula.

In applied calculations, the problem is essentially simplified if C is assumed constant within the entire nebula. Anyhow, this assumption is fully justified in qualitative analysis.

The average value of C can be determined by one of the following two methods. An "infinite" nebular model can be assumed, in which the material density approaches a constant value n_0 at infinite distance from the center, *i.e.*, $n(\infty) = n_0$ as $r \to \infty$. Then,

$$C = 2n_0kT. \qquad (21)$$

Alternatively, for a "finite" nebular model the basic assumption is the constancy of masses of the "perturbed" and the "unperturbed" nebulae. Thus,

$$C = 2n_0kT + \frac{1}{8\pi V} \int_V |H(x, y, z)|^2\, dV, \qquad (22)$$

where the integral is taken over the entire volume of the nebula.

Quantitative analysis shows that the difference between "finite" and "infinite" nebulae is very slight. In the following we therefore consider only the "finite" nebular model.

The equation (18) is the basis of all subsequent calculations. In every individual case the distribution of gas density, and therefore the distribution of brightness, over the nebula are determined by the nature and the magnitude of the magnetic field $H(x, y, z)$. The apparent shape and structure assumed by the nebula is governed by this factor.

Our problem then, is to find the kind of magnetic field that will account for the observed bipolar structure of planetary nebulae. The magnetic field of an extended dipole is found to fit these requirements very closely. However, this only goes as far as the similarity in the *distribution of the magnitude* of the field strength over ϕ, the coordinate between the two poles. Indeed, as we have noted in the preceding, the assumption that the magnetic lines of force in a nebula are parallel to one another does not apply to an extended dipole. In other words, relations of the form $H = f(\phi, r)$, derived for point and extended dipoles, are used in the following as an *interpolation* formula for determining the field strength at a given angular distance from the magnetic axis (or the magnetic equator) of the nebula.

Despite the artificial character of these assumptions, they account satisfactorily for a number of facts pertaining to structure and apparent shape of planetaries.

We now proceed with a discussion of the relationship between the nebular structure and the magnetic field of point and extended dipoles, remembering that this approach is, to a certain degree, purely formal.

4. A Point Dipole

A magnetic dipole is a system comprising two different 'charges" of opposite sign at a distance I from each other. Normally in physics the term dipole is used only when I is much less than the distance r of the "charges" from the field point, *i.e.*, when $r \gg I/2$. In planetary nebulae, however, we are dealing with dipoles whose size is comparable with the radius of the nebula, *i.e.*, $r \sim I/2$. We therefore speak of a "point" dipole (Figure 9-2) when $r \gg I/2$, *i.e.*, when the dipole is in fact situated at the center of the nebula, and of an "extended" dipole (Figure 9-3), when $r \sim I/2$, *i.e.*, when the dipole "occupies" the entire nebula or, more precisely, when the "charges" or the poles of the dipole are at the two opposite ends of the nebula. The line through the "charges" (or the S and N poles) is the dipole axis or the magnetic axis of the nebula.

In this section we will consider the effect of a point dipole on a nebula and show that a dipole magnetic field does indeed produce bright "beads" situated symmetrically about the nucleus.

Consider a full spherical nebula of radius R with constant gas concentration (electrons and protons) n_0 at all points. Take a certain section through the center O of the nebula ("a central section") and place a point dipole of moment a at the origin. The field strength along a given line of force at a point (r, ϕ) due to the point magnetic dipole is (in plane polar coordinates)

$$H(r, \phi) = a\eta(r, \phi), \tag{23}$$

where

$$\eta(r, \phi) = \frac{1}{r^3}\sqrt{1 + 3\sin^2\phi}. \tag{24}$$

The field vector is directed along the tangent to the line of force and makes an angle

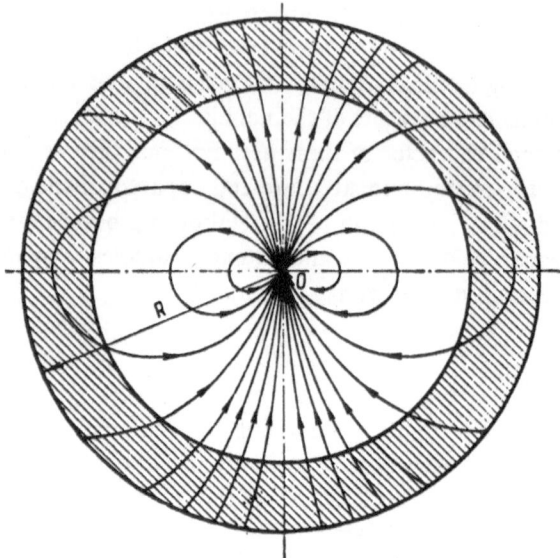

Figure 9-2
A point dipole in a ring nebula.

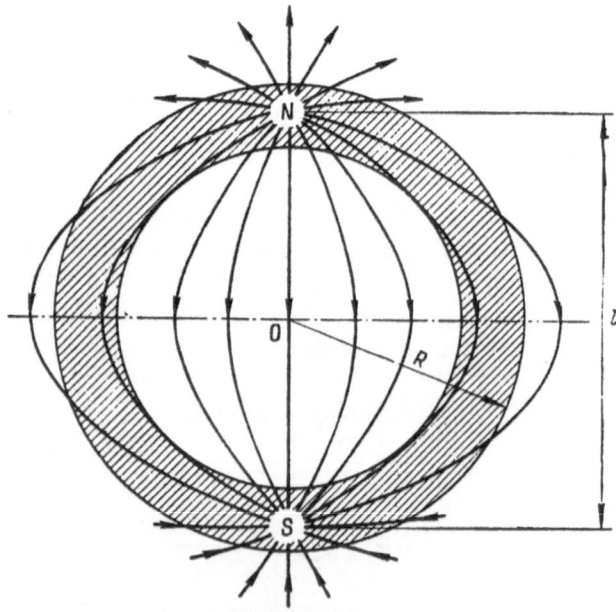

Figure 9-3
An extended dipole in a ring nebula.

α with the radius-vector, which is defined by

$$\tan \alpha = \frac{1}{2} \cot \phi. \tag{25}$$

Note that α is constant for all lines of force, being independent of r. This is a characteristic feature of the point dipole, not observed for extended dipoles.

Expression (23) shows that the magnetic field in the central section has gradients in both coordinates. The field gradient sets up a magnetic pressure gradient which, as we have noted in the preceding, destroys the initially uniform gas density distribution in the nebula.

Inserting for $H(r, \phi)$ in (18) its expression from (23), we obtain an equation for the relative concentration of gas at the point (r, ϕ):

$$\frac{n(r, \phi)}{n_0} = 1 - \sigma \eta^2 (r, \phi), \tag{26}$$

where

$$\sigma = \frac{a^2}{16 \pi k T n_0}. \tag{27}$$

It follows from this relation that, first, the density distribution is symmetric about the magnetic axis of the nebula and, second, the concentration at a given distance from the center has a minimum in the direction of the magnetic axis ($\phi = 90°$) and a maximum in the direction of the magnetic equator ($\phi = 0°$). Moreover, the concentration of ions is relatively low near the center, increasing further out.

The volume emissivity of the nebula is proportional to the square of the electron or ion concentration. Therefore even a slight difference between the concentration at different points in the nebula may lead to a substantial difference in the surface brightness at these points. The largest difference occurs between the brightness in the direction of the magnetic axis of the nebula (region of minimum brightness) and that on its equatorial axis (region of maximum brightness). The contrast is preserved when the actual space distribution obtained by revolving the central section about the magnetic axis is projected onto the image plane perpendicular to the magnetic axis of the nebula. The apparent image is a nebula with two brightness maxima at points symmetrically located about the center, *i.e.*, a *bipolar* nebula.

For a given value of σ, which is a function of the magnetic moment a and the initial concentration n_0, the problem of gas equilibrium is meaningless for certain values r_0 and ϕ_0, *i.e.*, within a certain region of "avoidance" in the nebula. The equation of the curve delimiting this region of avoidance is formally obtained from the condition $n(r_0, \phi_0) = 0$. In the following we will therefore consider the distribution of gas only in the principal volume of the nebula, its envelope.

The hypothesis of a point dipole at the center of the nebula thus explains qualitatively the formation of bipolar planetaries. The results, however, are quantitatively inadmissible. If the actual size of nebulae is compared with the geometrical
extent of a dipole, $l = (a/H)^{1/3}$, the magnetic field at the dipole is found to be of the
order of 10^6 gauss, assuming a nebular field of 10^{-4}–10^{-5} gauss and a dipole
length of the order of a stellar radius.

The difficulty can be eliminated if the hypothesis of a point dipole is replaced
by one involving an extended dipole of size comparable to the size of the nebula.

5. An Extended Dipole

The magnetic field strength in a nebula at a point (r, ϕ) arising from an extended
dipole is

$$H(r, \phi) = a\eta_1(r, \phi), \tag{28}$$

where $\eta_1(r, \phi)$ has the form [167]

$$\eta_1(r, \phi) = 2^{5/2}\kappa^2 \frac{\{(1 + \kappa^2)^2 + 4\kappa^2 \sin^2\phi - (1 - \kappa^2)[(1 + \kappa^2)^2 - 4\kappa^2 \sin^2\phi]^{1/2}\}^{1/2}}{(1 + \kappa^2)^2 - 4\kappa^2 \sin^2\phi}, \tag{29}$$

and $\kappa = l/2r$.

This is a general formula applicable to large and small distances from the origin.
In particular, for $l/r \ll 1$, the expression (29) reduces to (24). In practice, the dipole
can be regarded as a point dipole for l/r of the order of 0.1. The function $\eta_1(r, \phi)$
is plotted against r/l in Figure 9-4 for various values of ϕ.

The distribution of the relative concentration of electrons in the central section
of the nebula is given, like in (26), by

$$\frac{n(r, \phi)}{n_0} = 1 - \sigma\eta_1^2(r, \phi). \tag{30}$$

In the case of an extended dipole the geometry of the central section and the shape
of the nebula as projected onto the sky are also markedly affected by the ratio l/R
of the dipole length to the outer radius of the nebula. This ratio also characterizes
the relative "depth" of the magnetic poles in the nebula. On the other hand, the
outer radius R of the nebula and the thickness h of the envelope enter the theory
as independent quantities, unrelated to the magnetic field. We are therefore dealing with two parameters l/r and l/h, which may take on different values in different
nebula.

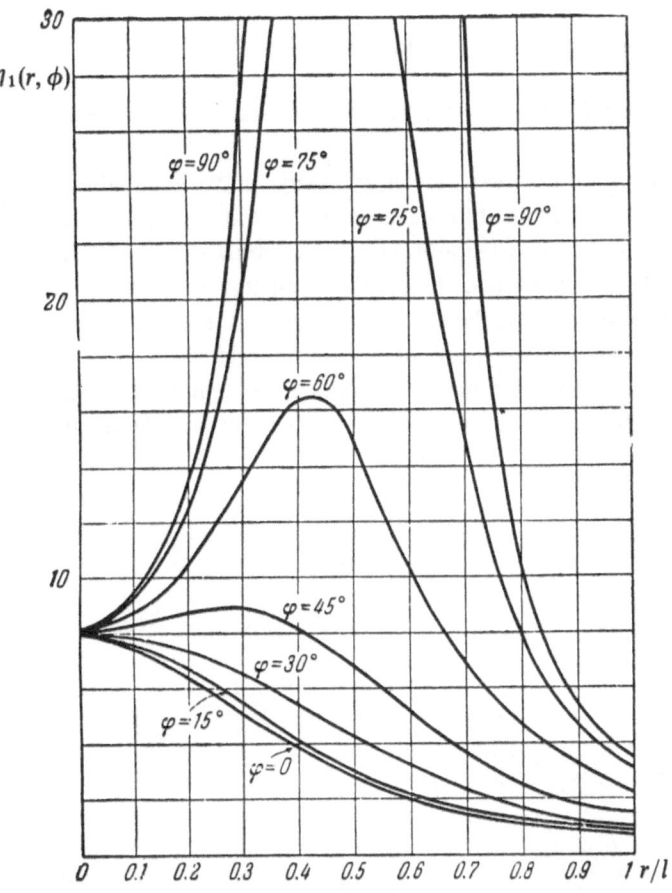

Figure 9-4

$\eta_1 (r, \phi)$ vs. r/l for various values of ϕ.

6. Brightness Distribution of a Nebula with a Magnetic Field

In this section the isophotes derived from the above theoretical calculations are compared with the observed isophotes or shapes of individual planetary nebulae.

We will consider the particular case of a nebula with the magnetic axis at right angles to the line of sight. The relative surface brightness $J (p, \psi)$ at the point (p, ψ) of the image of the nebula projected onto the celestial sphere is then given by

$$J (p, \psi) = C_1 \int_p^{\sqrt{R^2 - p^2 \cos^2\psi}} \frac{n^2 (r, \psi)\, r\, dr}{\sqrt{r^2 - p^2}}, \quad (0 \leq \psi \leq 90°), \qquad (31)$$

where C_1 is a constant and $n(r, \psi)$ is given by (30), with $\sin \phi$ replaced with

$$\sin \phi = \frac{p}{r} \sin \psi.$$

In (31) it is implied that the nebula is transparent to the visual radiation that it emits.

The integral in (31) is evaluated graphically for various values of the radius R of the nebula, the thickness of the envelope h (both in units of l), and the parameter σ.

Figure 9-5 is a plot of the isophotes for a full nebula with $R = 2.25$ (*i.e.*, $R/l = 2.25$) and $\sigma = 1$. The numbers near the isophotes in this and following figures give the intensity in arbitrary units; the dashed line marks the limits of the region of avoidance. The two "beads"—regions of maximum brightness in the equatorial plane of the nebula arranged symmetrically about the center—are very prominent on this drawing. The regions of minimum brightness are on the magnetic axis.

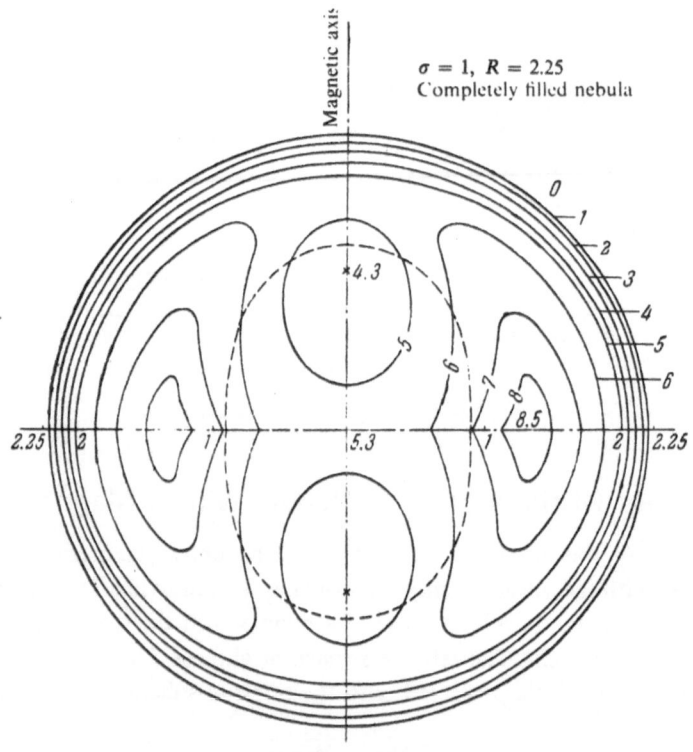

Figure 9-5

Theoretical isophotes of a "full" planetary nebula with $R = 2.25$ and $\sigma = 1$.

The surface brightness distribution of NGC 3587 (the Owl Nebula, see Plate III) is very similar to the theoretical distribution of Figure 9-5. The observed isophotes of the Owl Nebula are shown in Figure 9-6, reproduced from Minkowski and Aller [156]. The similarity is both qualitative and quantitative. The ratio of the maximum (in the equatorial section of the nebula) to the minimum (on the magnetic axis) intensity for NGC 3587 (the numbers in Figure 9-6 are log intensities in relative units) is approximately 1.9. The theoretical ratio, according to Figure 9-5, is 2.

Since the theoretical and the observed isophotes are identical, we can estimate the magnetic field in various parts of NGC 3587. In this case $\sigma = 1$, and we have from (27)

$$\sigma = \frac{a^2}{16\pi k n_0 T_e} = 1. \tag{32}$$

According to Table 4-15 the electron concentration of NGC 3587 is $n_0 = 150 \, \text{cm}^{-3}$ at $T_e = 10,000°$. Using these numerical data, we find (32) $a \approx 10^{-4}$ gauss cm^3. For the field strength at any point in the nebula we have from (28)

$$H(r, \phi) = a\eta_1(r, \phi) = 10^{-4}\eta_1(r, \phi). \tag{33}$$

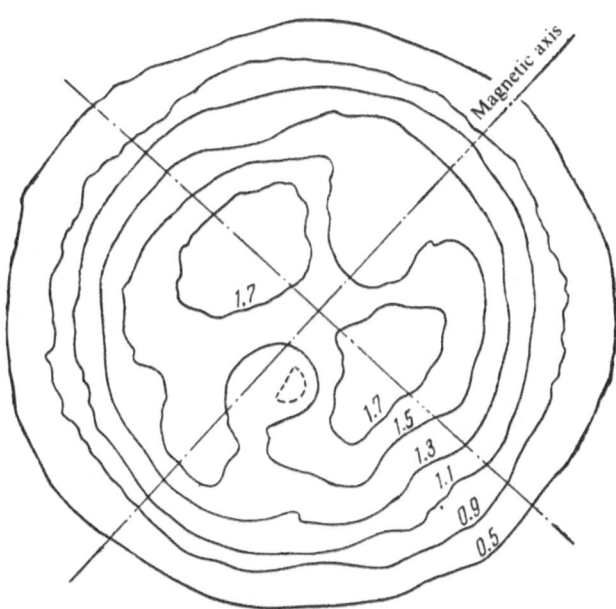

Figure 9-6
The observed isophotes of NGC 3587 (Owl Nebula) in red light (Minkowski and Aller).

For $r = 1$ (approximately half the radius) we obtain from (29) or from Figure 9.4, $\eta_1 (1, 0°) = 0.175$ and $\eta_1 (1, 90°) = 3.56$. Hence, $H (1, 0°) = 0.7 \times 10^{-4}$ gauss and $H (1, 90°) \approx 4 \times 10^{-4}$ gauss. For $r = 2$ (near the edge of the nebula) we have $\eta_1 (2, 0°) \approx 10^{-5}$ gauss, $H (2, 90°) \approx 3 \times 10^{-5}$ gauss. At the center of the nebula $(r = 0)$, $\eta_1(0) = 8$, so that $H \approx 0.8 \times 10^{-3}$ gauss.

We note that the magnetic field strength is not overly sensitive to the electron concentration and temperature: $H \sim (n_e T_e)^{1/2}$.

Figure 9-7 shows the theoretical isophotes for a full nebula with $\sigma = 1$, as before, but with $R = 1.67$. Here the "beads" are peaked on the inside, and the regions of maximum intensity are, on the whole, more rounded. The result is a "dumbbell" nebula, similar to NGC 3195 and anon. $16^h 13^m.3$. A good example of this kind is provided by A 70 (see Plate II).

Figures 9-8 and 9-9 give the theoretical isophotes for an envelope of thickness h (a hollow nebula), which constitutes respectively 1/4 and 1/8 of the nebular radius. The outer radius of the nebula is $R = 1.34$ and $R = 1.17$, respectively. Both figures are plotted with $\sigma = 7.9 \times 10^{-2}$.

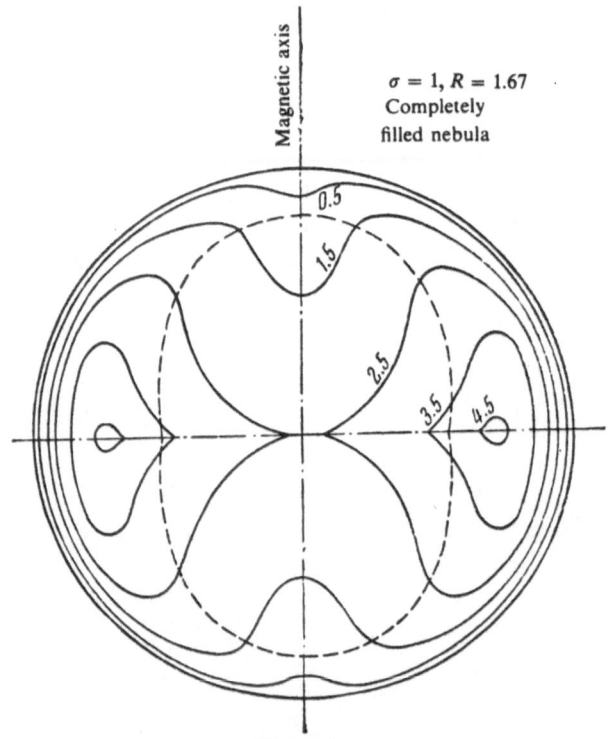

Figure 9-7

Theoretical isophotes of a "full" planetary nebula with $R = 1.67$ and $\sigma = 1$.

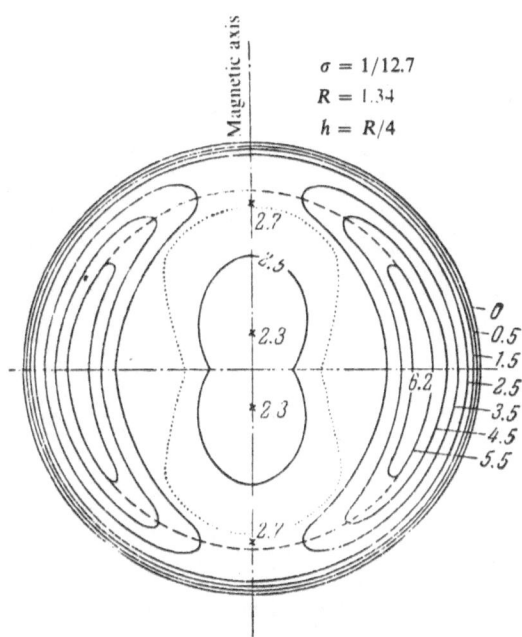

Figure 9-8

Theoretical isophotes of a hollow planetary nebula with $R = 1.34$, $h = R/4$, and $\sigma = 1/12.7$.

The isophotes of Figure 9-8 are structurally similar to a ring nebula, where, although no distinct "beads" are observed, the brightness increases at the ends of the minor axis (NGC 6720, NGC 7293, *etc.*). From the observed isophotes of NGC 6720, the ratio of the maximum intensity to that at the center of the nebula is about 2.5–3 (see, *e.g.*, [8], p. 247). An almost identical ratio is obtained from the isophotes of Figure 9-8. The numerical values used in the theoretical calculations above therefore appear to fit the actual conditions for NGC 6720. We therefore find $H \approx 10^{-4}$ gauss for the magnetic field strength at a distance $r \sim 1-1.3$ in the region of enhanced brightness ($\phi = 0°$), $H \approx 5 \times 10^{-4}$ gauss at the magnetic poles ($\phi = 90°$), and $H \approx 10^{-3}$ gauss at the center of the nebula.

The isophotes of Figure 9-9 are in satisfactory qualitative agreement with the structure and the shape of anon. $16^h 10^m.5$ [166] and NGC 7662 [34]. The "beads" of these nebulae are "banana-shaped".

Figure 9-10 is still another example of theoretical isophotes. It corresponds to $\sigma = 10$, $R = 2$, and $h = R/10$. Here the contrast between the "beads" and the central brightness. is even greater. This brightness distribution is observed, *e.g.*, in NGC 40, NGC 6058, and other planetary nebulae. Remarkable representatives of this class are the planetary nebulae A 66 (Plate II) and A 19.

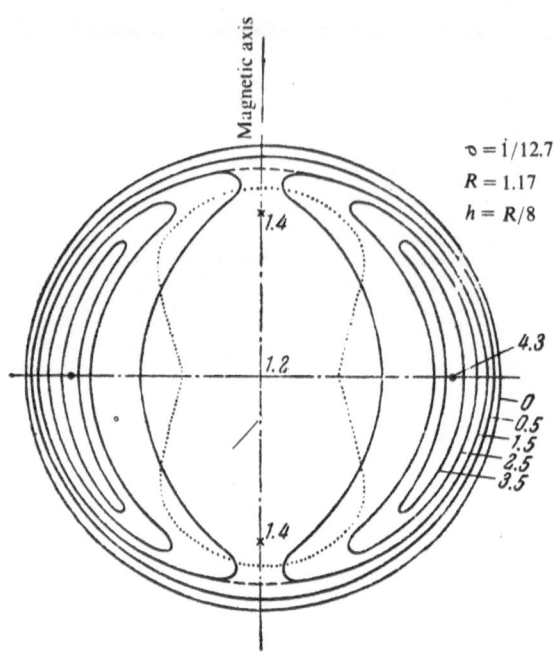

Figure 9-9

Theoretical isophotes of a hollow planetary nebula with $R = 1.17$, $h = R/8$, and $\sigma = 1/12.7$.

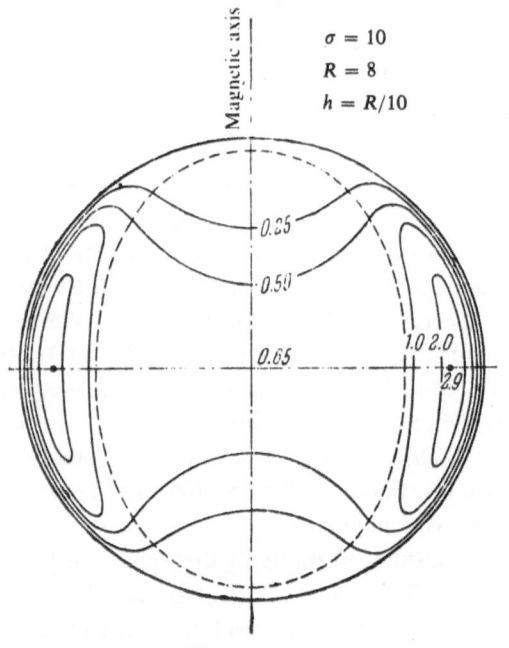

Figure 9-10

Theoretical isophotes of a hollow planetary nebula with $R = 2$, $h = R/10$, and $\sigma = 10$.

In the above examples the magnetic axis of the nebulae was assumed to lie perpendicular or nearly perpendicular to the line of sight. The apparent shape of the nebula will clearly change with the orientation of the magnetic axis relative to the observer. For example, a nebula with its magnetic axis along the line of sight is projected onto the celestial sphere in the form of a ring with uniform brightness in all directions. The bright parts of nebulae with intermediate orientation of the magnetic axis appear in projection in the form of elliptical annular structures, although the basic spatial structure is clearly spherical. The bipolar structure is nevertheless apparent even if the magnetic axis is far from perpendicular to the line of sight.

The various forms of bipolar structure and of some structural features of planetary nebulae can thus be understood if the magnitude of the magnetic field strength in the nebula is assumed to follow the pattern of the field of an extended dipole.

7. Elongated Nebulae

The intrinsic magnetic fields of the planetary nebulae also affect their external configuration, *i.e.*, their apparent shape. The question of the mechanical equilibrium of a nebula (or a gaseous sphere) is of crucial significance in this respect.

The effect of a magnetic field on the figure of a conducting fluid sphere in mechanical equilibrium (a star) was first considered by Chandrasekhar and Fermi [200]. They showed that a dipole magnetic field acting on the outermost parts of the sphere and a uniform magnetic field acting in the interior would flatten a gaseous sphere in the direction of the magnetic axis, transforming it to an oblate body. The relative flattening (eccentricity) in this case is proportional to the square of the magnetic field strength at the poles (see also [201]).

The situation is entirely different in gaseous spheres which are not in a state of mechanical equilibrium, such as planetary nebulae. It has been shown (Sec. 2) that a magnetic field causes a nebula to expand, and imparts additional kinetic energy to it. In the presence of a magnetic field a nebula will therefore expand somewhat more rapidly than without fields. If the magnetic field is homogeneous, its effect is isotropic and the external appearance of the nebula will not change: an initially spherical nebula will remain spherical while expanding.

The results of the previous sections show, however, that the magnetic fields in planetary nebulae are not homogeneous: the field strength along the magnetic axis is larger than in the direction of the magnetic equator. The nebula will therefore expand at different rates in different directions, and its initial spherical shape will be distorted.

Suppose that the force (unknown to us) which separated or expelled the nebula from the nucleus and initiated its expansion acts uniformly in all directions. This isotropic force causes uniform expansion of the nebula in all directions, *i.e.*, the gas retains its spherical shape. The intrinsic kinetic energy T_0 per unit volume of the nebula due to this expansion will remain constant in all directions at every stage of expansion.

If now a dipole magnetic field is superimposed on the nebula, the total energy per unit volume is modified by the magnetic energy and is

$$T_0 + \mathfrak{M}(r, \phi), \tag{34}$$

where $\mathfrak{M}(r, \phi)$ is the magnetic energy per unit volume; it is a function of direction, since the dipole field is not homogeneous. The additional energy will increase the rate of expansion $v(r, \phi)$ of each volume element in a given direction. We may therefore write

$$\frac{\rho v^2(r, \phi)}{2} = T_0 + \mathfrak{M}(r, \phi), \tag{35}$$

where ρ is the gas density.

The magnetic field has its maximum at the poles and its minimum on the equator. The rate of expansion of a nebula with a dipole magnetic field is therefore largest along the polar axis and smallest in the direction of the equator. An initially spherical nebula is thus gradually stretched in the direction of the magnetic axis. This prolateness naturally increases with the ratio of the magnetic energy to the intrinsic kinetic energy.

In this way, dipole magnetic fields stretch the nebulae along the magnetic axis distorting their initial spherical shape. This conclusion is consistent with observational results; the great majority of planetaries are nonspherical and elongated to a certain degree.

We have discussed only the qualitative aspects of the phenomenon. An exact quantitative treatment is impossible without writing and solving the equations of magnetohydrodynamics for a moving medium. It is a little premature to venture in this direction, however, since we do not know how the geometry of the magnetic field and the dipole size vary as the nebula expands (to give only the most obvious reason). We therefore restrict our analysis to an approximate solution of the problem which ignores higher order changes of the field elements and the material parameters, the aim being to derive the various possible configurations of planetary nebulae with magnetic fields.

The magnetic energy density at the outer boundary of a nebula is

$$\mathfrak{M}(r, \phi) = \frac{a^2}{8\pi} \eta_1^2(r, \phi).$$

Substituting this expression into (35), we find

$$\left[\frac{v(r, \phi)}{v_0} \right]^2 = 1 + \frac{a^2}{8\pi T_0} \eta_1^2(r, \phi), \tag{36}$$

where v_0 is the rate of expansion without a magnetic field. At the center of the nebula

$$\mathfrak{M}_0 = \frac{a^2}{8\pi}\, \eta_1^2\,(0,0),$$

and from (36) we obtain

$$\left[\frac{v\,(r, \phi)}{v_0}\right]^2 = 1 + \frac{\mathfrak{M}_0}{T_0}\, \frac{\eta_1^2\,(r, \phi)}{\eta_1^2\,(0, 0)},\tag{37}$$

or, putting

$$\frac{\mathfrak{M}_0}{T_0}\, \frac{1}{\eta_1^2\,(0, 0)} = q,\tag{38}$$

we find for the relative rate of expansion of the nebula in the direction ϕ,

$$\frac{v\,(\phi, r)}{v_0} = [\,1 + q\eta_1^2\,(r, \phi)\,]^{1/2}.\tag{39}$$

The extent of the nebula (more precisely, of some thin lamina a distance r from the center) in the given direction, $R\,(\phi)$, is clearly proportional to $v\,(\phi, r)$. We may therefore write for the outermost contour of the nebula

$$R\,(\phi) = [\,1 + q\eta_1^2\,(r, \phi)\,]^{1/2}.\tag{40}$$

The shape of the nebula in the presence of a dipole magnetic field is thus a function of q, *i.e.*, in the final analysis it depends on the ratio of the magnetic energy density at the center to the intrinsic kinetic energy and on the "depth" at which the magnetic poles are located, *i.e.*, on the function $\eta_1\,(r, \phi)$.

Figure 9-11 shows schematically the outlines of planetary nebulae according to (40) and (29) for various values of q (from $q = 0$ to $q = \infty$) and three values of the ratio r/l (0.5, 1, and 2) corresponding to various "depths" of the magnetic poles in the nebula. The case $q = \infty$ implies that the magnetic field is entirely responsible for the shape of the nebula. For $q = 0$, there is no magnetic field and the nebula is spherical. An increase in r/l signifies that the magnetic poles are situated deeper in the interior of the nebula (reckoning from the outer boundary, the radius of which is r), and the ratio of the two diameters is correspondingly smaller. The minimum diameter ratio for $q = \infty$ is 2; it is obtained in the limit $r/l \rightarrow \infty$ (a point dipole).

Dipole magnetic fields are also responsible for the appearance of regions of enhanced brightness. These "beads" are situated on the equator symmetrically relative to the nucleus. In Figure 9-11 the equator is represented by the horizontal, minor axis; the magnetic, major axis is along the vertical. The brightness at the ends of the major axis falls off gradually, and no distinct boundary should be observed.

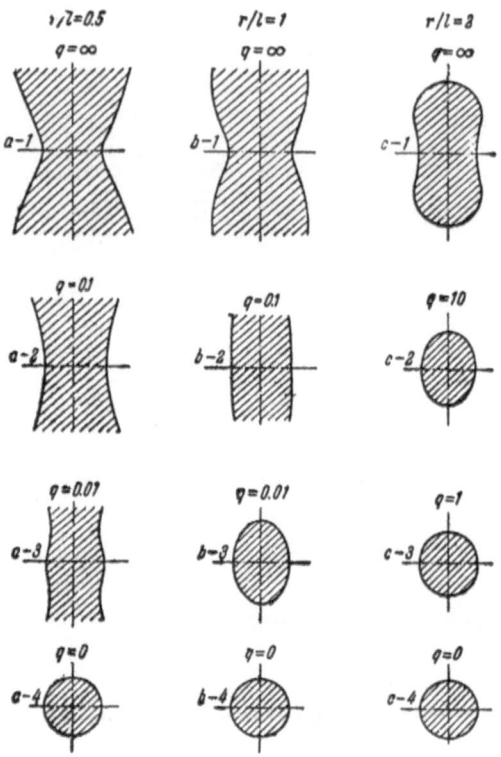

Figure 9-11
Various theoretical forms of planetary nebulae with dipole magnetic fields.

At the ends of the minor axis, conversely, the brightness is at a maximum and the outer rim of the nebula (the edge of the "beads") is very sharp. It would therefore appear at a first glance that some restraining or arresting force operates in the direction of the minor axis, preventing the expansion of the nebula. In fact, however, the nebula expands at a normal rate in this direction, as we have seen before, the rate of expansion increasing in other directions with the distance from the equator.

Figure 9-11 clearly shows that the configuration of planetary nebulae with dipole magnetic fields may vary between wide limits depending on the field parameters. Though our calculations are highly approximate, it seems improbable that further improvement of the theory will modify this conclusion radically. The planetary nebulae may assume a variety of shapes depending on the geometry and the relative strength of the field. They include "hour-glass" nebulae (types a-1, a-2, b-1), "dumbbells" (type c-1), "rectangular" (b-2), ellipsoidal (b-3, c-2) and, finally, almost spherical (c-3) nebulae.

Planetary nebulae with these shapes are invariably bipolar. We thus conclude that bipolar planetaries may range from almost spherical to almost rectangular. Examples of these extreme types are NGC 2474–5 (a spherical nebula, see Plate II borrowed from [156]) and IC 4406 (a rectangular nebula; see below).

The shape of most planetary nebulae corresponds to one of the types in Figure 9-11. Some of these objects are discussed in the following.

8. "Rectangular" Nebulae

A remarkable representative of this type is the planetary nebula IC 4406 (Plate V). Its shape is almost rectangular; more precisely, it has straight edges parallel to the longitudinal axis. Figure 9-12a shows the isophotes of this nebula (in red light) according to Evans [169]. The numbers indicate intensities in relative units.

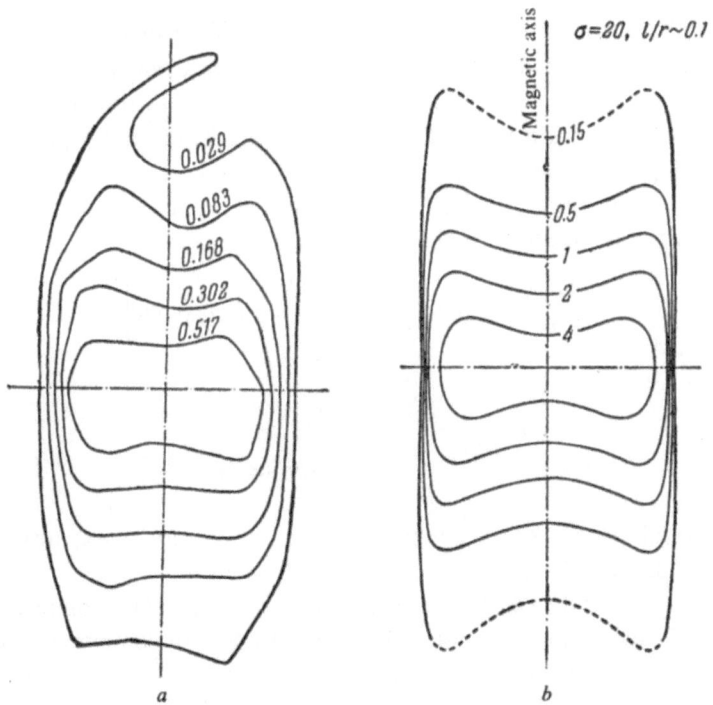

Figure 9-12

a) Isophotes of IC 4406 in red light (Evans). *b*) Theoretical isophotes of a "rectangular" nebula with $l/r \sim 0.1$, $\sigma = 20$.

According to its shape, IC 4406 is classified as type b-2 of Figure 9-11 (the ends of the major axis are not rounded; here the light is so faint because of the extreme thinness of the gas that these parts do not appear with ordinary exposures). This configuration is therefore of electromagnetic origin, arising from a dipole magnetic field of a certain type with the dipole axis pointing along the major axis of the nebula. We will try to plot the approximate theoretical isophotes for this "rectangular" nebula. We again start with equation (18), introducing into the left-hand side an additional term which allows for the nonuniform expansion in various directions, so that (18) takes the form

$$\frac{H^2(r, \phi)}{8\pi} + 2nkT + \frac{\rho v_\phi^2}{2} = C, \qquad (41)$$

where v_ϕ is the translational velocity in the direction ϕ and ρ the mean gas density in that direction ($\rho = nm_H$).

In principle, we should insert for v_ϕ in (41) its expression from (39). However, as we are dealing with a rectangular nebula, a simpler function can be used. The equations of the straight edges parallel to one of the coordinate axes can be written in polar coordinates as $R(\phi) \sim \sec \phi$ in place of (40). Comparison with (39) gives for the outermost parts of the nebula $v_\phi = v_0 \sec \phi$. We may use this simple relation to describe the expansion of a rectangular nebula in the desired approximation.

Substituting this expression for v_ϕ into (41) we find, proceeding along the same lines as in (30), an expression for the relative gas concentration in a central section of a nonuniformly expanding nebula:

$$\frac{n(r, \phi)}{n_0} = \delta(\phi) \left[1 + \left(\frac{v_0}{w}\right)^2 - \sigma\eta_1^2(r, \phi) \right], \qquad (42)$$

where

$$\delta(\phi) = \frac{1}{1 + (v_0/w)^2 \sec^2\phi},$$

w is the termal velocity of the gas particles, and v_0 the rate of expansion in the direction $\phi = 0$.

Using (42) we find the gas density at any point of a central section of a rectangular nebula and hence, using a formula similar to (31), we can compute the distribution of brightness over the entire projection of the nebula onto the celestial sphere.

Figure 9-12b shows the isophotic contours calculated by this method for a rectangular nebula with $w = 10$ km/sec, $v_0 = 20$ km/sec, and $\sigma = 20$. A point dipole was assumed ($l/r \sim 0.1$).

We see from the figures that despite the considerable simplifications introduced in the theory, the two isophote plots are fairly similar. The similarity extends beyond the general shape of the isophotes to the almost identical brightness gradients, especially in the direction of the magnetic axis. Calculations show that the magnetic field strength near the center of the nebula should be of the order of 3×10^{-3} gauss (for $n_0 = 5 \times 10^3$ cm^3, $T = 10^{4\circ}$K).

The very odd and unusual shape of IC 4406 is thus clearly a product of a dipole magnetic field. The magnetic field of this nebula is perpendicular, or nearly so, to the line of sight. The nebula is shaped like a circular axisymmetric cylinder with its axis directed along the magnetic axis. There is thus nothing anomalous about the "anomalous" nebula IC 4406; this is one of the simplest configurations of planetary nebulae which are favorably oriented in space in relation to the observer.

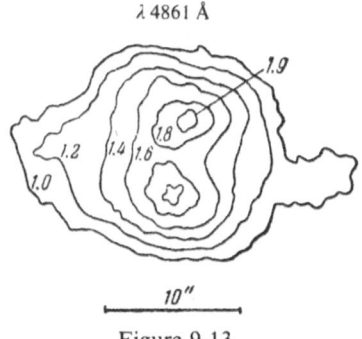

λ 4861 Å

Figure 9-13

Monochromatic isophotes of NGC 2440 in λ 4861 Å (Aller).

Despite the low probability of observing rectangular planetary nebulae, a number of nebulae are known to have a partially rectangular or almost regular rectangular shape. The bipolar nebula NGC 2440 is very similar to IC 4406, and its isophotes are shown in Figure 9-13 [8]. NGC 7026 has a very pronounced rectangular shape (see Figure 9-21). NGC 6772 and CD $-$ 29°13998 (Figure 9-22) are also noticeably rectangular. A 1, like IC 4406, has almost straight edges parallel to the magnetic axis; the second envelope of NGC 6781 is apparently also a regular rectangle. The effect of the magnetic field is very noticeable on the isophotes of NGC 40 [8].

"Hour-glass" planetary nebulae are less frequent. There is nevertheless one known nebula of rare beauty in this class. It is anon. $\alpha = 07^h07^m$, $\delta = 00°43'.4$, shown on Plate V. Note the narrow equatorial "waist" and the wide "jets" at the ends of the major axis. The rim of the nebula is very sharp at the ends of the minor axis and almost indistinguishable at the ends of the major axis.

Other planetary nebulae also display a certain likeness to the "hour-glass" configuration, being somewhat constricted in the direction of the minor axis (*e.g.*, NGC 6772).

9. The Magnetic Field of NGC 7293

An attempt was made to determine the actual functional dependence of the magnetic field on the coordinate ϕ in one of the typical bipolar planetaries, NGC 7293 [208]. The detailed isophotes were plotted from the photographs of this nebula taken in H_α light with the 40″ Schmidt camera of the Burakan Observatory. These isophotes are shown in Figure 9-14, where the numbers indicate intensities in relative units. The line I–I through the two "beads" is the magnetic equator, and II–II is the magnetic axis of the nebula.

The fundamental assumption in this analysis is that the sum of the kinetic and magnetic energy is constant at all points of the envelope, i.e.,

$$\frac{H_\phi^2}{8\pi} + 2n_\phi kT_e = \text{const.} \tag{43}$$

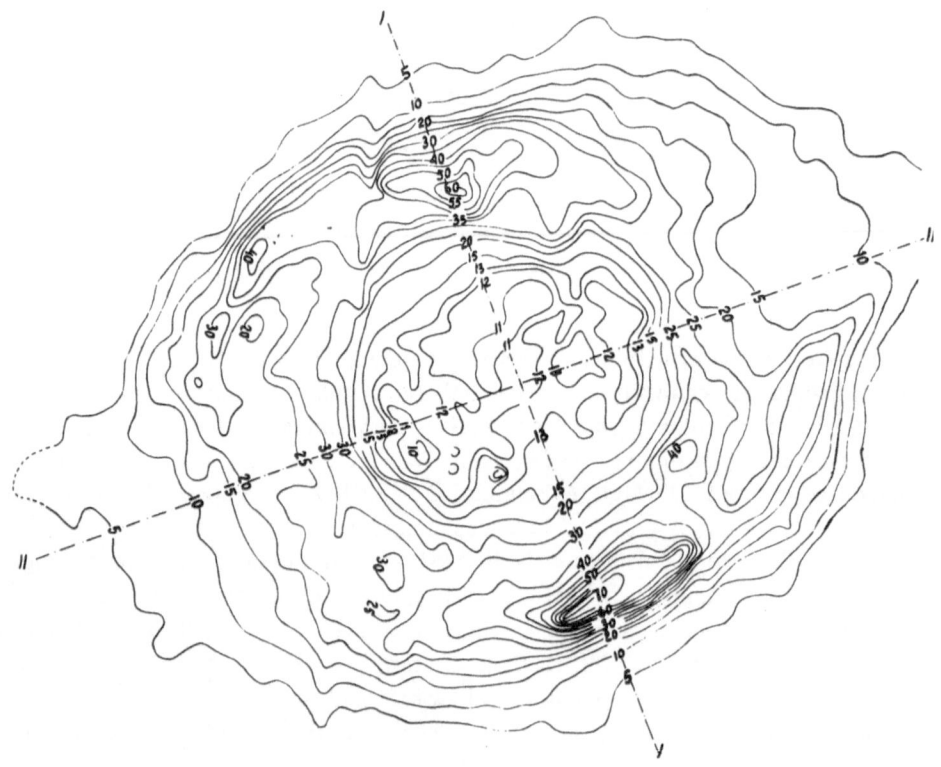

Figure 9-14
Isophotes of NGC 7293 in H_α + [N II]. Figures indicate intensity in arbitrary units.
I − I is the equatorial axis and II − II is the magnetic axis of the nebula.

The problem is simplified if we concentrate on the condition for equilibrium of the gas at the points of a fixed circle through the maxima of the "beads".

Substituting for the constant in (43) the corresponding expression with $\phi = 0$, we obtain

$$\frac{H_\phi}{H_0} = \left[1 + \frac{\gamma_0}{H_0^2} \left(1 - \frac{n_\phi}{n_0} \right) \right]^{1/2}, \tag{44}$$

where H_0 and n_0 are the field strength and the electron concentration in the "beads" ($\phi = 0$), and

$$\gamma_0 = 16\pi n_0 k T_e. \tag{45}$$

The actual field distribution along the circle through the two "beads" can be determined with fair accuracy from (44) if the distribution of electron concentration n_ϕ along the circle is known from observations. The exact distribution of the electron concentration is unfortunately not known, but a rough estimate is quite acceptable for our qualitative purposes. If the nebula is assumed to have a constant line-of-sight thickness everywhere on this circle, n_ϕ can be determined from the distribution of the surface brightness i_ϕ on the circle. Figure 9-15 shows the distribution of the surface brightness i_ϕ of NGC 7293 (averaged over the entire northern hemisphere) from the equator ($\phi = 0°$) to the magnetic pole ($\phi = 90°$); the curve is normalized to $i_0 = 1$. The top curve in the same figure gives n_ϕ/n_0 vs. ϕ, assuming $n \sim i_\phi^{1/2}$.

However, the knowledge of n_ϕ/n_0 is insufficient for calculating the distribution of H_ϕ from (44): the field strength H_0 in the "beads" is also required. In other words, an additional relation is needed, and we have to make use of the fact that the ratio H_{90}/H_0 is equal to 2 for a point dipole and is greater than 2 for an extended dipole.

Writing (44) for $\phi = 90°$ and substituting $n_{90}/n_0 = 0.67$ (from Figure 9-15) and $H_{90}/H_0 = 2$, we find $H_0 = 3.5 \times 10^{-5}$ gauss ($\gamma_0 = 1.14 \times 10^{-18}$ for $n_0 = 160 \, \mathrm{cm}^{-3}$ and $T_e = 10,000°\mathrm{K}$). The field at the pole is $H_{90} = 7 \times 10^{-5}$ gauss. With the aid of these numerical data we derive from (44) H_ϕ/H_0 as a function of ϕ, i.e., the distribution of the magnetic field strength with the coordinate ϕ along the envelope of NGC 7293. This distribution is plotted by the solid curve in Figure 9-16. The same figure also gives the field distribution for $H_{90}/H_0 = 5$ (an extended or distorted dipole). Dashed lines are the theoretical H_ϕ vs. ϕ curves for an extended dipole with $r/l = 1$ and $r/l = 0.2$ (calculated using (29)).

It follows from these curves that even for $H_{90}/H_0 = 2$, the actual magnetic field geometry in the interior of the nebula is markedly different from the field of the classical dipole. The difference is manifest in the field strength distribution for $0 < \phi < 90°$ and in the ratio of the field strengths for the two main axes of the nebula, the magnetic and the equatorial.

Once the variation of the electron concentration n_ϕ (Figure 9-15) and the variation of the magnetic field strength H_ϕ (Figure 9-16) are known, we can find the ratio

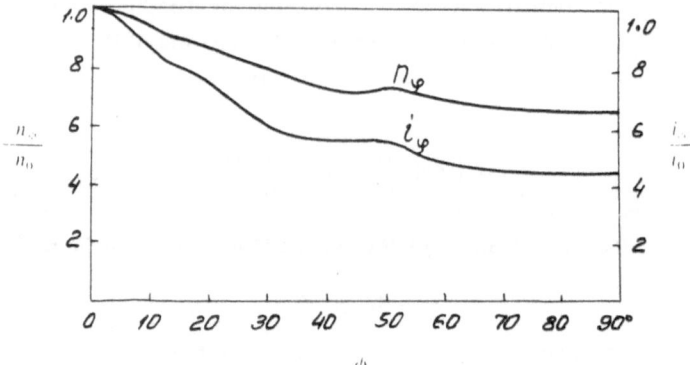

Figure 9-15

Relative distributions with direction of the emission coefficient i_ϕ and the electron concentration η_ϕ in the envelope of NGC 7293 ($\phi = 0$ on the equatorial section, $\phi = 90$ in the direction of the magnetic axis).

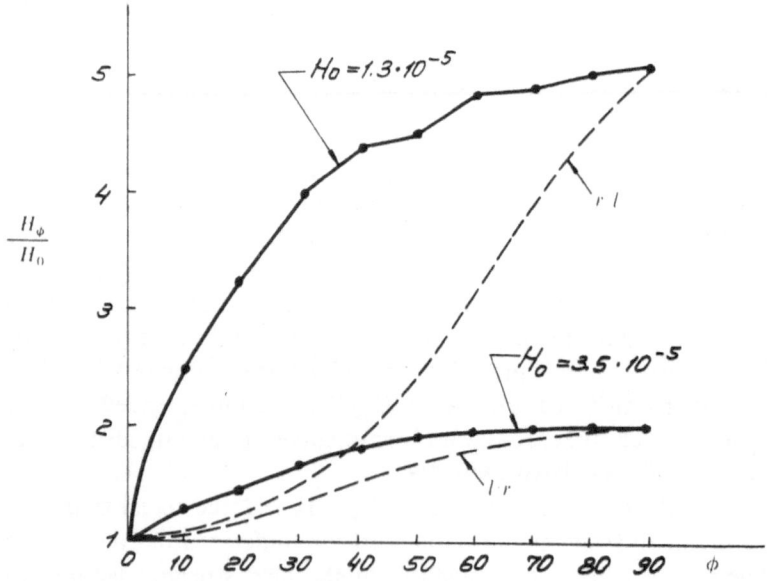

Figure 9-16

Distribution of absolute magnitude of magnetic field strength in the envelope of NGC 7293 with $H_0 = 1.3 \times 10^{-5}$ and $H_0 = 3.5 \times 10^{-5}$ gauss (solid lines). Dashed lines give theoretical field strengths for $r/l = 1$ and $r/l = 0.2$.

of the kinetic energy density to the magnetic energy density at any point in the nebula using the relation

$$\frac{E_{kin}}{E_{mag}} = 16\pi k T_e \frac{n_\phi}{H_\phi^2}. \tag{46}$$

E_{kin}/E_{mag} is plotted in Figure 9-17 for two values of H_0 taking $T_e = 10{,}000°K$. Near the equator ("beads") the kinetic energy density of NGC 7293 is several time higher than the magnetic energy density. In the direction of the magnetic axis ($\phi \sim 90°$), the two energies are comparable.

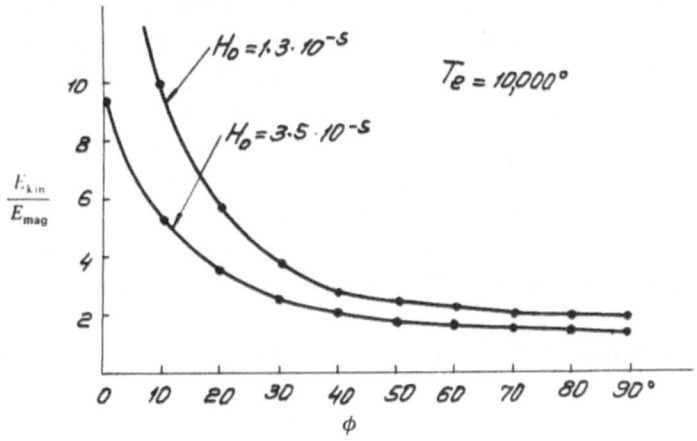

Figure 9-17

Variation of the specific magnetic energy through the envelope of NGC 7293 for two values of H_0.

10. Spiral Planetary Nebulae

Spiral planetary nebulae have two spiral arms arranged symmetrically about the core. Typical representatives of this class are NGC 4361, NGC 6210, NGC 7026, A 65. Plates II and IV show some of them.

Spiral structure has by now been established for the following planetary nebulae:

NGC 650–1	NGC 6309
J 320	NGC 7009
NGC 4361	NGC 7026
NGC 6210	A 65
NGC 4634	CD −29°13998.

Some spiral planetary nebulae are also bipolar (NGC 650–1, NGC 7026) or possess two envelopes (NGC 4361, NGC 7009). The spiral arms vary in their configurations and prominence. In some planetary nebulae the spiral arms have distinct outlines and are fairly well developed (NGC 4361, NGC 6210), while in others they are thin and hardly visible (NGC 6309). In some cases the spiral arms reduce to mere projections (NGC 7026, Figure 9-20) or "pincers" (CD − 29°13998, NGC 650–1, Figures 9-21 and 9-22), but they are always arranged symmetrically about the nebula. In one instance (A 65) the spiral arms are so pronounced that the planetary nebula can easily be mistaken for a galaxy, whereas in another case, although no typical spiral arms are apparent, if the spatial orientation of the nebula relative to the observer is allowed for, certain features (*e.g.*, the conical projections on both sides of NGC 7009) can be interpreted as manifestations of spiral structure.

Though the known spiral planetary nebulae are few, they are nevertheless of certain interest. The spiral arms of the planetary nebulae are apparently of *electromagnetic* origin, probably resulting from the combined action of the inhomogeneous intrinsic fields of the nebula and the general regular magnetic field of the Galaxy. Our discussion of this phenomenon is highly schematic and purely qualitative, and we again assume a dipole-like magnetic field distribution in the interior of the nebulae.

An expanding planetary nebula with a dipole magnetic field is inevitably stretched in the direction of the magnetic axis, as we have seen above. One gets the impression in several cases that gaseous matter escapes from the poles of the nebula in a wide jet, forming projections on the two sides.

Now suppose that these projections have grown away from the magnetic poles of the nebula into a region where the dipole field strength is comparable with the general magnetic field of the Galaxy, *i.e.*, of the order of 10^{-5}–10^{-6} gauss. The general Galactic field can be regarded as homogeneous within the space occupied by the nebula. The effect of the Galactic magnetic field can no longer be ignored at this stage, and the structure of the nebula at the two ends of its magnetic axis, where the projections are found, is effectively determined by the combined action of the nebular *dipole* field and the *homogeneous* galactic field.

We will consider only one half of a central section with one pair of magnetic lines of force, as shown in Figure 9-18. The magnetic axis of the nebula is assumed not to be parallel to the local Galactic magnetic field. For simplicity we take it to be perpendicular to the Galactic field.

Consider two points A and B within the projections, which are symmetric about the magnetic axis and are thus at equal distance from the pole N. The magnetic line of force of the dipole field through the point A is clearly a mirror image of the magnetic line of force through B. The dipole field vectors at points A and B are therefore equal in magnitude and make the same angle with the radius-vectors to A and B. The magnitude of the field vector for an extended dipole is given by (31).

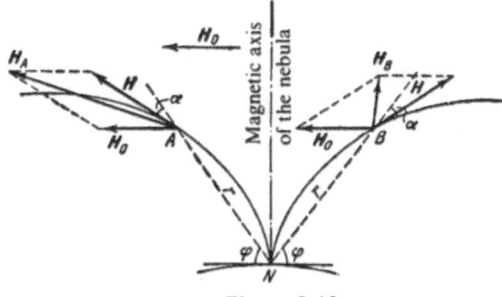

Figure 9-18

Illustrating the theory of spiral planetary nebulae.

The Galactic magnetic field H_0 is now superimposed at right angles to the magnetic axis of the nebula; its magnitude is comparable to the dipole field H_r (H in the figure) in the projections, *i.e.*, at points A and B. The resultant field at these points is thus

$$H_A = H_r + H_0, \qquad H_B = H_r + H_0.$$

Clearly $H_A > H_B$ (see Figure 9-18). The strength of this inequality varies with the distance from the pole. Near the pole, where $H_r \gg H_0$, we have $H_A \sim H_B \sim H_r$. Far from the pole, $H_r \ll H_0$ and therefore $H_A \sim H_B \sim H_0$. Left and right of the magnetic axis there are consequently closed regions with $H_A < H_B$; this clearly corresponds to such distances from the pole where $H_r \sim H_0$.

The density of the ionized gas at the point (r, ϕ) is again determined from the condition of constancy of the sum of the magnetic and the kinetic thermal energy, $H^2/8\pi + 2nkT = C$; hence $n = C/2kT - H^2/16\pi kT$.

According to this expression, the concentration of gas at B is greater than the concentration at A, *i.e.*, $\rho_B > \rho_A$ for certain closed regions around A and B.

This conclusion should be interpreted as follows. Without the Galactic magnetic field the density is symmetrically distributed about the magnetic axis, but this is no longer so if the Galactic field cannot be ignored, as the density near the rim of the nebula on one side of the magnetic axis is higher than on the other side. As the volume emissivity is a quadratic function of the concentration of ions or electrons, a certain difference may arise between the surface brightness on the two sides of the magnetic axis. In some cases the difference may become large enough to be noticeable, and the observer will perceive the projections as spiral arms. The entire pattern is repeated (as in a mirror reflection) in the southern half of the nebula, and the resultant effect is therefore a nebula with symmetric spiral arms on both sides. Figure 9-19 is a schematic diagram of a prolate (elongated) nebula without the galactic magnetic field (a) and in a superposition of the nebular dipole field and the homogeneous Galactic field (b).

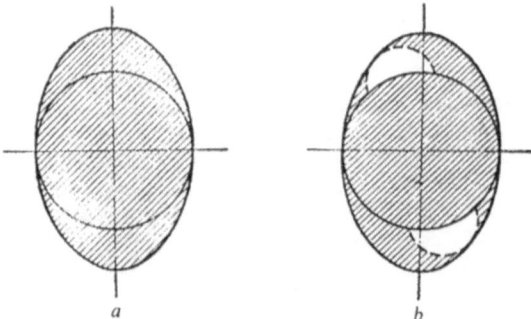

Figure 9-19

Theoretical forms of planetary nebulae: (a) without the galactic magnetic field; (b) with
the galactic magnetic field superimposed on the dipole field of the nebula.

In distinction from the earlier hypothesis [51], which attributes the formation of spiral planetaries to ejection of gaseous matter at two antipodal points and the axial rotation of the nebula, our approach requires neither rotation nor outflow of gases. The spiral arms are in fact the locus of points where the magnetic field strength is at a minimum. Moreover, these are stationary fixtures, ornaments "adorning" the nebula.

NGC 7026 was listed among the spiral planetaries. Plate II contains a photograph of this nebula taken with the 200-inch telescope. Figure 9-20 shows the main isophotes of this nebula (the figure is borrowed from [8]).

The existence of symmetric spiral arms on the two sides of the nebula is more than obvious in this case. That the nebula indeed carries the dipole magnetic field necessary for the formation of spiral arms can be inferred from the following facts. First, this is a bipolar nebula, and second, it is stretched in the direction of the equatorial diameter, which accounts for its nearly rectangular shape (like IC 4406). The dashed line in Figure 9-20 is the expected direction of the magnetic axis, arrived at from structural considerations; the solid line is the plane of the Galactic equator and presumably the direction of the Galactic magnetic lines of force.

Figures 9-21 and 9-22 show the isophotes of other spiral planetary nebulae, CD-29°13998 and NGC 650−1 [8]. The most prominent feature is the highly flattened, almost rectangular shape of these nebulae, clear evidence of intrinsic magnetic fields. The spiral arms at the ends of the magnetic axis (the dashed line) are pincer-shaped.

A fairly distinct spiral structure is also observed in the second envelope of anon. $16^h10^m.5$, one of the most remarkable bipolar planetary nebulae (Plate III).

In the double-envelope nebula NGC 7009 the spiral arms are oriented differently in relation to the observer: the plane of the spiral arms is nearly perpendicular to

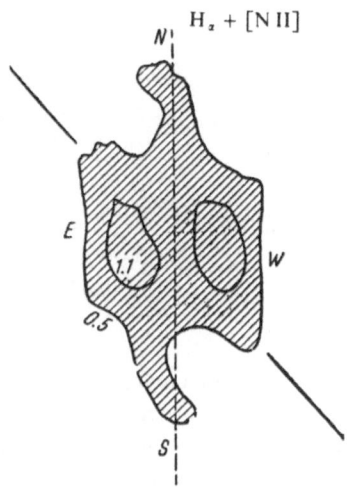

Figure 9-20
The silhouette of NGC 7026. The numerical values are those of log J. The dashed
line is the magnetic axis, the solid line is a direction of a plane, parallel to the galactic
equator.

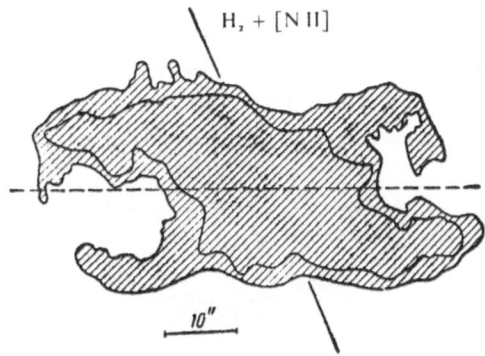

Figure 9-21
The silhouette of CD −29°13998. The dashed line is the magnetic axis, the solid line
is a direction of a plane, parallel to the galactic equator (Aller).

the image plane. The Galactic magnetic field vector is also in a plane perpendicular
to the image plane, but judging from the symmetric dark spots in the second envelope, we conclude that it is slightly inclined (∼ 30°) to the magnetic axis of the nebula; the magnetic axis follows the line through the center and the two prominences.

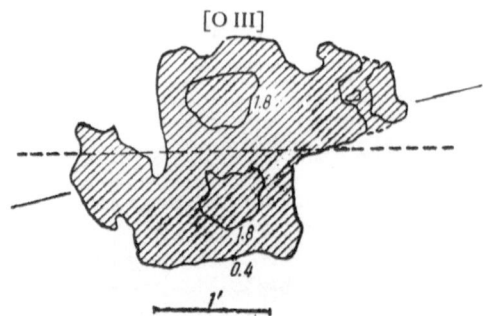

Figure 9-22

The silhouette of NGC 650-1. The numerical values are those of log J. The dashed line is the magnetic axis, the solid line is a direction of a plane parallel to the galactic equator (Aller).

Signs of spiral structure were also observed for NGC 7293. Figure 9-23 is a silhouette of this nebula borrowed from [208].

Z-shaped nebulae [51] are a dynamic variety of spiral planetary nebulae. These are essentially bipolar nebulae with "beads" joined by a diagonal crosspiece of marked brightness. There are very few planetary nebulae of this kind: NGC 2452, NGC 6778, NGC 6853, and apparently anon. $16^h30^m.2$ (a photograph of the last nebula is given in [174]). Nevertheless, they cannot be regarded as a freak phenomenon. The Z-shaped configuration of some planetary nebulae is probably also of electromagnetic origin and can be interpreted as a particular case of the magnetic field geometry in the nebular interior. This question is treated in some detail in [167].

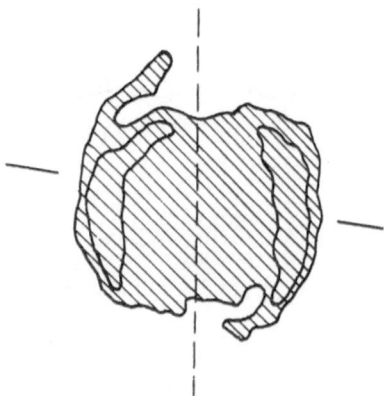

Figure 9-23

The silhouette of NGC 7293. The dashed line is the magnetic axis, the solid line is a direction of a plane parallel to the galactic equator.

In addition to spiral and Z-shaped planetary nebulae, there are several objects with odd shapes and structures which apparently also can be attributed to electromagnetic factors. Curtis' early photographs show one such anomalous snail-shaped planetary, NGC 6543, which in the more recent photographs taken taken with the 200-inch telescope appears in the form of two interlaced rings (see Plate V). It is remarkable that the central star is not situated at the center of either ring but at the center of the nebula as a whole. Even if we assume that these rings originated in two successive explosions of the nucleus, it is not clear why they should be displaced in such a way relative to one another. The drag of the interstellar medium is of no consequence in this case, since the linear dimensions of this nebula are apparently fairly small. Note also that this nebula has the most extended second envelope among all known planetary nebulae.

The configuration of NGC 6543 may apparently be attributed to electromagnetic factors. This is evident from the existence of two relatively faint spiral arms, which are noticeable, upon careful examination, in the direction of the major axis.

There are two possibilities in this respect. The first is that the real (three-dimensional) shape of NGC 6543 is one of the known configurations of bipolar planetary nebulae (*e.g.*, an "hour-glass" or a rectangular nebula, etc.), and it is only because of its peculiar orientation relative to the observer that the nebula has such an anomalous apparent shape in projection (interlacing rings). The second possibility is that NGC 6543 is really made up of two interlacing rings or loops, which are in turn the product of hitherto unknown electromagnetic processes.

NGC 1514 is apparently similar in certain respects to NGC 6543.

Anomalous planetary nebulae are very rare. Besides NGC 6543 and NGC 1514, we can mention only two or three other curiously shaped objects (NGC 6573, MH_α 362). Future studies will enable us to understand their odd structure.

The principal conclusion that follows from the discussion of this section is that spiral arms, as well as any other symmetrical formations and elements such as crosspieces, snails, *etc.*, should be interpreted as signs of inhomogeneous but symmetrical magnetic fields in planetary nebulae.

11. Dissipation of Magnetic Energy

Let S be the ratio of the magnetic energy density E_{mag} to the kinetic energy density E_{kin} at a given point in the nebula, *i.e.*,

$$S = \frac{E_{mag}}{E_{kin}} = \frac{H^2}{4\pi\rho v^2}.$$ (47)

In the central regions and near the magnetic poles of a bipolar nebula, S is considerably greater than unity. Near the "beads" S is of the order of unity. This implies that the internal motions are predominantly governed by the magnetic field every-where in the nebula, although in the "beads" the kinetic energy makes a comparable contribution. The nebular structure is not influenced by the magnetic field only when S is less than unity.

The average ratio $\bar{S}$ is defined by

$$\bar{S} = \frac{1}{V} \int_{(V)} S dV, \tag{48}$$

where V is the volume of the nebula. The variation of $\bar{S}$ during the expansion of the nebula is clearly indicative of the relative significance of the magnetic field at various stages of the nebula's life. Let M be the total magnetic energy of the nebula at the time of its seperation from the nucleus. As a starting point we assume that this energy remains constant as the nebula expands. The total kinetic energy $\mathscr{E}$ of the nebula, however, behaves differently.

In the most general case $\mathscr{E}$ is a sum of three components: the kinetic energy of expansion, the kinetic energy of the ionized component (due to the thermal motion of gas particles at a temperature of $10^4°$K), and the kinetic energy of the nonionized component of the nebula (thermal motion of much colder gas particles, at about $100°$K). The rate of expansion is fairly constant regardless of the size of the nebula, and the first component is virtually constant during the nebula's life. The third com-ponent is small and ignorable. The second component is determined by the amount of the ionized matter relative to the total mass of the nebula. The total optical thick-ness τ_c in the ultraviolet region of a small nebula is markedly greater than unity, and therefore only a small fraction of the nebula is heated to $10^4°$K. This fraction, however, increases as the nebula expands, and $\mathscr{E}$ increases correspondingly. In small nebulae, when $\tau_c \gg 1$, $\bar{S}$ (clearly, $\bar{S} = M/E$) will therefore decrease as the nebula expands until τ_c has dropped to about unity.

In double-envelope planetary nebulae, τ_c is less than unity. Many of these nebulae are bipolar, and their structure is therefore governed by magnetic fields. If we as-sume $\bar{S} \sim 1$ for double-envelope planetary nebulae, then $\bar{S}$ should be much greater than unity in small nebulae (when $\tau_c \gg 1$). In other words, the significance of mag-netic fields in the dynamics of planetaries is greatest for small nebulae, and decreases with expansion.

Let us now consider how the situation develops when τ_c becomes less than unity. In accordance with the preceding arguments, $\bar{S}$ remains constant, so that the rela-tive contribution of the magnetic and inertial forces does not change until the very moment when the nebula breaks up and dissolves entirely. This conclusion, how-ever, does not correspond to what really happens in planetary nebulae at this stage.

As the nebula expands, the magnetic energy is inevitably dissipated. At least two sources for the dissipation of magnetic energy are apparent. The first is con-nected with the expansion: although the dipole remains a dipole during the expansion, its size changes, the concentration of the lines of force decreases, and magnetic energy is dissipated as mechanical work done in stretching the magnetic lines of force and pullling them apart [170]. The second source of magnetic energy dissipation is the partial conversion of magnetic energy to Joule heat (j^2/σ per unit volume).

There is, however, a third source of dissipation, namely turbulence. Although turbulence increases the field strength and thus the magnetic energy, the dissipative action more than offsets this effect. Dissipation by turbulence is not pronounced in small nebulae, where only small-scale turbulence is possible. Turbulence apparently sets in at a much later stage in the life of the nebula, when it becomes relatively large.

Irrespective of the rate of dissipation of magnetic energy, $\bar{S}$ is clearly much less than unity at this stage in the life of a nebula. The magnetic lines of force are stretched to the ultimate limit and "snap", so that the field breaks up into small fragments. The dipole field ceases to exist, and it is replaced by local magnetic fields. Magnetic energy fluctuations arise, and this in turn causes redistribution of the ionized nebular matter in small pockets. In other words, fluctuations of surface brightness should be observed. Indeed, the outer envelopes of some double-envelope planetary nebulae (NGC 2392, NGC 6720, NGC 7662, and others) are distinguished by considerable fluctuations in surface brightness, whereas the inner envelopes do not show any such effect. This clearly supports our hypothesis.

It should be noted, however, that there exist certain bipolar and double-envelope planetary nebulae (NGC 7354, NGC 3587, *etc.*) in which neither envelope shows brightness fluctuations. $\bar{S}$ in both envelopes is greater than unity and magnetic energy dissipation has not yet terminated.

On the other hand, certain planetary nebulae (NGC 40, NGC 2461, NGC 1501, NGC 1514, A 56, *etc.*) are entirely covered by spots of nonuniform brightness. The dipole magnetic fields of these nebulae apparently have long since ceased to exist (if they ever existed), and only local, fragmentary fields remain. The bright spots are situated between these magnetic pockets.

Back in Chapter I we pointed to one of the most remarkable facts, established by Wilson [34], namely that planetary nebulae with large brightness fluctuations do not show any turbulence. We now see that this "anomaly" is readily explained away if a relationship is established between brightness fluctuations (nonuniform gas density in the nebula) and the magnetic field geometry in the nebula.

An interesting example is provided by the diffuse Orion Nebula. Radial velocity measurements at 85 points of this nebula, carried out by Campbell and Moore [5], showed a considerable scatter about some average. This scatter was naturally attributed to turbulence of the nebular gas. Collateral evidence in favor of this hypothesis was the clearly nonuniform distribution of brightness over the nebula, with some large fluctuations. But if brightness fluctuations and radial velocity fluctuations are indeed caused by turbulence, they should obey Kolmogoroff's well-known law of locally isotropic turbulence, which sets in at high Reynolds numbers. This law can be stated as follows: the relative velocity v of two points a distance l from each other is on the average proportional to $l^{1/3}$ (l being sufficiently large). More precisely, putting ρ for the gas density, we write

$$v \sim \left(\frac{l}{\rho}\right)^{1/3}. \tag{49}$$

Von Hoerner [172] compared this law with Campbell and Moore's observations and came to the conclusion that Kolmogoroff's relation was satisfied for the Orion Nebula. On a later occasion, Münch [173] analyzed a much more extensive collection of material (50,000 radial velocity measurements, to mention one set of data only!) and established that Kolmogoroff's law did not apply to the Orion Nebula. The deviations from this law were seen to increase as l decreases. It therefore follows that brightness fluctuations and radial velocity fluctuations are not entirely dependent on turbulence and that some other factor also makes a contribution. Münch suggests that the velocity distribution deviates from the expected relation because of the existence of velocity discontinuities, i.e., a system of compression waves which propagate in all directions throughout the nebula and are caused by the interaction of hot and cold gases. This may be so, but the role of local magnetic fields in the nebula should not be disregarded.

Another factor should be kept in mind when discussing the variation of the magnetic field during the expansion of a planetary nebula. If the total magnetic energy of an average-sized planetary nebula is attributed to the gaseous mass detached from the nucleus and subsequently formed into a nebula, the magnetic field strength of the nucleus at the time of the catastrophe is found to be at least 10^5 gauss. We are therefore led to the conclusion that either planetary nebulae are formed from objects with exceptionally strong magnetic fields, or they possess an independent mechanism for generating or amplifying magnetic fields which operates in the nebula during its formation and evolution. A similar difficulty is encountered with other celestial objects, e.g., the Crab Nebula, which, being a powerful source of radio waves, is undoubtedly endowed with relatively strong magnetic fields.

12. The Interstellar Magnetic Field of the Intermediate Galactic System

An expanding planetary nebula eventually dissolves into interstellar space. The gaseous matter of the planetary carries off the nebular magnetic field and it is dispersed in the interstellar space, provided that the magnetic energy has not been fully dissipated by the time the nebula is finally dissolved. The magnetic fields of nebulae in the final stages of their life are highly irregular and even random, as we have seen. The fields of dissolved planetary nebulae therefore combine to form a field of *random* structure. Moreover, even if the individual planetary nebulae retain a dipole field to the very end, the general magnetic field of the Galaxy is nevertheless random, because the magnetic axis of each nebula is arbitrarily oriented in space.

In view of the relatively high rate of production of planetary nebulae (of the order of one nebula annually) and their short lifetime (a few tens of millennia), the planetary nebulae apparently make a certain contribution to the interstellar magnetic fields of the intermediate subsystem of the Galaxy.

Let one nebula be created in 10^9 years in each volume of V parsec^3 in the Galaxy. Assuming that no magnetic energy is dissipated in the interstellar medium, we have for the average magnetic field $\mathscr{H}$ in the intermediate system

$$\mathscr{H} = \mathscr{H}_0 \left(\frac{r}{r_0}\right)^3,$$

where $\mathscr{H}_0$ and r_0 are the mean field strength and the mean radius of a planetary nebula, and $r \approx V^{1/3}$. Calculations give $\mathscr{H} \sim 10^{-6} - 10^{-7}$ gauss for $V = 125$ pc^3 and $\mathscr{H} \sim 10^{-8} - 10^{-9}$ gauss for $V = 125{,}000$ pc^3. In both cases, $r_0 = 10{,}000$ a.u. and $\mathscr{H}_0 \sim 10^{-3} - 10^{-4}$ gauss have been assumed.

13. Some Conclusions

An analysis of the shape and structure of bipolar, spiral, and other planetary nebulae inevitably points to the existence of nonhomogeneous magnetic fields in the nebulae. It has been firmly established that planetaries have a magnetic axis and a magnetic equator, and that a nonzero magnetic field gradient exists between the magnetic axis and the magnetic equator. The magnetic fields in planetary nebulae are symmetric and dipole-like. The likeness to a dipole field is observed only in the distribution of the *magnitude* of the field vector. The geometry of the magnetic lines of force in a planetary nebula is essentially different from that of a dipole field. The magnetic lines of force in the main part of the nebula, in its envelope, are virtually parallel to one another, extending along the envelope.

The field geometry in the region of the magnetic poles is not at all clear. The field in the central parts of the nebula, if it exists at all, is definitely not a dipole field. This is evident from the structure of the monochromatic images of the central regions of planetary nebulae. For example, NGC 6720, NGC 7662, and IC 2165 show distinct bipolar structure in the main emission lines, whereas the central regions of these planetaries, when seen in $\lambda 4684$ Å He II, $\lambda 3869$ Å [Ne III], and $\lambda 3476$ Å [Ne V], are disk- or ring-shaped, without any clear signs of bipolarity.

Chapter X

The Origin of Planetary Nebulae

1. The Ejection Hypothesis

According to the older notions, planetary nebulae resulted from explosions of novae and supernovae. This hypothesis is based on the obvious fact that the explosion of a nova or a supernova is accompanied by the ejections of a gaseous envelope. It is unacceptable, however, for a variety of reasons. In view of the fundamental importance of this problem, we will list the objections to this hypothesis, some of which are by no means new, in some detail.

1. The mass ejected by an exploding nova is of the order of 10^{-4}–10^{-6} solar masses. This is three to four orders of magnitude less than the average mass of planetary nebulae.

This objection does not apply to supernovae, which eject masses at least comparable to the mass of planetary nebulae.

2. The gaseous envelopes produced by nova or supernova explosions expand with a velocity of a few thousand kilometers per second. This is two orders of magnitude larger than the observed rate of expansion of planetary nebulae. The suggestion that the initially enormous velocities of nova or supernova shells are eventually slowed down to a convenient level had to be rejected. As we have seen in Chapter IV, none of the known deceleration mechanisms, radiation pressure included, is sufficient to reduce the initial ejection velocities to the rates of expansion observed for planetary nebulae. We can think of no other, more powerful decelerating mechanisms.

That a substantial slowing down of the gaseous envelopes is impossible also follows from observational data. All known gaseous envelopes having been ejected by exploding novae and supernovae expand at tremendous velocities even though they have reached a size comparable to that of planetary nebulae. For example, the gaseous envelope of N Persei (1901) is currently expanding at a rate of 1200 km/sec, though its diameter is $D = 22,000$ a.u.; the envelope of N Aquilae (1918) is expanding at a rate of 1700 km/sec ($D = 11,000$ a.u.); the Crab Nebula, a remnant of a supernova explosion in 1054, is currently expanding at a rate of 1300 km/sec, and its diameter is about 1.5 parsecs.

3. The hypothesis attributing the origin of planetary nebulae to supernova explosions meets with another, statistical difficulty. Supernovae are very rare phenomena; only three supernovae were recorded in our Galaxy during the last millennium, *i.e.*, an average of one explosion every three hundred years. Planetary nebulae, on the other hand, are created at an average rate of one nebula annually (in our Galaxy).

4. The system of planetary nebulae is markedly different from the system of novae; the latter display a fairly strong galactic concentration, whereas the planetaries occupy an intermediate position between the disk-like and spherical galactic subsystems.

5. The ejection hypothesis is unacceptable for dynamic reasons as well: the gaseous envelopes ejected by exploding novae and supernovae are unstable formations which break up fairly rapidly.

The above arguments are sufficiently convincing to make us reject any possible relation between nova and supernova explosions and the phenomena culminating in the formation of planetary nebulae.

Continuous ejection of gaseous matter by some types of nonstationary stars, *e.g.*, Wolf-Rayet stars, cannot explain the formation of planetary nebulae, either. Stability of the gaseous envelope and its deceleration cannot be accounted for within the framework of this mechanism.

2. The "Residue" Theory

As we have seen, all known mechanisms for the ejection of gaseous matter by stars —either with a catastrophic explosion (resulting in speeds of the order of 1000 km/sec) or through continuous ejection—cannot produce a planetary nebula. The expansion of these nebulae indicates that the gaseous envelopes separated from the central stars with a small initial velocity, of the order of tens of kilometers per second. But these velocities are significantly smaller than the escape velocity at the stellar surface. This implies that the gaseous envelope could not have been ejected from the star's surface directly. If, however, the forces acting in the interior of the nucleus are assumed to build up gradually over a long period (and not increase instantaneously), the envelope can swell without separating from the central star and recede to larger distances.

Let the stellar forces be such that they impart to the envelope a velocity of a few tens of kilometers per second. As the gas moves away from the surface of the central star, the escape velocity decreases in inverse proportion to the square of the distance. Very soon the envelope reaches a region where the escape velocity is less than the rate of expansion ([51], p. 143).

This seemingly plausible argument meets with a serious obstacle, however: any expansion of photosphere of a hot star, whether rapid (catastrophic explosion)

or gradual (swelling), inevitably entails an increase in brightness. We should therefore expect a relatively large number of stars to increase in brightness up to a certain limit, without subsequently growing fainter, as is the case with real variables.
As far as we know, no such phenomenon is observed.

It may be objected that these brightness variations are difficult to observe because
of their extreme slowness. In the case of N Pictoris, however, in which gaseous matter was ejected at a velocity of up to 70 km/sec during the "explosion", a variation
of brightness was in fact observed. There is another remarkable star, 17 Leporis,
which periodically ejects a gaseous envelope with a speed of about 50 km/sec.
The variation of its brightness, as well as all the complex changes in its spectrum
over a fairly long period, have been observed repeatedly (this star has been observed
for 14 years at the Yerkes Observatory, in which time it exploded 9 times; several
less pronounced eruptions were also recorded [176]).

This proves that the swelling of a star, if it did occur, could not go by unnoticed
because of the inevitable variation in brightness.

We thus come to the apparently paradoxical conclusion: a planetary nebula
remains "latent", being invisible as long as its size is less than a few thousands of
astronomical units (when it is recognized by the characteristic spectrum of a planetary nebula), and then it is abruptly "exposed", becoming fully visible.

Any attempt to relate the origin of planetary nebulae to stars of certain types or
to certain gas ejection mechanisms thus inevitably leads to fundamental difficulties.
These obstacles, however, are easily overcome if we assume that *planetary nebulae
are not of stellar origin.*

The origin and the formation of planetary nebulae can be related to the formation
of certain types of stars that make up the intermediate galactic subsystem. The nebular gas is the residual matter from which a star has formed. Without making any
assumptions concerning the sources of stellar energy and its release, we can naturally
agree that the temperature starts increasing at the star's center. In other words,
the outermost layers are initially cold. The creation of stars is inherently associated
—and this is the cornerstone of the theory—with the shedding of a certain amount
of primordial matter, since otherwise the star will not reach an equilibrium state.

Indeed, if the prestellar body is in equilibrium, a certain relation exists between
its fundamental parameters, namely the mass $\mathfrak{M}_0$, the radius R_0, the central temperature T_0, and the chemical composition A_0. This relationship can be written in the
form $F_1(\mathfrak{M}_0, R_0, T_0, A_0) = 0$.

As the central temperature increases, the prestellar body becomes nonstationary,
i.e., all of its parameters change. In the second state of equilibrium, when the star
is fully formed, a different relationship is therefore obtained between the final mass
M_*, radius R_*, temperature T_*, and composition A_*, this relation being
$F_2(\mathfrak{M}_*, R_*, T_*, A_*) = 0$. Here clearly $T_* > T_0$, whereas R_* may be either greater
or smaller than R_0. The composition A_* is also different from A_0. As regards $\mathfrak{M}_*$,

three cases are conceivable. First, $\mathfrak{M}_* > \mathfrak{M}_0$, *i.e.*, the final mass (the mass of the star) is greater than the initial mass (the mass of the prestellar body). This is feasible if mass is received from "external" sources during the formative stages. We will not consider this possibility. Second, $\mathfrak{M}_* = \mathfrak{M}_0$, *i.e.*, the rates of change of R_0, T_0, and A_0 and their final values are interrelated in such a way that the initial mass has not changed at all and is fully incorporated in the final form of the star. This clearly presents fairly rigid requirements which are difficult to satisfy. This case, though not to be ruled out *a priori*, constitutes an essentially "unstable equilibrium" of sorts. A slight deviation from the "equilibrium" state due to internal disturbances will either stop the process of star formation (since the system will have to pass from $\mathfrak{M}_* = \mathfrak{M}_0$ to $\mathfrak{M}_* > \mathfrak{M}_0$) or will allow it to go on with $\mathfrak{M}_* < \mathfrak{M}_0$ (the third case). The mass difference $\mathfrak{M}_0 - \mathfrak{M}_*$ is discarded by the star in the process of formation, and remains outside the stellar volume. This mass deficit is large in some cases and small in others, but the essential point is that a certain quantity of gaseous matter remains after the star is formed.

Formation of stars is not an instantaneous process. The temperature and pressure in the interior of an evolving star should increase gradually. The extra matter is therefore also shed gradually, and this is the swelling that we have described above. In the initial formative stages, when the stellar temperature is low, the outermost layers of the star (the swelling shell) are not luminous. The dark envelope recedes to a considerable distance during this period and becomes highly rarefied, with a possible gradual change in its physical state. The stage is thus set for the envelope to become self-luminous. We will thus observe a kind of a combination spectrum, with the characteristic nebular spectrum (the emission of the gaseous envelope) being superimposed on the spectrum of a young, newly-formed star. The twelve early-type stars discovered by Merrill back in 1932, whose spectra show numerous characteristic lines of planetary nebulae (N_1, N_2, $\lambda\,4686$ He II, $\lambda\,4363$ [O III], *etc.*), may possibly be identified with these transitional objects [177]. We should also mention Minkowski's discovery in 1946 [12] of small (stellar) planetary nebulae emitting only weak hydrogen lines with a steep Balmer decrement and very weak forbidden lines (in some objects even these were not observed). At the same time the spectra of these objects was reminiscent of those observed in certain cases for B and Be stars. This also may be an example of the early stage of formation of planetary nebulae.

We conclude from the preceding that planetary nebulae are not formed from any existing stars in the Galaxy but rather from prestellar objects which evolve into proper stars producing planetary nebulae as a byproduct. In other words, planetary nebulae constitute a transition phase, in which developing stars pass from a nonstationary to a stationary state.

Young stars are characteristically nonstationary. Their atmospheres are unstable and very extended, and they lose gaseous matter, *etc.* All this is observed in varying degrees, *e.g.*, in Wolf-Rayet stars, P Cygni and T Tauri stars, *etc.* Herbig-Haro objects occupy a special place in this category [179, 180].

Herbig-Haro objects constitute the earliest stage in the evolution of T Tauri stars [178]; they are surrounded by a very small nebulous envelope with a spectrum of intense emission lines. We are witnessing here the *simultaneous formation of dwarf stars and planetary nebulae.* It can hardly be doubted that the nebula in this case is formed from the "excess" mass discarded by the "newborn" star.

In the above examples the nonstationary state is attributable to the necessity of "getting rid" of the extra matter by any means. The different cases differ only in the mechanism and intensity of "ejection", the process reaching its apogee, so to speak, in the case of planetary nebulae, when a tremendous quantity of gaseous matter is "ejected" at one time in the form of a more or less regular gaseous envelope.

The age of planetary nebulae is of the order of a few tens of millennia. In view of our hypothesis, their central stars should be of the same age. The stars in O associations, as we know, are also young, but their age is estimated as a few million years. It thus seems that the central stars of planetary nebulae are among the youngest, if not the youngest, stars in our Galaxy. This provides additional confirmation of Ambartsumyan's notion that stars continue to form in our Galaxy to this very day.

The existence of gaseous envelopes around these hot stars should thus be interpreted as a sign of the extreme youth of the star itself.

3. The Early Age of the Nuclei

The conclusion of the last section, namely, that the central stars of planetary nebulae are extremely young, recently formed stars, is additionally corroborated by observational data, some of which are listed in the following.

1. We have already noted (Chapter I) that the nuclei of some planetary nebulae are Wolf-Rayet (WR) stars. The existence of WR central stars is by now established at least for 20 planetary nebulae.

The spectra of WR stars are characteristically varied. There are hardly two stars of this class with identical spectra. Nevertheless, if the secondary spectral features are ignored, we are left with essentially two series of classes of WR stars as outlined by Beals and others. Carbon lines predominate in the spectra of one series (WC, carbon stars), and nitrogen lines in the spectra of the other (WN, nitrogen stars). Almost all known WR stars, with very few exceptions, fall into one of the two series.

We know that the majority (some 60%) of WR stars are members of O associations. Finally, it is generally accepted that WR stars, being typical nonstationary objects, constitute an early stage of stellar formation and evolution.

Now suppose that we have succeeded in locating two WR stars with sufficiently similar spectra, one of which is a member of some O association and the other the central star of a planetary nebula. If the former WR star is assigned a certain age ᴉ virtue of its membership in the O association, it would be illogical to assign a different age to the other WR star which, though the nucleus of a planetary nebula, is in the same physical state. The two stars should probably have the same age (in orders of magnitude).

We should now return to a certain point raised back in Chapter I. The WR stars which are nuclei of planetary nebulae are markedly different from ordinary WR stars in their luminosity and hence in their size. They are different objects in this respect, and there may conceivably be other differences between them. For example, Swings points to an increased He/H ratio in nebular WR stars [181]. As far as the physical state of the two stars is concerned, however, we can assume that they are in all probability identical.

Satisfactory photometric data on the intensities of the emission lines of nebular WR stars are unfortanetly very scanty. Table 10-1 lists the relative intensities of the most characteristic emission bands for two central stars (CD +30°3639 and NGC 40) and three ordinary WR stars (HD 192103, HD 192641, and HD 193793), according to measurements of Beals, Aller, and Andrillat [27, 183, 182]. The nebular nuclei and the WR stars in this table all belong to the carbon series: their spectra show virtually no nitrogen lines.

It follows from the table that, within the normal scatter of relative line intensities for this WR series, the central stars of CD +30°3639 and NGC 40 are sufficiently close to the other WR stars in Table 10-1 for our purpose. We should emphasize that the central star of NGC 40, and especially that of CD +30°3639, is very rich in emission lines, this being a characteristic property of WR stars.

2. The spectra of a very small number of WR stars constitute a transitional type between WN and WC; the carbon and nitrogen lines in these spectra are of comparable intensities. Typical representatives of these WC + WN stars are HD 45166,

TABLE 10-1

Relative Intensities of Some Emission Lines in the WR Central Stars of
Planetary Nebulae and in Ordinary WR Stars

Relative intensities	WR nuclei		WR stars		
	CD 30° 3639	NGC 40	HD 192103	HD 192641	HD 193793
$\dfrac{4323\ C\ III}{4686\ He\ II}$	0.77	0.91	0.30	0.36	0.18
$\dfrac{4650\ C\ III}{4686\ He\ I}$	$\begin{cases} 7 \\ 4.75 \\ 9 \end{cases}$	4.36	$\begin{cases} 4.25 \\ 3.24 \end{cases}$	4.75	$\begin{cases} 4.10 \\ 3.20 \end{cases}$
$\dfrac{3889\ He\ I}{4686\ He\ II}$	1.52	0.27	0.24	0.1	0.07

HD 62910, HD 90657 [6]. The nuclei of NGC 6543, NGC 6572, NGC 6826 and some other planetary nebulae are also WC + WN stars. In view of the overall scarcity of WC + WN stars, their occurrence as ordinary stars and as nuclei of planetary nebulae acquires special significance.

3. Closely adjoining to WR stars are the Of stars with spectra characterized by the following features [24, 29]: (1) intense emission lines of λ 4643 Å N III, λ 4640 Å N III, and λ 4686 Å He II; (2) the lines are relatively narrow, corresponding to gas ejection velocities of 100 km/sec from the surface; (3) the profiles and the equivalent widths of the emission lines vary in time.

According to the painstaking spectrophotometric studies of Aller and Wilson [31] and of Oke [24], some planetary nebulae have Of central stars. These are the planetary nebulae NGC 2392, IC 4593, NGC 6210, and apparently IC 418, although Swings, e.g., identifies the central star of this nebula as a transitional type from WR to Of [29]. Aller and Wilson [31] have published spectrograms of the central star of NGC 2392 and two WR stars, HD 14947 and HD 16691; these spectra are virtually identical, the emission lines in the nebular nucleus being only slightly narrower.

Of particular significance is the variability of the line profiles and equivalent widths in the spectra of the above nebular nuclei. These variations are in general similar to those observed in ordinary Of stars. This apparently constitutes direct proof that, if not all, then at least some nuclei of planetary nebulae are *typical nonstationary stars*.

The currently available data are insufficient to evaluate the amplitude and the period of these variations. The data on hand nevertheless indicate that in the spectrum of the nucleus of NGC 2392 the intensity of the emission line λ 4686 Å He II may vary by a factor of two, the intensity of λ 4634 Å N II by a factor of four–five; the emission lines λ 4634 Å N III and λ 4651 Å C III in the spectrum of the IC 4593 nucleus sometimes vanish entirely. Figures 10-1 and 10-2 show the profiles of these and other emission lines in the spectra of the central stars of IC 418, IC 4593, and NGC 2392, obtained at different times [31].

Thus, by careful examination we can always find WR, WC + WN, and Of stars with spectra fairly similar to those of the central stars of some planetary nebulae. Similar correspondences can also be identified among ordinary absorption O stars, provided that the sample is limited mainly to stars with emission lines, in which the nonstationary behavior is the most pronounced.

We will now try to determine whether or not these stars occur in certain O associations. Table 10-2 is a brief list of WR, WC + WN, and Of stars whose spectra resemble those of certain nebular nuclei (see note at the bottom of the table) and which are also members of various O associations (the last column of the table). The members of stellar associations are generally regarded as young formations, and we are therefore led to the conclusion that the nuclei of the planetary nebulae listed

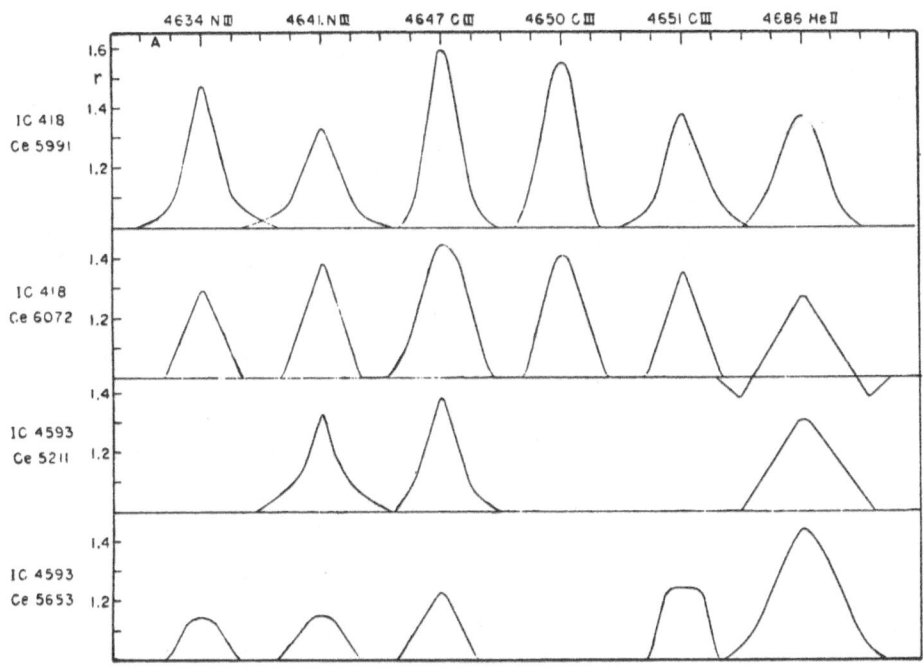

Figure 10-1

Spectral line profiles of the central stars of IC 418 and IC 4593. The spectrograms were taken on different occasions (Aller and Wilson).

in the table are also young stars (at least as far as stellar atmospheric phenomena are concerned).

4. In speaking of the nonstationary behavior of the nebular nuclei, we cannot ignore the anomalous phenomena observed in the continuous spectrum of some planetaries. These phenomena were discussed in some detail in Chapter 5. The main anomalies are the following:

(a) The ratio of the nebular continuum intensity in the visual region to the intensity of Balmer lines is different for different planetaries. If, however, the continuous spectrum has the usual, *i.e.*, thermal, origin, this ratio should be constant or nearly constant for all planetary nebulae (certain differences may arise through the variation of electron concentration from one nebula to another).

(b) The ratio of the continuum to the Balmer lines intensities sometimes varies even within one nebula, an inexplicable effect if the radiation is of thermal origin.

(c) There are indications of variability of the continuum intensity in planetary nebulae.

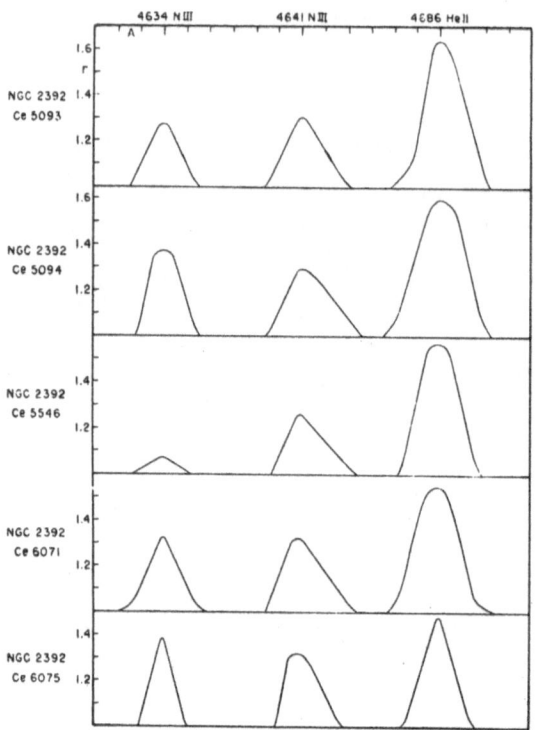

Figure 10-2

Spectral line profiles of the central star of NGC 2392. The spectrograms were taken
on different occasions (Aller and Wilson).

The above anomalies are difficult to reconcile with the thermal emission mechanism.
We are therefore led to postulate the existence of some other, nonthermal mechanism
contributing to the continuous spectra of planetary nebulae. Since all physical processes in a nebula are governed by the activity of the central star, the agent exciting the intense nebular emission is also of stellar origin. The intensity of this agent
on the surface of the central star should be sufficiently high to produce a noticeable effect over the entire expanse of the nebula.

A star capable of engendering such a prominently nonequilibrium phenomenon
clearly cannot be classified as a normal star; it must be a very young star, or even
a partially formed star.

We are thus again led to the conclusion that the nuclei of planetary nebulae are
newborn stars. Indeed, if the nuclei rapidly evolve into normal galactic stars, we are
acquainted with the final stage of their evolution. In the course of their "life" as the
nucleus of a nebula, they are typical nonstationary stars, and we thus recognize

TABLE 10-2

Nonstationary Stars in Stellar Associations Similar to
Central Stars of Planetary Nebulae

HD	α	δ	m_p	Spectral type	Association
14947 (1)	$02^h19^m.5$	$+ 58°25'$	8.04	Of	Cassiopeia VI
16691 (1)	02 35 .5	$+ 58\ 28$	8.4	Of	Cassiopeia VI
45166 (2)	06 20 .8	$+ 08\ 13$	9.6	WC + WN	Monoceros II
62910 (2)	07 41 .1	$- 31\ 41$	10.0	WC + WN	
90657 (2)	10 22 .9	$- 51\ 08$	9.8	WC + WN	Carina
151804 (3)	16 44 .5	$- 41\ 04$	5.37	Of	Scorpius
152408 (3)	16 48 .0	$- 41\ 00$	6.03	Of	Scorpius
192103 (4)	20 08 .1	$+ 35\ 54$	8.25	WC	Cygnus
192641 (4)	20 10 .8	$+ 36\ 21$	8.16	WC	Cygnus
193793 (4)	20 17 .1	$+ 43\ 32$	6.8	WC	Cygnus

Note: The spectra of the stars listed in the first column are identical to the spectra of the central stars in the following planetary nebulae:

(1) IC 418, NGC 2392, 6210, IC 4593.

(2) NGC 40, BD + 30° 3639.

(3) IC 418 (see [26]).

(4) NGC 6543, 6572, 6826, IC 4997.

the intermediate stage of their evolution. The question now is, what is the prenebular stage of this star's life? It cannot be a normal star: given the tremendous scale of nonstationarity, which encompasses a considerable proportion of the galactic objects, we conclude that the prenebular stage of the central stars of planetary nebulae is of necessity a prestellar stage. The salient point is thus not the formation and growth of nebulae, but the formation and growth of new stars, the nebula being a mere byproduct.

4. The Evolution of Nuclei

There are virtually no normal stationary stars whose spectra match those of nebular nuclei. In particular, no normal Galactic star (with the exception of some white dwarfs and similar objects) has a pure continuous spectrum. This indicates that long before the planetary nebula expands to such an extent as to become invisible, the nucleus undergoes marked evolutionary changes. In other words, the central star manages to change its spectral type during a few tens of millennia. It eventually evolves into a normal star.

Once the "excess" mass, *i.e.*, the nebula, has receded from the surface of the new-born star, the hot interior is exposed. At this stage the nucleus should have the charac-

teristic spectrum of hot stars. It is not very probable that once the "excess" mass
has been cast off, all ejection of gaseous matter from the nucleus will cease. The
resulting stellar object is therefore a kind of a Wolf-Rayet star, *i.e.*, a very hot star
with an extended atmosphere. Hence the great similarity in their spectra. The only
difference is in the intensities of the individual bands, which reflect the relative a-
bundances of the various elements, *i.e.*, the composition of the nucleus, and the rate
of ejection. When the ejection of gaseous matter has stopped and the nucleus is deprived of its extended atmosphere, it gives the continuous spectrum of a "bare"
hot star, without any lines.

As the nebula continues to grow, the nucleus is "cooled" and a temperature gradient is set up in the photospheric layers, whose continuous emission just reaches
the observer. The necessary condition for the appearance of absorption lines is thus
satisfied, and the spectrum of the nucleus now corresponds to the spectrum of a
very hot star with absorption lines.

If all of this is true, certain differences should be expected between the spectra
of the central stars of young type I nebulae (planet-like nebulae) and those of the
nuclei of older type II and III nebulae (ring and double-envelope nebulae). The
observational data are unfortunately insufficient to check this prediction. There are
certain indications, however, that such really is the case. Vorontsov-Vel'yaminov
([50], p. 81) has noted an interesting relationship between the nebular structure
and the spectra of the central stars. Most small and stellar planetary nebulae have
Wolf-Rayet central stars. Ring nebulae mainly have nuclei with a continuous spectrum. Finally, the nuclei of irregularly-shaped nebulae (transition type between
planetary and diffuse nebulae) show absorption spectra with shallow lines.

Table 10-3 gives the number distribution of planetary nebulae of different types
according to the spectral type of their central stars. Planet-like nebulae indeed have
mainly WR nuclei, whereas the nuclei of ring and double-envelope nebulae are
C-type (continuous spectrum) and later-type stars.

In certain cases available data seem to reveal a significant change in the spectral
class of the nucleus. The planetary nebula NGC 1514, for example, is fairly well
developed and is almost at the last stage of its life; this is a double-envelope object
with intense brightness fluctuations in the principal envelope. The nucleus of this

TABLE 10-3

Frequency of Nuclear Spectra in Planetary Nebulae
of Different Types

Type of nebula	WR	C	O	B
Planet-like (I)	10	3	—	—
Ring and double-envelope (II)	3	5	4	1

nebula should therefore be a later-type star than the nuclei of younger planetary nebulae (*e.g.*, WR stars). This is apparently so, as McLaughlin's findings [33] indicate that the nucleus of NGC 1514 is a B 9 star.

Another example is discussed in Chapter 1. This is NGC 246, a nebula of very large apparent size ($\sim 4'$), apparently also in the last stage of its life. The nucleus of this planetary nebula is a binary with an O7 star for one of its components, the other component being a late F or G [8] star.

A very large nebula ($\sim 6'$) in the southern sky, previously regarded as a diffuse nebula but now known to be a planetary, has been studied by Henize [174]. It has a distinct Z shape, with two bright "beads" joined by a faint diagonal "crosspiece". The nebula ($\alpha = 16^h 30^m.2$, $\delta = -48°$) is apparently in the last stage of its life, and its nucleus should be a star of a relatively late type. This is indeed so: the nucleus of this planetary is a B0 star (HD 148937, $m_{pv} = 6.9$).

5. The Temperature of IC 4997

A remarkable case of reduction or variation of nuclear temperature during the lifetime of a planetary nebula leading to a change in the spectral type of the central star is provided by the small nebula IC 4997. In Harvard and Lick spectrograms taken in 1913–1916 the forbidden 4363 Å line of doubly ionized oxygen was stronger than the H_y line (λ 4340 Å). In 1938–1939, 4363 [O III] was 30% brighter than H_y, and in 1940 the two lines had equal intensities. Mount Wilson observations in 1955–1956 revealed a continued reduction in the intensity of λ 4363 Å, which was then 77% of the H_y intensity [184]. Over a period of some forty years the intensity of 4363 Å was almost halved. However, we must not ignore the possibility that this is a case of slow but periodic (with a period of a few decades) variation of the nebular spectrum, caused by a periodic variation in the activity of the central star.

The apparent size of its nebula is small, less than $2''$. Its linear dimensions in all probability are also small, as is evident from the relatively very high density of IC 4997 (see Sec. 5, Chapter IV). The decrease of I_{4363}/I_{H_y} is therefore attributable at a first glance to a reduction in the electron concentration as a result of the expansion of the nebula.

A more detailed analysis, however, shows that the decrease in I_{4363}/I_{H_y} cannot be caused by a decrease in n_e. The intensity of the forbidden line λ 4363 Å [O III] is proportional to n_e, while the intensity of H_y is proportional to n_e^2. To a first approximation, I_{4363}/I_{H_y} is therefore inversely proportional to n_e, and an opposite effect is indicated: the ratio I_{4363}/I_{H_y} should increase as the electron concentration decreases.

Let us derive a more exact relation for the ratio I_{4363}/I_{H_y}. According to [185]

$$\frac{I_{4363}}{I_{H_y}} \sim \frac{N^{++}}{n_e} T_e^{-1/2} e^{-\varepsilon/kT_e}, \tag{1}$$

where N^{++} is the concentration of doubly ionized oxygen atoms, T_e the electron temperature of the nebula, and ε is the excitation potential of the $^1D_2$ metastable level of O^{++}.

For planetary nebulae with a constant degree of ionization throughout the entire thickness, the following condition is observed in the course of expansion:

$$\frac{N^{++}}{n_e} \approx \frac{N_0}{n_e} \approx const, \tag{2}$$

where N_0 is the total concentration of oxygen atoms. In very large nebulae (*e.g.*, diffuse nebulae), when the degree of ionization cannot be taken as constant across the nebula, relation (2) is replaced by

$$\frac{N^{++}}{n_e} \approx \frac{N_0}{n_e^2} \frac{1}{r} \sim \frac{N_0}{n_e} r, \tag{3}$$

where r is the radius of the nebula.

From (1) and (2) or (1) and (3) it follows that for a constant electron temperature I_{4363}/I_{H_γ} should remain almost constant in the former case and increase approximately in proportion to the radius of the nebula in the latter.

The observed decrease of I_{4363}/I_{H_γ} in IC 4997 is thus not caused by variation of electron concentration. The decisive factor seems to be the decrease in the electron temperature of the nebula. Indeed, taking N^{++}/n_e to be constant in (1), we see that I_{4363}/I_{H_γ} should decrease with the electron temperature.

The exact electron temperature of IC 4997 is not known. The methods described in Chapter IV for the determination of electron temperature require the knowledge of electron concentration in the nebula. The angular size of the object is unfortunately too small for the only independent method—the method of surface brightness—to give a reliable estimate of electron concentration. We will therefore assume fairly arbitrarily that in 1938–1939, when the ratio I_{4363}/I_{H_γ} was 1.3, the electron temperature of IC 4997 was $T_e = 20,000°K$. This value of T_e corresponds to an electron concentration $n_e \approx 10^6$ cm^{-3} (see Sec. 5, Chapter IV).

Assuming that the rate of cooling of IC 4997 (due to collisions of O^+, O^{++}, and other ions with the free electrons) has remained constant during the relevant period, we can obtain an approximate estimate for its electron temperature as a function of time by writing (1) in the form

$$\frac{I_{4363}}{I_{H_\gamma}} \sim T_e^{-1/2} e^{-\varepsilon/kT_e}. \tag{4}$$

The results are given in Table 10-4. It appears that the electron temperature has decreased nearly 20% in twenty years.

TABLE 10-4

Electron Temperature of IC 4997 in
Different Years

Years	I_{4363}/I_{H_γ}	T_e
1913–1916	> 1	. . . . .
1938–1939	1.3	20,000°K
1949	1	18,000
1956	0.77	16,500

At high electron concentrations, collisions of the second kind make a considerable contribution in nebulae. The prevalence of collisions of the second kind implies that only a fraction of the free-electron energy excites forbidden lines. The mean energy of free electrons is therefore somewhat higher than in nebulae with relatively low values of the electron concentration. It is no longer true that the electron temperature of the nebula is entirely independent of the temperature of the central star, as is the case in normal nebulae. A certain correlation should exist between the electron temperature of the nebula and the temperature of its nucleus. This dependence becomes more pronounced as the proportion of collisions of the second kind increases. Electron temperature fluctuations in relatively dense nebulae should therefore be interpreted to a certain extent as fluctuations in the temperature of the central star. The reduction in the temperature of IC 4997 thus reflects a certain reduction in the temperature of its nucleus.

The cooling of the nebular electrons should also reduce the relative intensities of other forbidden lines, in particular N_1 and N_2. Calculations show, however, that the corresponding effect is approximately one fifth of the effect observed for $\lambda 4363$ Å; it can be detected by precise measurements, but simple comparison of spectrograms apparently is inadequate. The hypothesis of nebular cooling or electron temperature fluctuations and, in the final analysis, the cooling of the central star or fluctuations of its temperature can be verified by observing the dependence of the variation in the relative intensity of forbidden lines on their excitation potential and comparing the observational data with the corresponding theoretical dependence.

The example of IC 4997 shows that present-day observational techniques are quite sufficient to detect the "evolution" *i.e.*, cooling of the nebular nuclei. The cooling is too slow to produce a marked change in the brightness of the nucleus, but it is nevertheless fast enough to cause a noticeable change in the relative intensities of certain nebular emission lines over a few decades. A systematic survey of the spectra of planetary nebulae should therefore be instituted, with spectrograms taken systematically at fixed intervals. Small ("planet-like") nebulae should be given priority in this project. A variation in the intensity of some weak lines in their spectra could be detected by "blinking". The largest change should be expected primarily in the intensities of forbidden lines associated with metastable levels of high excitation potential, *e.g.*, $\lambda\,4363\,\text{Å}$ [O III], and in the intensities of lines (both forbidden and allowed) with high ionization potentials, such as $\lambda\,4684\,\text{Å}$ He II, Ne V lines, *etc.*

No data are as yet available on the variation in the intensity of $\lambda\,4363\,\text{Å}$ [O III] in other planetaries. There is a suspicion of such variation in two nebulae, No. 506 and No. 510 in the catalog [209], but it is based on one series of observations only and has not been verified.

6. The Contribution of Continuous Ejection (Continued)

The conception of planetary nebulae as the residue of star-formation processes, and not as a product of a catastrophic explosion or continuous ejection of gaseous matter from central stars, receives collateral support from the following remarkable fact. In certain cases, the central star of a planetary nebula is seen to eject gaseous matter at a velocity of 1000 km/sec, whereas the expansion velocity of the nebula is a few tens of kilometers per second. This point is of particular significance, since it is clear that such tremendous ejection velocities can in no way be slowed down by two orders of magnitude in the nebulae.

Table 10-5 lists the expansion velocities v_m of some planetary nebulae and the ejection velocities of gaseous matter from their central stars, v_0. The values of v_m are from Table 1-6; the values of v_0 are collected from various sources, as indicated by the numbers brackets in the last column.

The data of Table 10-5 reveal a distinct gap between the rates of expansion of the nebulae v_m and the velocities of gas ejection by the central stars, v_0. Characteris-

TABLE 10-5

Expansion Velocities of Planetary Nebulae (v_m) and Gas Ejection
Velocities from the Nuclei (v_0)

Planetary nebula	Expansion velocity of nebula v_m (km/sec)	Ejection velocity from nucleus v_0 (km/sec)
J 351	24.5	1000 (1)
IC 418	—	80 (3)
NGC 2392	56.3	100 (2)
NGC 6210	21.1	100 (2)
NGC 6543	12	100 (3)
NGC 6572	16	500 (3)
NGC 6751	—	1000 (3)

TABLE 10-5 (continued)

Planetary nebula	Expansion velocity of nebula v_m (km/sec)	Ejection velocity from nucleus v_0 (km/sec)
NGC 6826	—	250 (1)
NGC 6905	—	1000 (3)
NGC 7026	40.8	1000 (1)
Anon. 18^h09^m	—	185 (1)
CD + 30° 3639	—	300 (3)
NGC 4593	—	100 (2)
IC 4997	—	100 (3)

(1) Aller, L.H., *Ap. J*, **97**, 135, 1943.
(2) Oke, J.B., *Ap. J*. **120**, 22, 1954.
(3) Swings, P., *Ap. J*. **95**, 112, 1942.

tically, there is no correlation between v_m and v_0; the order of magnitude of v_0 for the nebulae in Table 10-5 ranges between fairly wide limits, whereas v_m is almost of the same order of magnitude for all these nebulae. We thus have indirect proof that the conditions corresponding to the shedding of the extra mass by the central object in the prestellar stage are radically different from the conditions causing ejection of gaseous matter from newly formed stars.

7. Rate of Ejection of Gaseous Matter by Wolf-Rayet Stars

In view of the special significance of Wolf-Rayet stars in the evolution of planetary nebulae, we should consider in some detail how much mass $\mathfrak{M}$ is ejected by one such star annually. The current estimates, which incidentally are based on highly arbitrary and unfounded assumptions concerning the physical conditions and the geometrical parameters of ordinary WR stars, give $\mathfrak{M} \sim 10^{-5}\ \mathfrak{M}_\odot$ per year.*

If WR stars in planetary nebulae eject gaseous matter at the same rate, the mass of the nebula will be at least doubled over 10^4 years (assuming an initial nebular

---

* The yearly loss of mass by a Wolf-Rayet star is in general described by the formula

$$\frac{d\mathfrak{M}}{dt} = -4\pi R_1^2 \rho_1 v \cdot 3 \cdot 10^7,$$

where ρ_1 and v are the density and the ejection velocity at a distance R_1 from the star center. The main difficulty is that R_1 is not known, even approximately, whereas the annual mass decrement is highly sensitive to this parameter, as we see from the above relation.

mass of 0.1 $\mathfrak{M}_\odot$), and the rate of expansion will reach 500 km/sec (assuming ejection velocities of 1000 km/sec by the WR star). In other words, if central WR stars actually eject matter at a rate of $\mathfrak{M} \sim 10^{-5} \mathfrak{M}_\odot$, the expansion rates of planetary nebulae should show a scatter of a few hundred kilometers per second, which is not so. Since the gas ejected by a WR nucleus virtually is not slowed down, the total momentum of the matter ejected over the entire lifetime of the nebula is necessarily less than the original momentum of the nebular mass. This is possible only if the total ejected mass is substantially less than the mass of the nebula.

The mass rate $\mathfrak{M}$ can be assessed if we assume that the momentum of a planetary nebula at the end of its life, *i.e.*, after 10^4-10^5 years, is twice its initial momentum, because of the continuous ejection by the WR nucleus (a dynamic estimate). This gives a value of $\mathfrak{M}$ of the order of $10^{-7}-10^{-8} \mathfrak{M}_\odot$ annually. Note that even the introduction of differential deceleration because of the radiation pressure of the nebular L_α radiation does not influence the order of magnitude of this estimate (Chapter 6).

There is yet another approach to the problem. A WR star ejects about $10^{-5} \mathfrak{M}_\odot$ annually, but the ejection goes on for much less than 10^4-10^5 years. This may be so in some cases, but according to much of the data a central WR star will continue to eject for the entire lifetime of the nebula.

The above dynamic estimate of the rate of ejection ($10^{-7}-10^{-8} \mathfrak{M}_\odot$ annually) applies to WR stars which are nuclei of planetary nebulae. There are certain indications, however, that this estimate is also true to a certain extent for WR stars in O associations. This suggestion is based on the request, voiced by various authors, to revise our current notions of the nature of WR stars and of the physical state of their outermost layers. On the basis of extensive observational material, Zanstra and Weenan [96] and Van Pelt [186] have established definite stratification of radiation in the atmospheres of WR stars, similar to that observed in planetary nebulae (Bowen's model). If this is a valid conclusion, we must assume that the radiation at the outer boundary of the envelope around a WR star is strongly diluted, *i.e.*, that the conditions in these stellar atmospheres are similiar to the conditions in planetary nebulae.

Johnson [187], on the other hand, cites certain data which rule out the existence of highly extended atmospheres around WR stars and the continuous ejection of enormous masses at tremendous velocities. The intense bands in the spectra of WR stars, according to Johnson, may be attributed to electric discharges (analogous to polar aurorae) presumably taking place in the comparitively dense region near the photosphere.

Some of the arguments and conclusions of these authors are clearly questionable and cannot be accepted without further convincing proof. This single example of a contradiction, however, indicates beyond all doubt that the current notion of the nature of WR stars is in need of revision.

8. Former Nuclei of Planetary Nebulae

The discussion in the previous sections indicates that the central stars of planetary nebulae evolve fairly rapidly, and that when a nebula disappears and dissolves in the interstellar space its nucleus, which has started as a WR star with a spectrum of emission lines, an O star with an absorption line spectrum, or a C star with a continuous spectrum, will have grown into a normal star of the intermediate spherical subsystem of the Galaxy. There are certain exceptions to this rule, however, when for certain reasons (*e.g.*, because of the so-called "visibility condition" [51]) the nebula vanishes before the central star has changed its spectral type. In other words, there is a certain probability that some of the ordinary WR stars and O stars scattered in the Galaxy are former nuclei of planetary nebulae which have vanished prematurely.

This hypothesis can be justified if we remember that WR and O stars mostly occur in groups, in O associations. The universal applicability of this rule explains the considerable confusion caused by the discovery of isolated WR and O stars in various parts of the Galaxy. The occurrence of these scattered "individuals" is a serious obstacle in the theory of stellar associations, although there is always a possibility that these are remnants of open star clusters or stars ejected with great velocities from still existing associations. If we now assume that at least a part of these stars are former nuclei of planetary nebulae, the above difficulty clearly does not arise.

It would be interesting to carry out a special study of the spectra of WR and O stars outside stellar associations, in order to identify stars with spectra similar to those of nebular nuclei (along the same lines as in Sec. 1 above).

When discussing the possible genetic relation of the free WR and O stars outside stellar associations to central stars of planetary nebulae, we must not ignore the statistics of their spatial, or at least apparent, distribution. The point is, that only 1% of all known WR stars have $|b| > 10°$. The corresponding percentage is even less for free WR stars outside stellar associations. On the other hand, almost half ($\sim 40\%$) of the planetary nebulae with WR nuclei occur at Galactic latitudes $|b| > 10°$ (see Table 10-6). This discrepancy would appear to rule out any possible relationship between the free WR stars and the nuclei of vanished planetary nebulae. However, this is not so and the facts can be reconciled in the following way.

We see from Table 10-6 that 24%, *i.e.*, about one quarter of the total population of planetaries, occur at Galactic latitudes $|b| > 10°$. If we assume equal probabilities for WR and O nuclei to occur in any two nebulae, one near the Galactic equator and the other at high Galactic latitudes, nebulae with WR and O nuclei should be uniformly distributed in the coordinate b, *i.e.*, 24% should occur above $|b| = 10°$, 76% below $|b| = 10°$. It follows from the second row in Table 10-6, however, that 40% of all the planetary nebulae with WR nuclei occur above $|b| = 10°$. An even higher percentage is observed for nebulae with O nuclei (76%). In view of the pre-

TABLE 10-6
Distribution of Planetary Nebulae, WR Stars and
O Stars in Galactic Latitude

Object	Total number	Percentage at Galactic latitude	
		$|b| < 10°$	$|b| > 10°$
All planetary nebulae	288	76	24
Planetary nebulae with WR nuclei	23	60	40
Planetary nebulae with O nuclei	25	24	76
O stars	193	89	11
WR stars	97	99	1

ceding assumption, this peculiar distribution can be explained only by a selection effect in the observations: the nuclei of both WR and O planetary nebulae were studied predominantly at high Galactic latitudes. It would be interesting to find the reason for this selectivity (or was it pure chance?). Meanwhile it is clear that if we concentrate on the nuclei of planetary nebulae at low latitudes, the percentage of nebulae with central WR stars in that part of the sky will increase.

There is an alternative approach to the problem, namely by postulating the existence of two kinematically different subsystems of WR (and O) stars in the Galaxy. The first disc-like subsystem includes the existing and decayed stellar association. The second, spherical (intermediate) subsystem comprises the former nuclei of planetary nebulae. The representatives of the first subsystem are also characterized by a low variance of their peculiar velocities and a relatively high lifetime in the WR stage (of the order of a million years). Stars of the second subsystem, conversely, show a high velocity variance and a short lifetime in the WR phase (a few tens or hundreds of millennia). The last point accounts for the almost total absence of free WR stars, without any nebular formations, at high galactic latitudes.

This is a very attractive hypothesis, but the data on hand are unfortunately insufficient to check its validity. Of the 23 planetaries with a WR nucleus, only 4 have a central star brighter than 12^m; the remaining nuclei are as faint at 15^m. No ordinary WR star fainter than twelfth magnitude has ever been studied.

To summarize: the statistical distribution is not inconsistent with the assumption that a fraction of the early-type stars (WR and O stars) outside stellar associations are the nuclei of vanished planetary nebulae. The corresponding percentage for O stars is greater than for WR stars.

9. Hot Stars at High Galactic Latitudes

The hypothesis that a fraction of the early-type stars are former nuclei of planetary nebulae is also supported by other findings. Some of the early-type stars are markedly different from normal representatives of this class in their spatial distribution and kinematic characteristics. Table 10-7 gives some data on these objects. They are all O stars outside stellar associations. These stars characteristically occur at high Galactic latitudes, and if they are regarded as high-luminosity objects (hot giants), we inevitably obtain for them very high values of the z-coordinate (Table 10-7), typical of the intermediate spherical subsystem of the Galaxy.

These and other similar objects are in all probability former nuclei of planetary nebulae.

This assumption is best verified by photographing the parts of the sky where these stars occur through a high-power telescope with narrow-band filters, for the purpose of detecting traces of vanished nebulae around at least some of them. If such traces are detected, the hypothesis will gain considerable support, but even if nothing of the kind is found we will not be able to reject the hypothesis: the nebulous matter may be too faint to show on any of the plates, although the star is actually a former nucleus of a planetary nebula.

In case this impasse is reached, valuable collateral evidence can be provided by the low age of these objects and by their low luminosity.

BD +28°4211. This is a very blue peculiar Op star [190], with an unusually low color index, CI = $- 0^m.62$ [223]. Greenstein estimated the distance to this star from the intensities of interstellar H and K lines and obtained for its luminosity $M_v = + 4^m$. Hall [225] determined the proper motion of this star and was led to suggest that it is not brighter than first mangitude ($M_v > + 1^m$).

BD +28°4211 is thus a subdwarf or an intermediate white dwarf.

TABLE 10-7
"High-Latitude" Hot Stars

Star	l	b	m	Spectrum	z (parsec)	Source
HD 60848	170°.3	+ 19°.0	7.2	08V	750	[191]
HD 93521	150°.1	+ 63°.6	6.9	$09V_p$	1480	[191]
HD 127493	299°.9	+ 33°.6	10.0	09	2700	
BD +28°4211	50°.2	− 19°.9	9.9	0_p	1700	[190]
BD +75°325	107°	+ 32°.0	8.9	0	3000	[222]
HD 128220	343°	+ 65	8.5	0_p	> 3000	[230]
BD 425 2354	352°	+ 86	10.5	0_p	> 3000	[230]
BD −11°162	44°	− 74	11.2	0_p	> 3000	[230]

On the other hand, the star is seen to behave in certain respects as a typical new-born, partly formed star. Its color index is anomalously low, and the spectrophoto-metric gradients are very small. According to Fringant [226], the absolute spectro-photometric gradients of this star are $\phi_b = 0.63$ and $\phi_{ul} = 0.48$. In this latter case, the Planck temperature in the ultraviolet region is about 90,000°K. According to other data [227], the spectrophotometric gradients of this star are variable, and there is even a suspicion of continuous emission of nonthermal origin in the ultraviolet.

BD +75°325. In many respects this star is similar to BD +28°4211. According to Elvius [222] it has the lowest color index among all known stars, CI = $- 0^m.80$. Although no direct distance and luminosity determinations are available, Gold, Herbig, and Morgan [228] regard it as a subdwarf, i.e., a low-luminosity object. These authors were the first to point out the similarity in the spectra of BD + 75°325 and BD +28°4211 and to stress the diffuse character of the helium lines in the spectrum of BD +75°325. To judge from the spectrophotometric gradients $\phi_b = 0.62$ and $\phi_{ul} = 0.39$ [226], the Planck temperature in the ultraviolet is even higher (> 100,000°K) than that of BD +28°4211.

HD 93521. This O star occurs at a very high Galactic latitude ($b > 60°$), which rules out any possible relation between it and the type I Galactic population.

Williams [229] reports an extremely intense line λ 4471 Å He I in its spectrum, a characteristic feature of B 3 dwarfs. Hence the conclusion that this is a low-lu-minosity star. An independent luminosity estimate based on the intensity of inter-stellar lines gives $M_v \sim - 1^m.3$.

Spectrophotometric observations reveal certain fluctuations with time in the spectrophotometric gradients [227]. The nonstationary behavior of this star is apparently less pronounced than that of the two just discussed.

We have described only three early-type stars with distinct signs of nonstationary behavior, all of which (and this is the essential point) occur at high Galactic latitudes. If these objects are high-luminosity stars, they originated in one of the stellar asso-ciations near the Galactic plane and then traveled to high Galactic latitudes, cover-ing a distance of about 1000 parsecs. A trip of this kind would take a few tens of mil-lions of years, and all signs of nonstationary behavior would have long since vanished. The assumption that these stars were once nuclei of planetary nebulae removes these difficulties. The short-lived nebula disappears relatively quickly, while the central star is still in the active stage of a partly formed, newborn stellar object, behaving like the young stars in stellar associations.

We should emphasize that only part of the nebular nuclei will remain nonstation-ary when the nebula has vanished. The majority of central stars apparently evolve fairly rapidly, turning into normal stars of the intermediate (spherical) subsystem by the time the nebula dissolves and vanishes.

10. Humason-Zwicky Objects

Humason and Zwicky discovered a great number of early-type low-luminosity stars ("faint blue stars") at high galactic latitudes, which apparently do not belong type I population [229]. Of particular interest are the objects of this type discovered near the north galactic pole. Two of the 31 objects in this part of the sky are white dwarfs, while the rest have normal spectra (mainly B type). They are all fainter than tenth stellar magnitude (up to $15^m.2$). The authors show convincingly that these stars are not hot giants and that they should be low-luminosity objects. Indeed, had they been ordinary hot giants, the faintest of them would have been located at the distance of the Andromeda Nebula. The absolute magnitude of these objects is therefore apparently about $+ 1^m$. Four objects also show intense blue color.

Luminosity determinations of blue stars at high galactic latitudes are discussed by Klemola [230]. He starts with a very comprehensive list of blue stars at high galactic latitudes, which includes some two hundred stars (seven of these are O stars). He then proceeds to measure the τ-component of proper motion for nearly half of these stars and obtains $M_v = + 2^m.1$ for B0 $-$ B3 stars, $M_v = + 0^m.9$ for B4–B7 stars, $M_v = 1^m.5$ for B8–A0 stars. The average luminosity is $M_v = + 1^m.4$ for B0–A0 stars, a value in good agreement with Greenstein's estimate ($M_v = + 1^m.3$) based on a much smaller sample [231] and with Luyten's estimate ($M_v = + 1^m.9$) derived from mean secular parallaxes [232].

Humason-Zwicky objects thus (a) belong to population II (the spherical component), (b) have low luminosity (dwarfs), and (c) possess high surface temperature (blue color). These are the characteristic properties of the central stars of planetary nebulae. Hence the conclusion is that, if not all, then at least *the majority of Humason-Zwicky objects may be former nuclei of planetary nebulae.*

If this conclusion is true, it implies that the nuclei of planetary nebulae are "early-type hot dwarfs", which lie 4^m–8^m below the hot giants and 10^m–15^m above the the white dwarfs on the spectral type-luminosity diagram.

The possible relation of the nuclei of planetary nebulae to low-luminosity early-type stars again indicates that the nebular nuclei are young, newly born stars. Any attempt to associate the origin of planetary nebulae with normal or old stars would imply that the stars are "rejuvenated" at the end of their lives.

11. The Origin of Diffuse Nebulae

The theory of "residues", being the most probable explanation for the mechanism of formation of planetary nebulae, can be applied to explain the origin of diffuse nebulae as well.

There is no doubt whatsoever that there is a genetic relationship between the typical diffuse nebulae and hot stars responsible for their excitation [193]. There is also a distinct genetic relationship between stellar associations and diffuse nebulae (with few exceptions). Nevertheless, the notion that diffuse nebulae result from a corpuscular outflow or ejection of gaseous matter from the nonstationary stars of the association (in particular, WR stars) is unacceptable. This problem was analyzed in some detail by Shaïn [194], who advanced the following weighty arguments against the hypothesis attributing the formation of diffuse nebula (with emission spectra) to gaseous matter ejected from *existing* stars:

1. Nearly 40% of the WR stars are not associated with diffuse nebulae. There are no traces of any nebulosity around these stars to indicate substantial ejection of gaseous matter from these stars.

2. The Perseus association, with over one hundred hot high-luminosity giants, has no diffuse nebulae. A very faint emission background has been established in the region of this association.

3. The mass of some bright diffuse nebulae is typically a hundred and occasionally a thousand solar masses. Generally, only a few O and WR stars are embedded in these nebulae and their total mass is much less than the mass of the nebula. On the other hand, these stars are the most intense sources of ejected gaseous matter. It is therefore not clear where the "extra" mass of the nebula comes from.

4. No marked difference is observed in the absolute stellar magnitudes and spectral characteristics of O stars which associate with diffuse nebulae and those which occur independently of them. This is also a serious argument against the hypothesis which attributes the formation of the nebula to the ejection of gaseous matter from stars, as this hypothesis fails to account for the absence of diffuse nebulae around "identical" O stars.

5. The exciting stars genetically related to diffuse nebulae are very often located on the rim of the nebula or outside the nebula. This is again inconsistent with the ejection hypothesis; the exterior location of the star is reconciled with the theory only with the aid of various artificial assumptions (asymmetric ejection of matter, loss of symmetry due to internal motions in the nebula).

6. There are several distinctly nonstationary stars of type O (BD +6°1309, AO Cassiopeiae) with periodically appearing and disappearing spectrographic gaseous envelopes or equatorial rings. No diffuse nebulae are observed around these stars.

7. O stars are mostly (in no fewer than 95% of the cases) characterized by constant brightness and spectra. This finding is again inconsistent with the ejection hypothesis, if ejection is regarded as a rare and brief catastrophic occurrence.

8. A mass will separate from a hot star if it is accelerated to an escape velocity of 1500 km/sec. We can think of no decelerating mechanism which will slow down the ejected gaseous mass to very low velocities (about 10–20 km/sec) in the immediate vicinity of the active star.

Shain was thus led to the conclusion that the observed diffuse emission nebulae, with very few exceptions, did not form by ejection of gaseous matter from hot stars which are located in or near these nebulae. Though unable to advance an acceptable alternative mechanism, he spoke of the stellar origin of diffuse nebulae.

The idea that nebulae are formed from the "extra" mass shed by a prestellar object radically changes the situation.

As we know, stellar associations contain tens and hundreds of hot giants and, in all probability, thousands of dwarfs. If all of these stars are approximately the same age, they began to shed the excess mass almost simultaneously (on the cosmogonic scale). We have seen that this extra matter "recedes" from the developing stars with virtually zero velocity, *i.e.*, with a velocity much less than the escape velocity at the star's surface. On the other hand, the spatial density of "newborn" stars within the association is very high. There is thus an eventual build-up of "extra" matter in the association. The initial temperature of this gas is fairly low, and it neither expands nor disperses. As new hot stars develop, the residual matter is heated, its temperature increases, and its starts to expand. The resulting formation is what we call a diffuse nebula. This incidentally explains the presence of diffuse nebulae in stellar associations.

If diffuse nebulae are actually formed in this way, there need not be any correlation between the nebular mass and the mass of the hot stars causing the excitation: the mass of the nebula is made up of the residue of a great multitude of stars, which have no direct effect on its luminosity. The overall mass of the nebula, however, may be related to the total mass of all stars in the association. It is also clear that the star providing the excitation need not be located in any fixed position relative to the nebula; all positions inside and outside the nebula are equally probable, as long as the star is within the association. The size of stellar associations is of course much greater than the size of the diffuse nebulae within them.

The residues of various stars, before merging into a single nebula, move in different directions with different velocities. The overall mass of these residues—the diffuse nebula—can therefore be expected to be in a state of turbulence. The velocity variance and the average eddy velocity cannot be large, since the residues are shed with a relatively low velocity as the star "swells". A random magnetic field should also be expected in the nebulae; each residual portion possibly carried a certain amount of magnetic energy.

The presence of dust in diffuse emission nebulae also indicates that the residues of various types of stars (giants, dwarfs) are initially in different states of aggregation. Corpuscular streams or ejection of gaseous matter from nearly-formed stars are apparently of no consequence in the evolution of the nebula: their contribution to the mass and to the physical processes in the nebula is insignificant.

Highly remarkable is the absence of a diffuse nebula in the Perseus association. This fact was noticed at an early stage in the study of stellar associations and is seemingly incompatible with the above hypothesis. It apparently can be explained as follows. The Perseus association is very large, about 150 parsecs in diameter, and is therefore presumably a comparatively old association, with an age of over 10^6 years, probably of the order of 10^7 years. If this is so, the diffuse nebula expanding at a natural (thermal) rate of 10 km/sec will have by now become too tenuous to be visible. In other words, the diffuse nebula vanishes by expansion long before the young stars of the association grow into normal stars and the association as such is terminated by evolution. This hypothesis can be justified by the following considerations.

The visibility of a diffuse nebula is determined by its surface brightness. This, in turn, is proportional to the square of the electron concentration and the line-of-sight extent of the nebula (the nebula is assumed transparent to visual light). The surface brightness $I(r)$ of a nebula decreases very rapidly on expansion, in inverse proportion to r^5, i.e.,

$$I(r) = I_0 \left(\frac{r_0}{r} \right)^5 ,\tag{5}$$

where I_0 is the surface brightness of the nebula when its radius is r_0. In (a) it is implied that the mass of the nebula is not increased appreciably during the expansion.

It follows from (5) that as the radius of the nebula is doubled, its surface brightness diminishes by more than a factor of thirty, i.e., almost by four stellar magnitudes. It further follows from these considerations that for normal electron concentrations $n_e \sim 100$ cm^{-3} (the typical concentration of most observed diffuse nebulae), a three-fold expansion of a nebula suffices to bring the electron concentration to the normal value of interstellar hydrogen, i.e., $n_e \sim 3$ cm^{-3}. The disappearance of a diffuse nebula is further accelerated as the gas gradually recedes beyond the zone of hydrogen ionization and becomes nonluminous.

The preceding arguments also apply to the stellar association in Scorpio, which contains no diffuse nebulae either.

The absence of gaseous nebulae around "isolated" WR stars can be explained in three ways: (a) some of these stars are former nuclei of planetary nebulae (see Sec. 8); (b) some of them are former nuclei of small diffuse nebulae or remnants of stellar associations; (c) some of these stars are actually newborn objects, but their "excess" mass is too small to be observable.

It thus seems plausible that diffuse nebulae, like planetary nebulae, originate as a byproduct of star-formation processes in stellar associations. These are clouds of primordial matter which have not been consumed in forming stars.

12. The Cosmogonic Significance of Planetary Nebulae

To determine the overall number of planetary nebulae in the Galaxy, we must know their distance from the Sun. Another important factor is the distribution of planetary nebulae in other parts of the Galaxy. Our knowledge of these two factors (especially the latter) is very inadequate, and the current estimates of the total number of planetaries are hardly reliable. Parenago [82], for instance, estimates the total number of planetary nebulae at 10^4, a value which is obviously too low. Vorontsov-Vel'yaminov's later estimate [16] is 10^5, or even higher. This is a more plausible figure, but there are again indications that it is too low. On the other hand, the average lifetime of a planetary nebula is 30,000–100,000 years. We thus conclude that in order to maintain a constant number of planetary nebulae in the Galaxy, no more than a few new nebulae should form every year. Suppose that planetaries form at a rate of one nebula annually. In 10^9-10^{10} years, 10^9-10^{10} planetaries will have formed, and as many stars. This number is comparable with the total stellar population of the intermediate system of our Galaxy. Hence we may conclude that, if not all, then at least *a substantial part of the stars constituting the intermediate galactic subsystem are former nuclei of planetary nebulae.* The formation of some stars of this subsystem is inevitably accompanied by release of a certain quantity of prestellar matter.

The total number of stars in the Galaxy is estimated at 10^{11}. Comparison of this figure with the total number of planetary nebulae which have formed during the entire lifetime of the Galaxy (10^9-10^{10} years) shows that on the average one in every fifty stars in the Galaxy is former nucleus of a planetary nebula.

Star-formation processes in the intermediate subsystem are clearly different from the corresponding processes in the disc-like subsystem. According to Ambartsumyan, stars of this subsystem are created in clusters in T and O associations. This is apparently the principal feature of star-formation processes in the disc-like subsystem.

In the intermediate subsystem, on the other hand, stars are apparently formed one by one, as planetary nebulae show no tendency to occur in groups. This point deserves special attention. It suggests that star-formation processes are different in different parts of the Galaxy, each subsystem having its own characteristic mechanism.

In the light of this conclusion the existence of two subsystems in the Galaxy acquires special significance. Until recently the subsystems were treated, according to Kukarkin [195], as regions containing distributions of stars with certain kinematic and physical characteristics. We now see that these subsystems also differ in star-formation processes and possibly in stellar evolutionary trends.

One of the fundamental properties of stellar associations, as noted in the preceding, is their relation to diffuse nebulae (with very few exceptions). There are no diffuse nebulae in the intermediate subsystem and not a single case of association of a planetary with a diffuse nebula.* Although in the case of O and T associations

---

* With the exception of NGC 7635 [196], which as a matter of fact is not a planetary nebula: it is rather a fragment of the diffuse nebula NGC 7638 (?) which began to fluoresce after the passage of a shock wave initiated by a nova.

the presence of diffuse nebulae gave rise to speculations as to whether stars formed from the mass of these nebulae or *vice versa*, we can now state with some certainty that *stars do not form directly from the familiar diffuse matter*.

Speculations on the exact nature of the prestellar bodies from which stars and nebulae form are beyond the scope of this book. These may be globules (Bok), superdense objects (Ambartsumyan), or some other, hitherto unknown formations. Although we have discussed the possibility of star-formation processes in the intermediate spherical subsystem of the Galaxy, it is nowhere implied that this subsystem should "produce" its own prestellar objects. The "source" of these prestellar objects may be somewhere near the Galactic plane or its center. They reach the high Galactic latitudes by gradual migration from the Galactic plane, remaining in a state of "suspended animation" for a long time before they develop into stars.

REFERENCES

1. V.A. Ambartsumyan; On the Radiative Equilibrium of a Planetary Nebula, *Izv. Glav. Astron. Obs.* **13**, No. 3, 1933.
2. I.S. Bowen; The Origin of the Chief Nebular Lines, *P.A.S.P.* **39**, 295, 1927; The Origin of the Nebular Lines and the Structure of Planetary Nebulae, *Ap. J.* **67**, 1, 1928.
3. S. Rosseland; *Theoretical Astrophysics,* Oxford University Press, 1936.
4. H. Zanstra; An Application of the Quantum Theory to the Luminosity of Diffuse Nebulae, *Ap. J.* **65**, 50, 1927.
5. W.W. Campbell and J.H. Moore; The Spectrographic Velocities of the Bright-Line Nebulae, *Publ. Lick Obs.* **13**, 77, 1917.
6. G.O. Abell; Globular Clusters and Planetary Nebulae Discovered on the National Geographic Society-Palomar Observatory Sky Survey, *P.A.S.P.* **67**, 258, 1955.
7. R.C. Williams and W.A. Hiltner; A Self-Recording Direct-Intensity Microphotometer, *Publ. Obs. Univ. Mich.* **8**, 45, 1940.
8. L.H. Aller; *Gaseous Nebulae,* Chapman and Hall, 1956.
9. B.A. Vorontsov-Vel'yaminov; Photometric and Spacial Structure of the Ring Nebula NGC 6720 in Lyra, *Astron. Zh.* **14**, 194, 1937.
10. B.A. Vorontsov-Vel'yaminov and O. Kramer; Photometric Study of the Planetary Nebula NGC 6853 (Dumbbell), *Astron. Zh.* **14**, 301, 1937.
11. O.C. Wilson and L.H. Aller; The Structure of the Planetary Nebula IC 418, *Ap. J.* **114**, 421, 1951.
12. R. Minkowski; New Emission Nebulae. I and II, *P.A.S.P.* **58**, 305, 1946; *ibid.* **59**, 257, 1947.
13. B.A. Vorontsov-Vel'yaminov; Catalog of Integrated Photographic Magnitudes of Planetary Nebulae, *Astron. Zh.* **8**, 206, 1937.
14. W. Liller; The Photoelectric Photometry of Planetary Nebulae, *Ap. J.* **122**, 240, 1955.
15. H.D. Curtis; The Planetary Nebulae, *Publ. Lick Obs.* **13**, 55, 1917.
16. B.A. Vorontsov-Vel'yaminov; Systems of Planetary Nebulae, *Astron. Zh.* **27**, 285, 1950.
17. I.S. Shklovskiĭ; The Nature of Planetary Nebulae and Their Nuclei, *Astron. Zh.* **33**, 315, 1956.
18. L. Berman; A Study of Galactic Rotation from the Data of Planetary Nebulae, *Bull. Lick Obs.* **18**, 57, 1937.
19. W. Liller and L.H. Aller; Photoelectric Spectrophotometry of Planetary Nebulae, *Ap. J.* **120**, 48, 1954.
20. L.H. Aller, I.S. Bowen and R. Minkowski; The Spectrum of NGC 7027, *Ap. J.* **122**, 62, 1955.
21. L.H. Aller; Spectrophotometry of Representative Planetary Nebulae; *Ap. J.* **113**, 125, 1953.

22. L.H.Aller; A Study of Emission-Line Intensities in Some Bright Northern Wolf-Rayet Stars, *Ap. J.* **97**, 135, 1943.

23. W.H.Wright; The Wave-Lengths of the Nebular Lines and General Observations of the Spectra of the Gaseous Nebulae, *Publ. Lick Obs.* **13**, 191, 1917.

24. J.B.Oke; A Study of the Atmospheres of Early O and Of Stars, *Ap. J.* **120**, 22, 1954.

25. O.C.Wilson; The Nuclear and Nebular Spectra of the Planetary Nebula NGC 2392, *Ap. J.* **108**, 201, 1948.

26. P.Swings and J.W.Swensson; Les Spectres de Treize Nébuleuses Planétaires et de Leurs Noyaux, *Ann. d'Ap.* **15**, 290, 1952.

27. C.S.Beals; The Wolf-Rayet Stars, *Publ. Dom. Astron. Obs.* **4**, 271, 1930; Spectrophotometric Studies of Wolf-Rayet Stars and Novae, *ibid* **6**, 93, 1934; The Spectra of P. Cygni Stars, *ibid.* **9**, 1, 1951; On the Nature of Wolf-Rayet Emission, *M.N.* **90**, 202, 1929; On the Temperatures of Wolf-Rayet Stars and Novae, *M.N.* **92**, 677, 1932.

28. A.Van Pelt; On the Relation of the Spectra of Planetary Nebulae and Their Wolf-Rayet Nuclei, *B.A.N.* **13**, No. 477, 285, 1957.

29. P.Swings; The Spectra of Wolf-Rayet Stars and Related Objects, *Ap. J.* **95**, 112, 1942.

30. L.H.Aller; The Absorption-Line Spectra of the Central Stars of the Planetary Nebulae, *Ap. J.* **108**, 462, 1948.

31. L.H.Aller and O.C.Wilson; Spectrophotometry of the Central Stars of Four Planetary Nebulae, *Ap. J.* **119**, 243, 1954.

32. H.Andrillat; *Pub. l' Obs. Lyon* **3**, 19, 1952.

33. D.B.McLaughlin; The Nucleus of the Planetary Nebula NGC 1514, *P.A.S.P.* **54**, 31, 1942.

34. O.C.Wilson; A Survey of the Internal Motions in the Planetary Nebulae, *Ap. J.* **111**, 279, 1950; Internal Kinematics of the Planetary Nebulae, *Rev. Mod. Phys.* **30**, 1025, 1958.

35. H.Zanstra; Untersuchungen über planetarische Nebel. II, Parallaxen. Expansion der Nebelhüllen, *Zs. f. Ap.* **2**, 329, 1931.

36. G.L.Camm; A Study of Galactic Rotation, Based on the Velocities of the Planetary Nebulae, *M.N.* **99**, 71, 1939.

37. V.V.Sobolev; The diffusion of L_α-Radiation in Nebulae and Stellar Envelopes, *Astron. Zh.* **34**, 694, 1957; translation: *Sov. Astron.* **1**, 678, 1957.

38. K.F.Ogorodnikov; On the Second Order Terms of Galactic Rotation in Radial Velocities of Planetary Nebulae, *Circ. Pulkova Obs.* No. 21, 15, 1937.

39. C.Wirtz; Die Radialbewegungen der Gasnebel, *Astron. Nach.* **215**, 281, 1922.

40. A.J.Cannon; Peculiar Spectra in the Magellanic Clouds, Harvard Bulletin **5**, No. 891, 1931.

41. K.G.Henize and F.D.Miller; H-Alpha Emission Objects in the Magellanic Clouds, *Publ. Obs. Univ. Mich.* **10**, 75, 1951.

42. V.McK.Nail, C.A.Whitney and C.M.Wade; The Nebulosities of the Small Cloud, *Proc. Natl. Acad. Sci.* **39**, 1168, 1953.

43. E.M.Lindsey; Emission Objects in the Small Magellanic Cloud Showing the N_1, N_2 Nebular Lines, *M.N.* **115**, 248, 1955; summary: *Observatory* **75**, 108, 1955.

44. D.Koelbloed; A Search for Planetary Nebulae in the Small Magellanic Cloud, *Observatory* **76**, 191, 1956.

45. W. Baade; Planetary Nebulae in M 31, *Astron. J.* **60**, 151, 1955.

46. F.G. Pease; A Planetary Nebula in the Globular Cluster Messier 15, *P.A.S.P.* **40**, 342, 1928.

47. A.J. Cannon; Spectra Having Bright Lines, *Harvard Annals* **76**, 20, 1916.

48. T. Page; Continuous Emission in the Spectra of Planetary Nebulae, *Ap. J.* **96**, 78, 1942.

49. P. Stoy; Proposed Classification of Planetary Nebulae, *Observatory* **56**, 269, 1933.

50. B.A. Vorontsov-Vel'yaminov; Gazovÿe Tumannostei ĭ novÿe zvezdÿ, (Gaseous Nebulae, and New Stars), Academy of Sciences, U.S.S.R., 1948.

51. G.A. Gurzadyan; Voprosÿ Dinamiki Planetarnÿkh Tumannostĭ, (Problems of the Dynamics of Planetary Nebulae), Erevan, 1954.

52. A.S. Eddington; Forbidden Lines in the Spectra of Nebulae, *M.N.* **88**, 134, 1927.

53. R.H. Garstang; Energy Levels and Transition Probabilities in p^2 and p^4 Configurations, *M.N.* **111**, 115, 1951; Multiplet Intensities for the Lines $^4S-^2D$ of S II, O II and N I, *Ap. J.* **115**, 506, 1952.

54. V.A. Ambartsumyan; The Excitation of Metastable States in Gaseous Nebulae, *Circ. Glav. Astron. Obs.* No. 6, 10, 1933.

55. D. Bohm and L.H. Aller; The Electron Velocity Distribution in Gaseous Nebulae and Stellar Envelopes, *Ap. J.* **105**, 131, 1947.

56. I.S. Bowen; The Spectrum and Composition of the Gaseous Nebulae, *Ap. J.* **81**, 1, 1935.

57. G.G. Cillié; The Hydrogen Emission in Gaseous Nebulae, *M.N.* **92**, 820, 1932; The Theoretical Capture Spectrum of Hydrogen, *M.N.* **96**, 771, 1936.

58. J.G. Baker and D.H. Menzel; Physical Processes in Gaseous Nebulae III. The Balmer Decrement, *Ap. J.* **88**, 52, 1938.

59. A. Burgess; The Hydrogen Recombination Spectrum, *M.N.* **118**, 477, 1958.

60. E.R. Ryndina; Balmer Decrement in the Spectra of Planetary Nebulae, *Uch. Zap. L.G.U.* No. 190, 18, 1957.

61. L. Searle; The Recombination Spectrum of Nebular Hydrogen, *Ap. J.* **128**, 489, 1958.

62. M.J. Seaton; The Solution of Capture-Cascade Equations for Hydrogen, *M.N.* **119**, 90, 1959.

63. J.A. Gaunt; Continuous Absorption, *Phil. Trans.* **229A**, 163, 1930.

64. M.J. Seaton; Radiative Recombination of Hydrogenic Ions, *M.N.* **119**, 81, 1959.

65. G.A. Shaĭn; Note on the Intensity Decrement of Balmer Lines in Gaseous Nebulae. *Circ. Glav. Astron. Obs.* No. 11, 8, 1934.

66. R. Minkowski and L.H. Aller; Spectrophotometry of Planetary Nebulae, *Ap. J.* **124**, 93, 1956.

67. L. Berman; The Effect of Space Reddening on the Balmer Decrement in Planetary Nebulae, *M.N.* **96**, 891, 1936.

68. V.A. Ambartsumyan, E.R. Mustel', A.B. Severnyi and V.V. Sobolev; *Teoreticheskaya Astrofizika*, Gostekhizdat 1952; translation: *Theoretical Astrophysics*, Pergamon, 1958.

69. S. Miyamoto; *Mem. Coll. Sci. Kyoto Imp. Univ.* Ser. A **21**, No. 6, 1938; **27**, No. 4, 1939.

70. J.W. Chamberlain; Collisional Excitation of Hydrogen in a Gaseous Nebula, *Ap. J.* **117**, 387, 1953.

71. S.A. Kaplan and S.I. Gopasyuk; The Excitation of Emission in Interstellar Hydrogen by Electron Impact, *Circ. Astron. Obs. Lvov. Univ.* No. 25, 5, 1953.

72. H.S.W. Massey and C.B.O. Mohr; The Collisions of Slow Electrons with Atoms. III The Excitation and Ionization of Helium by Electrons of Moderate Velocity, *Proc. Roy. Soc.* **A140**, 613, 1933.
73. S. Geltman; Theory of Ionization Probability Near Threshold, *Phys. Rev.* **102**, 171, 1956.
74. L.H. Aller and R. Minkowski; The Interpretation of the Spectrum of NGC 7027, *Ap. J.* **124**, 110, 1956.
75. V.V. Sobolev; Determination of the Electron Temperatures of Planetary Nebulae and an Accurate Method "Nebulium" for the Determination of the Temperatures of their Nuclei, *Uch. Zap. L.G.U.* No. 82, 3, 1941.
76. V.V. Sobolev; Physics of Planetary Nebulae, *Voprosy Kosmogonii* **6**, 112, 1958.
77. D.R. Bates, A. Fundaminsky and H.S.W. Massey; Excitation and Ionization of Atoms by Electron Impact—The Born and Oppenheimer Approximations, *Phil. Trans.* **A243**, 93, 1950.
78. I.S. Bowen and A.B. Wyse; The Spectra and Chemical Composition of the Gaseous Nebulae NGC 6572, 7027, 7662, *Bull. Lick Obs.* **19**, 1, 1939.
79. L.H. Aller; *The Abundance of the Elements,* Interscience, 1965.
80. L.H. Aller; The Composition of the Planetary Nebula NGC 7027, *Ap. J.* **120**, 401, 1954.
81. A. Van Maanen; Stellar Parallaxes from Photographs, *Astron. J.* **44**, 9, 1934.
82. P.P. Parenago; Motions and Distribution in Space of Planetary Nebulae, *Astron. Zh.* **23**, 69, 1946.
83. See [35].
84. J.H. Oort; Reports on the Progress of Astronomy: Stellar Motions, *M.N.* **99**, 369, 1939.
85. B.A. Vorontsov-Vel'yamonov; General Catalogue of Planetary Nebulae with a Statistical Discussion, *Astron. Zh.* **11**, 40, 1934.
86. A.E. Whitford; An Extension of the Interstellar Absorption-Curve, *Ap. J.* **107**, 102, 1948.
87. H. Zanstra; Untersuchungen über planetarische Nebel I. Der Leuchtprozess planetarischen Nebel und die Temperatur der Zentralsterre, *Zs. f. Ap.* **2**, 1, 1930; Luminosity of Planetary Nebulae and Stellar Temperature, *Publ. Victoria Obs.* **4**, No. 15, 209, 1931.
88. D.E. Osterbrock; Electron Densities in Planetary Nebulae, *Ap. J.* **131**, 541, 1960.
89. K.H. Bohm and B. Schlender; Tabelle von Integralen über die Kirchoff-Planck-Funktion, *Zs. f. Ap.* **43**, 95, 1957.
90. V.V. Sobolev; *Dvizhushchiesya Obolochki Zvezd,* Leningrad, 1947; English translation: *Moving Envelopes of Stars,* Harvard Univ. Press, 1960.
91. V.A. Ambartsumyan; On the Temperature of the Nuclei of Planetary Nebulae, *Circ. Glav. Astron. Obs.* No. 4, 8, 1932.
92. L.H. Aller; Physical Processes in Gaseous Nebulae XIV. Spectrophotometry of Some Typical Planetary Nebulae, *Ap. J.* **93**, 236, 1941.
93. G.A. Gurzadyan; Spectrophotometry of the Orion Nebula, *Soob. Burakan Obs.* No. 16, 3, 1955.

94. G.A.Gurzadyan; On the Temperature of the Nuclei of Planetary Nebulae, *Soob. Burakan Obs.*, No. 18, 15, 1956.

95. D.H. Menzel and L.H. Aller; Physical Processes in Gaseous Nebulae. XII The Electron Densities of Some Bright Planetary Nebulae, *Ap. J.* **93**, 195, 1941.

96. H.Zanstra and J.Weenen; On Physical Processes in the Wolf-Rayet Stars I. Wolf-Rayet Stars and Beal's Hypothesis of Pure Recombination, *B.A.N.* **11**, No. 411, 165, 1950.

97. J.G. Baker, D.H. Menzel and L.H. Aller; Physical Processes in Gaseous Nebulae V. Electron Temperatures, *Ap. J.* **88**, 422, 1938.

98. V.A. Ambartsumyan; *Teoreticheskaya Astrofizika*, (Theoretical Astrophysics), Gostekhizdat, 1939.

99. M.J.Seaton; Electron Excitation of Forbidden Lines Occurring in Gaseous Nebulae, *Proc. Roy. Soc.*, **A218**, 400, 1953; The Hartree—Fock Equations for Continuous States with Applications to Electron Excitation of the Ground Configuration Terms of O I, *Phil. Trans.* **A245**, 469, 1953.

100. M.J.Seaton; Electron Temperatures and Electron Densities in Planetary Nebulae, *M.N.* **114**, 154, 1954.

101. H.Andrillat; Intensités Relatives des Raies et Témperatures Électroniques de 24 Nébuleuses Planétaires, *C.R.* **238**, 1781, 1954.

102. M.J.Seaton; Continuum Intensities in Planetary Nebulae, *M.N.* **115**, 279, 1955.

103. L.H.Aller; Thermal Effects of Collisional Line Excitation in Gaseous Nebulae, *Ap. J.* **118**, 547, 1953.

104. H.Zanstra; Dynamics of Radiation Pressure for a Diffuse Nebula, *M.N.* **97**, 37, 1936; Radiation-Pressure in an Expanding Nebula, *M.N.* **95**, 84, 1934.

105. M.J.Seaton; Relative Line Intensities for [O II] and [S II] $^2D-^4S$ in Planetary Nebulae, *Ann. d' Ap.* **17**, 74, 1954.

106. D.H. Menzel, L.H. Aller and M.H. Hebb; Physical Processes in Gaseous Nebulae XIII. The Electron Temperatures of Some Typical Planetary Nebulae, *Ap. J.* **93**, 230, 1941.

107. T.L. Page; The Continuous Spectra of Certain Planetary Nebula: A Photometric Study *M.N.* **96**, 604, 1936.

108. See [48].

109. D. Barbier and H. Andrillat; Mesure de la Répartition d'Énergie dans le Spectre Continu de Six Nébuleuses Planétaires, *C.R.* **238**, 1099, 1954.

110. A.Ya.Kipper; Sbornik *"O Razvitii Sovetskoi Nauki v E'stonckoi S.S.R."*, (On the Development of Soviet Science in the Estonean S.S.R.), p. 316, Tallin, 1950; Theory of the Double Emission of Light Quanta for Atomic Hydrogen, *Publ. Tartu Obs.* **32**, No.2, 63, 1952.

111. L.Spitzer and J.L.Greenstein; Continuous Emission from Planetary Nebulae, *Ap. J.* **114**, 407, 1951.

112. E.M.Purcell; The Lifetime of the $2^2S_{\frac{1}{2}}$ State of Hydrogen in an Ionized Atmosphere, *Ap. J.* **116**, 457, 1952.

113. B.Yada; The Effects of Two-Photon Emission on the Radiation Field of Planetary Nebulae. I and II, *P.A.S.J.* **5**, 128, 1954; *ibid* **6**, 76, 1954.

114. G.Breit and E.Teller; Metastability of Hydrogen and Helium Levels, *Ap. J.* **91**, 215, 1940.

115. M. Goppert-Mayer; Über die Wahrscheinlichkeit des Zusammenwirkens zwei Licht-quanten in einen Elementarakt, *Naturwiss.* **17**, 932, 1929; Über Elementarakte mit zwei Quantensprüngen, *Ann. d. Phys.* **9**, 273, 1931.

116. A. Ya. Kipper and V. M. Tiit; Processes of Disintegration of Light Quanta and Their Significance for the Physics of Gaseous Nebulae, *Voprosy Kosmogonii* **6**, 1958.

117. R. Minkowski; The Electron Temperature in the Planetary Nebula IC 418, *P.A.S.P.* **65**, 161, 1953.

118. H. Zanstra; On the Formation of Condensations in a Gaseous Nebula, *Vistas in Astronomy* **1**, 256, Pergamon, 1955.

119. V. A. Ambartsumyan; The Phenomenon of Continuous Emission and the Sources of Stellar Energy, *Soob. Burakan Obs.* **13**, 3, 1954.

120. G. M. Garibyan and I. I. Gol'man; The Polarization of the Radiation of the Relativistic Electrons in the Magnetic Fields of Stars and Nebulae, *Izv. A.N.Ar.S.S.R.* **7**, 31, 1954.

121. G. A. Gurzadyan and N. A. Razmadze; Polarimetric Investigation of the Planetary Nebula NGC 7026, *Soob. Burakan Obs.* **26**, 1959.

122. V. V. Vladimirskiĭ; Influence of the Terrestrial Magnetic Field on Large Auger Showers, *J.E.T.P.* **18**, 392, 1948.

123. D. Ivanenko and A. Sokolov; *Klassicheskaya Teoriya Polya.* (Classical Theory of Fields), Gostekhezdat, 1949.

124. V. L. Ginzburg; The Origin of Cosmic Rays, *Usp. Nauk.* **62**, 37, 1952.

125. G. A. Gurzadyan; On the Luminosity of Cometary Nebulae, *Soob. Burakan Obs.* **27**, 73, 1960.

126. G. A. Gurzadyan; Synchrotron Radiation in Cometary Nebulae, *D.A.N.* **130**, 47, 1960; translation: *Sov. Physics Doklady* **5**, 7, 1960.

127. Yu. N. Pariĭskiĭ: On the Connection Between Hydrogen Line Radiation and Radio Emission of Gaseous Nebulae. A New Method of Determining Distances of Nebulae, *Izv. G.A.O.* **21**, No. 5, 54, 1960.

128. C. R. Lynds; Observations of Planetary Nebulae at Centimeter Wavelengths, *Publ. Nat. Rad. Astr. Obs.* **1**, No. 5, 1961.

129. V. V. Sobolev; Radiation Pressure in Expanding Nebulae., *Astron. Zh.* **21**, 143, 1944.

130. H. Zanstra; On Scattering with Redistribution and Radiation Pressure in a Stationary Nebula, *B.A.N.,* **11**, No. 401, 1, 1949; On Radiative Equilibrium and Radiation Pressure in a Stationary Nebula, *ibid* **11**, No. 429, 359, 1951.

131. D. Koelbloed; An Accurate Solution of the Integral Equation for the Lyman Alpha Emission in a Stationary Nebula, *B.A.N.* **12**, No. 465, 341, 1956.

132. S. Miyamoto; On the Radiation Field of the Planetary Nebulae, *P.A.S.J.* **2**, 23, 1950.

133. L. G. Henyey; The Doppler Effect in Resonance Lines, *Proc. Natl. Acad. Sci.* **26**, 50, 1940.

134. W. Unno; On the Radiation Pressure in a Planetary Nebula. I and II, *P.A.S.J.* **2**, 53, 1950; *ibid* **3**, 158, 1951; Note on the Zanstra Redistribution in Planetary Nebulae, *ibid* **4**, 100, 1952.

135. V. V. Sobolev; Diffusion of Radiation with Redistribution in Frequency. I and II; *Vest. Leningrad Univ.,* No. 5, 85, 1955; *ibid,* No. 11, 99, 1955.

136. H. Zanstra; On the Acceleration by Radiation Pressure in a Nebula, *B.A.N.* **12**, No. 456, 349, 1956.

137. I.N.Minin; The Radiation Pressure and Dynamics of Planetary Nebulae, *Trudy IV Soveshchaniya po Voprosam Kosmogonii,* 214, Moscow, 1955.

138. G.A.Gurzadyan; Planetary Ring Nebulae, *Astron. Zh.* **34**, 820, 1957; translation: *Sov. Astron.* **1**, 796, 1957.

139. R.J.Trumpler; Preliminary Results on the Distances, Dimensions and Space Distribution of Open Star Clusters, *Bull. Lick Obs.* **14**, No. 420, 154, 1957.

140. E.A.Kreiken; The Apparent Diameters of Clusters and Faint Nebular Objects, *Zs. f. Ap.* **14**, 109, 1937.

141. G.A.Gurzadyan; *Radioastrofizika,* (Radio Astrophysics), Erevan, 1956.

142. L.Spitzer; Physical Properties of Interstellar Gas, p. 31 in *Problems of Cosmical Aerodynamics* (ed. J.M.Burgers and H.C.van de Hulst), Central Air Documents Office, 1951.

143. G.A.Gurzadyan; On the Stability of Gaseous Envelopes Ejected from Stars, *Izv. A.N.A.S.S.R.,* Ser. Phys-Math, **5**, No. 2, 1953; On the Hydrodynamics of Stellar Gaseous Envelopes, *ibid.* No. 13, 1953.

144. D.Lewis; The Instability of Liquid Surfaces When Accelerated in a Direction Perpendicular to their Planes II, *Proc. Roy. Soc.* **A202**, 81, 1950.

145. G.Taylor; The Instability of Liquid Surfaces When Accelerated in a Direction Perpendicular to Their Planes I, *Proc. Roy. Soc.* **A201**, 192, 1950.

146. L.Spitzer; Behavior of Matter in Space, *Ap. J.* **120**, 1, 1954.

147. E.A.Frieman; On "Elephant-Trunk" Structures in the Region of O Associations, *Ap. J.* **120**, 18, 1954.

148. C.Payne-Gaposchkin and S.Gaposchkin; A Spectrophotometric Study of Five Bright Novae, *Harvard Circular* No. 445, 1942.

149. M.Humason; The Spectrum of the Nebulosity Emitted by Nova Persei No. 2, *P.A.S.P.* **46**, 229, 1934.

150. J.Stebbins, C.M.Huffer and A.E.Whitford; The Colors of 1332 B Stars, *Ap. J.* **91**, 20, 1940. **30**, 383, 1953.

151. E.Hubble and J.C.Duncan; The Nebulous Envelope Around Nova Aquilae No. 3, *Ap. J.* **66**, 59, 1927.

152. H.Lamb; *Hydrodynamics,* 6th ed., Dover, 1945.

153. J.H.Oort; Some Phenomena Connected with Interstellar Matter, *M.N.* **106**, 159, 1946.

154. G.A.Gurzadyan; On the Nature of Double-Envelope Planetary Nebulae, *Astron. Zh.* **30**, 383, 1953.

155. G.A.Gurzadyan; The Dynamics of Planetary Nebulae, *Voprosy Kosmogonii* **6**, 157, 1958.

156. R.Minkowski and L.H.Aller; Structure of the Owl Nebula, *Ap. J.* **120**, 261, 1954.

157. J.C.Duncan; Discovery of a Faint Envelope Around the Bright Planetary Nebula NGC 6826 Cygni, *Publ. Amer. Astron. Soc.* **8**, 241, 1936; Photographic Studies of Nebulae. Fifth Paper, *Ap. J.* **86**, 496, 1937.

158. G.A.Shaĭn and B.F.Gaze; On the Connection of the Fibrous Structure of Nebulae with Motion, *Astron. Zh.* **30**, 127, 1953.

159. G.A.Gurzadyan; New Double-Envelope Planetary Nebulae; *D.A.N.* **133**, 1053, 1960; translation: *Sov. Physics Doklady* **5**, 651, 1960.

160. J.C.Duncan; Photographic Studies of Planetary Nebulae, *Publ. Amer. Astron. Soc.* **9**, 37, 1938.

161. I.N. Minin; Radiation Pressure and the Dynamics of Planetary Nebulae, *Voprosȳ Kosmogonii* **6**, 211, 1958.

162. M.J. Seaton; Thermal Inelastic Collision Processes, *Rev. Mod. Phys.* **30**, 979, 1958.

163. K. Wurm and O. Singer, Optische Dicke und He II-H-Linienintensitäten in planetari-schen Nebeln, *Zs. f. Ap.* **30**, 153, 1952.

164. A. Hattori, I. Kawaguchi, S. Miyamoto and T. Saigusa; Note on the Optical Thickness and the Temperature of the Planetary Nebulae, *P.A.S.J.* **4**, 152, 1953.

165. G.A. Gurzadyan; Magnetic Fields in Planetary Nebulae, *D.A.N.* **113**, 1231, 1957.

166. D.S. Evans and A.D. Thackeray; A Photographic Survey of Bright Southern Planetary Nebulae, *M.N.* **110**, 429, 1950.

167. G.A. Gurzadyan; On the Nature of Magnetic Fields in Planetary Nebulae, *Soob. Burakan Obs.* **24**, 31, 1958; On the Electromagnetic Character of Spiral Planetary Nebulae; *ibid.* **24**, 59, 1958.

168. T.G. Cowling; The Electrical Conductivity of An Ionized Gas in a Magnetic Field, with Application to the Solar Atmosphere and the Ionosphere, *Proc. Roy. Soc.* **A183**, 453, 1945.

169. D. Evans; IC 4406: A Double Nucleus Planetary Nebula, *M.N.* **110**, 37, 1950.

170. T.G. Cowling, *Magnetohydrodynamics,* Interscience, 1957.

171. M.P. Savedoff and J. Greene; Expanding H II Region, *Ap. J.* **122**, 477, 1955.

172. S. von Hoerner; Eine Methode zur Untersuchung der Turbulenz der interstellaren Materie, *Zs. f. Ap.* **30**, 17, 1951.

173. G. Münch; Internal Motions in the Orion Nebula, *Rev. Mod. Phys.* **30**, 1035, 1958.

174. K.G. Henize; Large Planetary Nebula in Norma, *Sky and Telescope* **18**, 315, 1959.

175. W.H. Bostick; Experimental Study of Ionized Matter Projected Across a Magnetic Field, *Phys. Rev.* **104**, 292, 1956; Experimental Study of Plasmoids, *ibid.* **106**, 404, 1957.

176. O. Struve; *Stellar Evolution,* Princeton, 1950.

177. P.W. Merrill; Objects Intermediate Between Planetary Nebulae and Stars, *P.A.S.P.* **44**, 123, 1932.

178. See [119].

179. G.B. Herbig; The Spectrum of the Nebulosity Surrounding T Tauri, *Ap. J.* **111**, 11, 1950; The Spectra of Two Nebulous Objects Near NGC 1999, *ibid.* **113**, 697, 1951.

180. G. Haro; Herbig's Nebulous Objects Near NGC 1999, *Ap. J.* **115**, 572, 1952.

181. P. Swings; Quelques Remarques Sur les Noyaux de Nebuleuses Planetaires: Rapport Introductif, *Mem. Soc. Roy. Liege* **20**, 36, 1958.

182. Y. Andraillat; Etude Spectrophotométrique des Etoiles de Wolf-Rayet, *Suppl. Aux. Ann. d'Ap.* No. 2, 1, 1957.

183. See [22].

184. W. Liller and L.H. Aller, Changes in a Planetary Nebula, *Sky and Telescope* **16**, 222, 1957

185. G.A. Gurzadyan; A Note On the Intensity Variation of the Line 4363 [O III] in the Spectrum of the Planetary Nebula IC 4997, *Astron. Zh.* **35**, 520, 1958; translation: *Sov. Astron.* **2**, 482, 1958.

186. J. Reynolds; Remarkable Planetary Nebula in Cassiopeia, *Observatory* **47**, 293, 1924.

187. M. Johnson; A Possible Clue to the Wolf-Rayet Atmosphere From Flare and Auroral Mechanisms, *Observatory* **74**, 124, 1954.

188. G.A. Gurzadyan; On the Problem of the Youth of the Nuclei of Planetary Nebulae, *Soob. Burakan Obs.* **25**, 101, 1958.

189. D.L. Harris III; The Color of BD + 28° 4211, *Ap. J.* **113**, 435, 1951.

190. D.A. MacRae, R. Fleischer and E.B. Weston; A Peculiar O star at High Galactic Latitude, *Ap. J.* **113**, 432, 1951.

191. W.W. Morgan; Some Astrometric Problems of Galactic Structure, *Astron. J.* **59**, 86, 1954.

192. J.L. Greenstein and J. Cuffey, A B-type Star of Population II., *P.A.S.P.* **66**, 187, 1954.

193. G.A. Gurzadyan; On the Problem of the Origin of Diffuse Nebulae, *Astron. Zh.* **29**, 121, 1952.

194. G.A. Shain; On the Nonstationarity of the Hot Stars and Diffuse Nebulae, *Trudy IV Soveshchaniya po Voprosam Kosmogonii*, 220, Moscow, 1955.

195. B.V. Kukarkin; *The Investigation of the Structure and Evolution of the Stellar Systems*, Moscow, 1949.

196. R. Minkowski and D. Osterbrock; Electron Densities in Two Planetary Nebulae, *Ap. J.* **131**, 537, 1960.

197. N.A. Razmadze; Spectrophotometry of Faint Planetary Nebulae, *Astron. Zh.* **37**, 1005, 1960; translation: *Sov. Astron.* **4**, 938, 1961.

198. A.I. Akhiezer and V.B. Berestetskii; *Kvantovaya Elektrodinamika*, 1st ed., p. 394, Fitzmatgiz, 1959: translation, *Quantum Electrodynamics*, Office of Technical Services, Dept. of Commerce, Washington, D.C.

199. See [153].

200. S. Chandrasekhar and E. Fermi; Problems of Gravitational Stability in the Presence of a Magnetic Field, *Ap. J.* **118**, 116, 1953.

201. V.C.A. Ferraro; On the Equilibrium of Magnetic Stars, *Ap. J.* **119**, 407, 1954.

202. S.B. Pikel'ner; Interstellar Gas and Magnetic Field, *Izv. Krym. Obs.* **10**, 74, 1953.

203. G.A. Gurzadyan; One Mechanism for L_α-Quantum "Splitting" in Gaseous Nebulae, *D.A.N.* **141**, 1061, 1961; translation, *Sov. Physics Doklady* **6**, 1031, 1962.

204. F.E. Obenshain and L.A. Page; A Measurement of the Effect of Static Electric Fields on the Formation of Positronium in Gases, *Bull. Am. Phys. Soc.* **3**, 228, 1958.

205. G. Feinberg; Effects of an Electric Dipole Moment of the Electron on the Hydrogen Energy Levels, *Phys. Rev.* **112**, 1637, 1958.

206. E.E. Salpeter; Some Atomic Effects of An Electronic Dipole Moment, *Phys. Rev.* **112**, 1642, 1958.

207. W.L. Fite, R.T. Brackmann, D.G. Hummer and R.F. Stebbings; Lifetime of the 2S State of Atomic Hydrogen, *Phys. Rev.* **116**, 363, 1959.

208. G.A. Gurzadyan; The Origin of Balmer Absorption Lines in Spectra of M 82 Type of Galaxies, *Soob. Burakan Obs.* **34**, 59, 1963.

209. B.A. Vorontsov-Vel'yaminov; A New Catalogue of Planetary Nebulae, *Soob. Gos. Astr. Inst. Stern.* No. 118, 3, 1962.

210. B.E. Westerlund; The Galaxy and the Magellanic Clouds, *IAU/IRSJ Symposium* No. 20, 1963.

211. S.N. Milford; Approximate Cross-Sections for Inelastic Collisions of Electrons with Atoms. I Allowed Transitions, *Ap. J.* **131**, 407, 1960.

212. D.G. Hummer; The Ionization Structure of Planetary Nebulae. II Collisional Cooling of Pure Hydrogen Nebulae, *M.N.* **125**, 461, 1963.

213. T. Osaki; The Thermal Equilibrium in a High-Excitation Planetary Nebula, *P.A.S.J.* **14**, 111, 1962.

214. C.R. O'Dell; A Distance Scale for Planetary Nebulae Based on Emission-Line Fluxes, *Ap. J.* **135**, 371, 1962.

215. M.J. Seaton and D.E. Osterbrock; Relative [O II] Intensities in Gaseous Nebulae, *Ap. J.* **125**, 66, 1957.

216. M.J. Seaton; The Forbidden Line Spectra of Gaseous Nebulae, I.A.U. Symposium No. 2, 75, 1955.

217. G.A. Gurzadyan; The Gradient of the Electron Temperature in Planetary Nebulae, *Astrofizika* **1**, 91, 1965.

218. C.E. Moore; *Atomic Energy Levels,* Natl. Bureau of Standards, 1949.

219. V.I. Pronik; Corpuscular Emission of the Nucleus and the Electron Temperature of the Planetary Nebula IC 418, *Izv. Krym. Obs.* **25**, 61, 1961.

220. G.A. Gurzadyan and M.A. Kazaryan; The Intensities of the Emission Lines of Some Planetary Nebulae, *Soob. Burakan Obs.* **36**, 23, 1964.

221. G.A. Gurzadyan; Photometric Data for Some Two-Envelope Planetary Nebulae, *ibid.* **35**, 59, 1964.

222. T. Elvius; Photometric and Spectrophotometric Investigation in Kapteyn's Selected Areas, *Stockholms Observatoriums Annaler* **19**, No. 3, 1956.

223. See [189].

224. J.L. Greenstein; The Luminosity of the Blue Star BD $+ 28° 4211$, *P.A.S.P.* **64**, 256, 1952.

225. R.G. Hall; The Proper Motion of BD $+ 28° 4211$, *P.A.S.P.* **65**, 154, 1953.

226. A.-M. Fringant; Étude Spectrophotometrique de deux Étoiles O tres bleues, *Jour. Observateurs* **41**, 98, 1958.

227. G.A. Gurzadyan and R.Kh. Ogannesyan; The Spectrophotometric Measurements of the Hypothetical Form of Nuclei of Planetary Nebulae, *Soob. Burakan Obs.* **35**, 43, 1964.

228. N.L. Gould, G.B. Herbig and W.W. Morgan; BD $+ 75° 325$: A Subluminous O-Type Star, *P.A.S.P.* **69**, 242, 1959.

229. M.L. Humason and F. Zwicky; A Search for Faint Blue Stars, *Ap. J.* **105**, 85, 1947.

230. A.R. Klemola; Mean Absolute Magnitude for the Blue Stars at High Galactic Latitude, *Astron. J.* **67**, 740, 1962.

231. J.L. Greenstein; The Spectra and Other Properties of Stars Lying Below the Normal Main Sequence, *Proc. 3rd Berkeley Symposium on Math. Statistics and Probability* **3**, 11, 1956.

232. W. Luyten and J.N. Anderson; *A Search for Faint Blue Stars,* **30**, Obs. of Univ. of Minnesota, 1962.

233. T.K. Menon and Y. Terzian; Radio Observations of Planetary Nebulae, *Ap. J.* **141**, 745, 1965.